Extruder Principles and Operation

Extruder Principles and Operation

M. J. STEVENS

*Formerly, Institute of Polymer Technology,
Loughborough University, UK*

ELSEVIER APPLIED SCIENCE
LONDON and NEW YORK

ELSEVIER APPLIED SCIENCE PUBLISHERS LTD
Crown House, Linton Road, Barking, Essex IG11 8JU, England

Sole Distributor in the USA and Canada
ELSEVIER SCIENCE PUBLISHING CO., INC.
52 Vanderbilt Avenue, New York, NY 10017, USA

British Library Cataloguing in Publication Data

Stevens, M. J.
  Extruder principles and operation.
  1. Polymers and polymerization—Extrusion
  I. Title
  668.4'13   TP1175.E9

ISBN 0-85334-336-5

WITH 92 ILLUSTRATIONS AND 21 TABLES

© ELSEVIER APPLIED SCIENCE PUBLISHERS LTD 1985

First Edition 1985
Reprinted 1986
Reprinted 1988

No responsibility is assumed by the publisher for any injury and/or damage to persons or property as a matter of products liability, negligence or otherwise, or from any use or operation of any methods, products, instructions or ideas contained in the material herein.

Special regulations for readers in the USA
This publication has been registered with the Copyright Clearance Center Inc. (CCC), Salem, Massachusetts. Information can be obtained from the CCC about conditions under which photocopies of parts of this publication may be made in the USA. All other copyright questions, including photocopying outside of the USA, should be referred to the publisher.

All rights reserved. No part of this publication may be reproduced, stored in a retrieval system, or transmitted in any form or by any means, electronic, mechanical, photocopying, recording, or otherwise, without the prior written permission of the publisher.

Filmset and printed in Northern Ireland by The Universities Press (Belfast) Ltd.

*To all those from whom I have learned,
including the students I essayed to teach.*

PREFACE

This book is intended to fill a gap between the theoretical studies and the practical experience of the processor in the extrusion of thermoplastic polymers. The former have provided a basis for numerical design of extruders and their components, but generally give scant attention to the practical performance, especially to the conflict between production rate and product quality. In practice extruders are frequently purchased to perform a range of duties; even so the operator may have to use a machine designed for another purpose and not necessarily suitable for the polymer, process or product in hand. The operator's experience enables him to make good product in unpromising circumstances, but a large number of variables and interactions often give apparently contradictory results. The hope is that this book will provide a logical background, based on both theory and experience, which will help the industrial processor to obtain the best performance from his equipment, to recognise its limitations, and to face new problems with confidence. Mathematics is used only to the extent that it clarifies effects which cannot easily be expressed in words; if it is passed over, at least a qualitative understanding should remain. The approximate theory will not satisfy the purist, but this seems to the author less important than a clear representation of the physical mechanisms on which so much of the polymer processing industry depends.

M. J. Stevens

ACKNOWLEDGEMENT

The author is responsible for statements and opinions expressed in this book, which are his own, but would like gratefully to acknowledge assistance over many years from Imperial Chemical Industries PLC, Petrochemicals and Plastics Division, and the staff of the Institute of Polymer Technology, Loughborough University of Technology.

CONTENTS

Preface vii

Acknowledgement viii

1 Introduction 1
    Scope and Limitations 1
    Method 2

2 Flow Behaviour Relevant to Extrusion 4
    2.1 Viscosity. 4
    2.2 Shear Flow 10
        Simple shear 10
        Circular capillary (Newtonian) 11
        Infinite slit (Newtonian). 13
        Strain energy in shear 15
    2.3 Extensional Flow and Elastic Effects 19
        Extensional flow 19
        Elastic effects 19
    2.4 Measurement of Viscosity and Elasticity . . . 20

3 Thermal and Energy Properties in Processing . . . 24
    3.1 Thermal Properties 24
    3.2 Thermal Conduction 25
    3.3 Non-isothermal Flow and Heat Transfer . . . 28
        Velocity distribution 28
        Surface heat transfer 29
    3.4 Mixing 31

Contents

4	**Practical Extrusion Processes**	34
4.1	Shaping Processes and Their Requirements	37
	Solid sections	37
	Hollow sections	37
	Wire covering	39
	Flat sheet	41
	Flat film	43
	Tapes	46
	Fibres and monofilaments	46
	Netting and mesh	47
	Tubular film (Blown film)	48
	Foams	50
4.2	Other Applications and Their Requirements	51
	Compounding	51
	Melt blending	53
	Degassing, drying	54
	Filtering	55
	Dewatering	55
	Polymerisation	55
	Blow moulding	56
	Injection moulding	56
	Comparison of requirements	60
4.3	Comparison of Machine Types	60
5	**Principles of Single-Screw Extrusion**	67
5.1	Functions of the Extruder	67
5.2	Derivation of Flow Equation	70
	Assumptions	71
	Drag flow	73
	Pressure flow	76
	Transverse flow	81
5.3	Leakage Flow	84
5.4	Output Equations and Longitudinal Pressure Profiles for Common Screw Types	89
	Analysis of variables	89
	Constant depth screw	91
	Double-parallel (stepped) screw	93
	Taper and taper-parallel screws	96
5.5	Graphical Representation of Output for Screw/Die Combinations, including Venting	101

		Graphical representation	.	.	.	.	.	101
		Venting or vacuum extraction	.	.	.	.	109
		Double-parallel screw .	.	.	.	.	.	115
		Screw extruder and gear pump	.	.	.	.	117
	5.6	Output Corrections .	.	.	.	.	.	120
	5.7	Pseudoplastic Flow	.	.	.	.	.	.	123
	5.8	Non-isothermal Flow .	.	.	.	.	.	129
	5.9	The Melting Process .	.	.	.	.	.	131
	5.10	Solids Conveying .	.	.	.	.	.	.	135

6 Principles of Energy Balance 142

- 6.1 Energy Balance and Efficiency 142
- 6.2 Experimental Determination of Energy Balance . . 147
  - Mechanical power input. 147
  - Heater energy 150
  - Heat content (enthalpy) of polymer 152
  - Energy losses 154
- 6.3 Power Consumption in the Screw—Newtonian Isothermal Case 155
  - Analysis of variables—power equation . . . 161
- 6.4 Pseudoplastic Isothermal Approximation . . . 163
- 6.5 Power in Non-isothermal Flow 164
- 6.6 Effect of Variables on Energy Balance 174

7 Operation of Single-Screw Extruders 188

- 7.1 Overall Performance of the Screw 188
  - Material flow 188
  - Pressure 193
  - Temperature 193
  - Energy transfer 196
  - Mixing 199
- 7.2 Effects of Controlled Variables 201
  - Screw speed 202
  - Back pressure 203
  - Heating 203
  - Temperature profile 206
  - Melt temperature 207
- 7.3 Polymer Properties 208
- 7.4 Screw Design 209
- 7.5 Operational Strategies 213

8 Extruder Operation as Part of a Total Process . . . 237
8.1 Quality 237
        Temperature variations 239
8.2 Stability 245
        Start-up and condition changes 250
8.3 Shear History 251
8.4 Control 253
        Control for specific processes 258
8.5 Scale-up 264

9 Practical Extruder Operation 272
9.1 Steady Operation 272
9.2 Colour and Grade Changing 275
9.3 Start-up and Shut-down 276
9.4 Dismantling and Cleaning 279
9.5 Waste Recovery 284

10 Application to the Individual Machine 287

Appendices 294
A Alternative Derivation of Flow Equation 294
B Pressure Gradients in a Stepped Screw 297
C Experimental Determination of Heat Losses . . . 301
D Stability of Melt Pumping Section 306
E Properties of Polymers for Heat and Flow . . . 308
F Temperature Variation in the Flight Clearance . . 317

References 324

Index 327

Chapter 1

INTRODUCTION

SCOPE AND LIMITATIONS

The objective of this book is to summarise the principles of single screw extrusion processes for plastics and rubbers, the practical performance and its theoretical explanation, and operating procedures to exploit this performance.

The scope of the work includes the principles and operation of the single screw extruder as a machine, as a combination with a die, and as part of a total fabrication process, including subsidiary functions such as compounding and venting. It includes the special requirements for the extruder of individual processes such as wire covering, tubular film blowing and blow moulding, but does not attempt to cover other aspects of these processes. In order that the objective may be adequately covered in a book of reasonable size, certain matters are excluded, and the reader's attention is directed to the references for further information. In particular, twin screw extruders, depending on their precise design, operate more or less as positive-displacement rather than drag pumps, leading to distinct design, constructional and operating characteristics. The principles of these machines are well described by Janssen.[1] The mechanical construction and adaptation for special purposes of single screw machines is governed partly by the process requirements described in Chapter 4, but also by standardisation, manufacturing methods and mechanical reliability. Mechanical construction is dealt with in Chapter 6 of Fisher[2] and Chapter 6 of Schenkel.[7] Commercial models and unconventional types of extruder are also covered by Fisher and Schenkel, as well as by periodic reviews such as *Plastics and Rubber Weekly*.[8] As this book is primarily concerned with operation, it does not purport to be a design manual for the precise dimensional design of machine, screw or die for a specific performance. This is covered by several authors including Fenner.[9]

METHOD

The method is to summarise the properties of plastics and rubber materials suitable for extrusion, especially the flow and thermal properties of polymer melts. The relevant general equations of mass and heat flow are presented and their application to extrusion problems developed. After a brief chapter on the process requirements of the single screw extruder, to provide an objective for the following chapters, a simplified extrusion theory is developed for both mass flow and energy consumption and presented in both algebraic and graphical form to demonstrate the effects of dimensional and operational variables on output, energy balance, melt temperature and product uniformity. These effects are used to propose strategies for operation and control to eliminate faults and optimise performance of the extruder, and of the system of which it is a part. In particular the problem of scaling up an extrusion process from the development laboratory to large-scale production is studied. Although the simplified theory is approximate, its predictions are borne out in practice. It is used in preference to more exact methods since the algebraic form of the solutions permits a ready identification and understanding of the effects of separate variables on extruder performance; this tends to be obscured in the more exact methods, which usually involve digital computation.

As a complement to the analysis of steady-state operation, recommendations for start-up, shut-down, dismantling and cleaning are given, based on practical experience.

The total commercial process of extrusion of plastics and rubbers covers the conversion of a raw polymer, usually in the form of powder or pellets, into a saleable finished or semi-finished product. This will include matters of management and plant organisation common to many other materials and processes. It will also include polymer handling and storage, product specification and quality control, plant layout, installation and maintenance, product finishing, assembly, packaging and distribution which are general to many plastics processes and only indirectly concerned with the actual operation of extrusion. The intention in this volume is to provide the specific requirements of extrusion, and their explanations, which form the technical background to the total process and its peripheral aspects.

In its simplest terms, extrusion consists of forcing a polymer in liquid form (usually molten) under pressure through a die to produce a continuous section or profile. This may then be sized, drawn down, corrugated, etc., to modify and control the shape and dimensions of the section and in some cases the properties (mechanical, optical, etc.). In the case of thermoplastics, the product or 'extrudate' must be cooled to retain its shape, while rubbers are chemically cross-linked ('cured') to achieve properties such as elasticity and resistance to chemicals and heat. In some processes, e.g. blow moulding, injection moulding and in-line vacuum forming, the extrudate may be shaped into a discontinuous or intermittent form before cooling or curing. Other subsidiary processes, such as printing, slitting and on-line testing, may precede coiling or cutting into handleable lengths. The detail of these operations after the die (collectively 'post-forming') is highly specific to the material and product, much of it confidential to each company, so space precludes exhaustive study here. However, it is just as important as the extruder to the 'quality', dimensions and surface finish of the product; an understanding of the effects of, say, the ratio of haul-off rate to extruder output is essential to economic production. The effects of changes after the die are usually visible, if not self-evident, but one cannot see what is going on inside the extruder, and this book aims to provide the basic mechanisms which help to explain the performance of the screw and die, and hence their interactions with post-forming operations. The book also introduces strategies for operation (Chapter 7) and control (Chapter 8), as well as some practical experience on selection of screws, start-up, cleaning and fault-finding (Chapters 8 and 9).

An important subject for development engineers is the prediction of the performance of full-scale production plant from that of a small laboratory process, since by its nature heavy moving machinery is expensive to modify or replace and, owing to special materials and manufacture, changes cause lengthy and costly production delays; this is given special attention. Literature references are given at the end of the book, numbered in order of first appearance. Mathematical equations are also numbered in order of appearance, but separately for each chapter. SI units are used throughout, with Imperial units in brackets where appropriate. Unless stated otherwise, numerical constants are dimensionless, and therefore the same in any consistent system of units.

Chapter 2

FLOW BEHAVIOUR RELEVANT TO EXTRUSION

In this chapter the principles and basic equations relating to fluid flow of polymers will be stated, with emphasis on those which are relevant to melt extrusion. In general, for proofs and other applications, reference should be made to standard works on rheology such as that by Brydson.[11] Design of extruder dies has received special attention[13,14] and will be mentioned further only where it is affected by extruder performance.

2.1. VISCOSITY

When a liquid is subjected to stress, it deforms continuously as long as the stress is applied, unlike a perfect solid which deforms a fixed amount, independent of time. On removal of stress, the liquid remains in its deformed state, whereas an elastic solid recovers its original dimensions instantaneously and completely (Fig. 2.1). Thus the stiffness or modulus, which is the relation between stress and deformation (strain), is unique for a solid, but for a liquid is dependent on time and so does not define a property of the material. However, under a constant stress, the *rate* of deformation with time is constant and the ratio is the viscosity. In direct stress σ (tension, compression), the strain ε is defined as extension/original length (Fig. 2.2) and the ratio stress/strain as the modulus of elasticity E, which is constant for a Hookean solid. For a liquid, the strain rate is:

$$d\varepsilon/dt = \dot{\varepsilon} = \frac{1}{L} \cdot \frac{dL}{dt} \qquad (2.1)$$

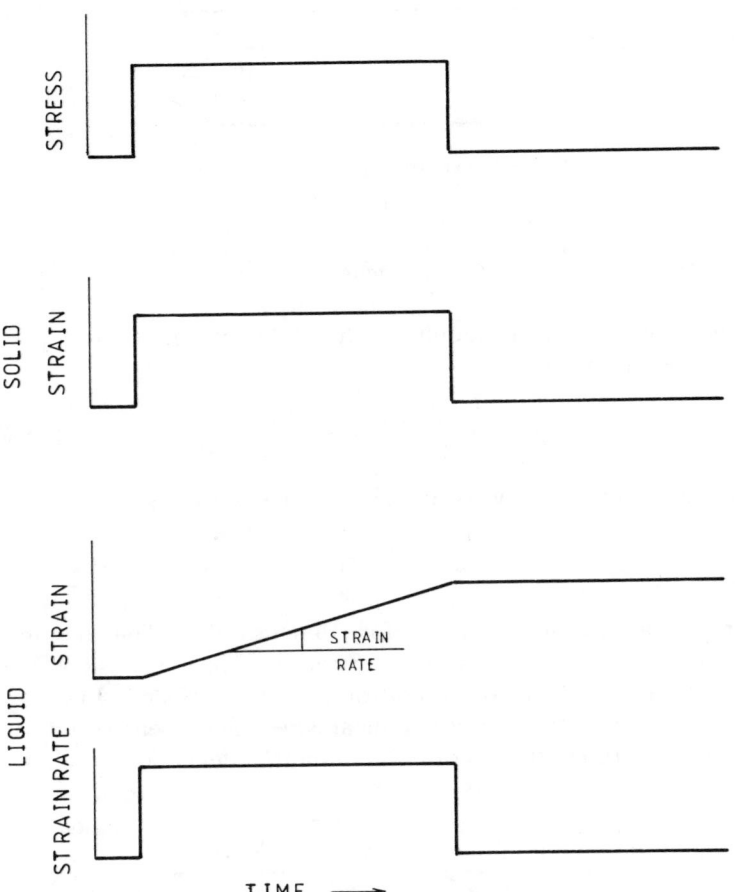

Fig. 2.1. Deformation under stress.

and the extensional viscosity:

$$\frac{\sigma}{\dot{\varepsilon}} = \lambda \qquad (2.2)$$

In shear, stress τ causes a deformation dx in the direction of the stress, but shear strain is defined (Fig. 2.3) as:

$$\frac{dx}{dy} = \tan \gamma \simeq \gamma \qquad (2.3)$$

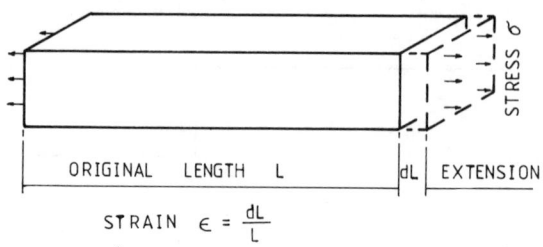

FIG. 2.2. Extensional strain.

for small strains. Shear modulus G (modulus of rigidity) is τ/γ. For a liquid, the strain rate is:

$$\frac{d\gamma}{dt} = \dot{\gamma} = \frac{d}{dt}\left(\frac{dx}{dy}\right) = \frac{d}{dy}\left(\frac{dx}{dt}\right) = \frac{dw}{dy} \tag{2.4}$$

where $w = dx/dt$ is the velocity. The shear viscosity is:

$$\frac{\tau}{\dot{\gamma}} = \eta \tag{2.5}$$

Strain is a pure ratio and therefore dimensionless. Thus the units of stress and modulus are both N/m^2. Strain rate has dimensions 1/time, expressed as 's^{-1}', and so the unit of viscosity is $N\,s/m^2$. In a Newtonian fluid, the relation between shear stress and shear strain rate is linear (Fig. 2.4), i.e. the viscosity is constant, independent of the values

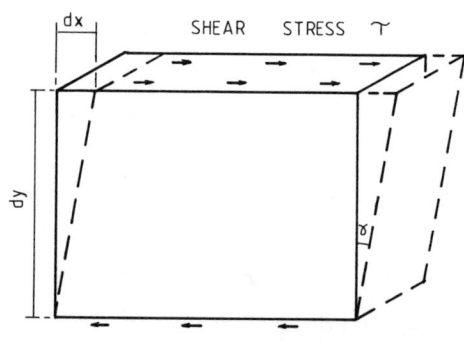

FIG. 2.3. Shear strain.

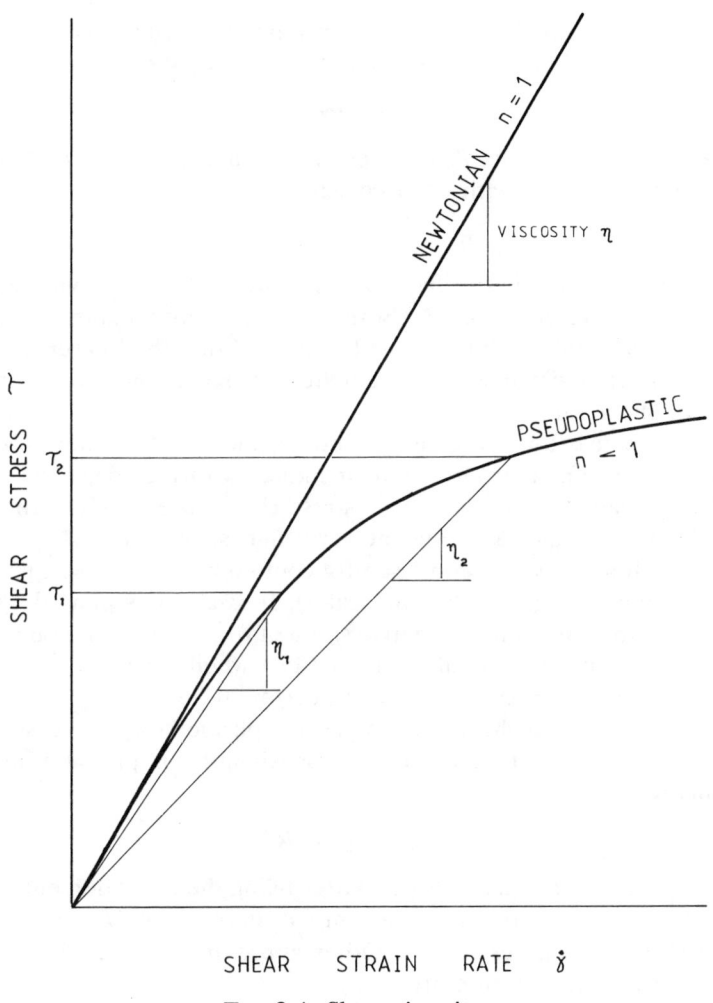

Fig. 2.4. Shear viscosity.

of stress and strain (shear) rate, but dependent on temperature. Many molten polymers are to a lesser or greater extent non-Newtonian, i.e. shear rate is no longer proportional to shear stress, the viscosity usually decreasing with increasing shear stress or shear rate, when they are said to be 'pseudoplastic' (Fig. 2.4). The definition of viscosity (eqn (2.5)) is still used, but this is now a function of shear stress or shear rate as well as temperature. An approximation which is frequently used expresses shear stress as a simple power of shear rate:

$$\tau = K(\dot{\gamma})^n \qquad (2.6)$$

where K is a constant for a specified material and temperature. Combining eqn (2.6) with (2.5) gives:

$$\eta \propto (\dot{\gamma})^{n-1} \propto (\tau)^{(n-1)/n} \qquad (2.7)$$

where for a pseudoplastic fluid, $n < 1$, and evidently Newtonian behaviour is equivalent to $n = 1$. Both of these approximations facilitate mathematical analysis, but it must be realised that the 'power-law' has little physical basis; it is a convenient approximation over a limited range of shear rates.

The index n is known as the 'pseudoplasticity index', and as values decrease from unity towards zero it indicates further departure from the Newtonian relation, i.e. shear stress rises more slowly with shear rate. Figure 2.4 indicates that at very high shear rates, shear stress tends to a limiting value and viscosity continuously decreases, giving a curve similar to Fig. 2.5 (log/log scales). If viscosity is plotted against shear rate, the curve has a negative slope of $n - 1$ (eqn (2.7)) on log/log scales, with a maximum value of -1 (45°) when $n = 0$. In both cases Newtonian flow is represented by a horizontal line.

Viscosity (at fixed shear rate for pseudoplastic flow) decreases with increase of temperature; the best approximation is probably an Arrhenius relation:

$$\eta = A \exp(E/RT) \qquad (2.8)$$

where A and E are constants, the latter being the activation energy for flow in J/mol, R is the gas constant in J/mol K and T the absolute temperature in degrees Kelvin. Other approximations used over limited ranges of temperature are:

$$\eta = B \exp(-bT) \qquad (2.9)$$

and

$$\eta = \eta_0[1 - \beta(T - T_0)] \qquad (2.10)$$

Equation (2.8) would imply that curves of $\log \eta$ v. $\log \dot{\gamma}$ were parallel, i.e. that the effect of temperature was independent of shear rate. For many real polymers this is not the case and either n must be varied with temperature or E must be a function of shear rate; this leads to

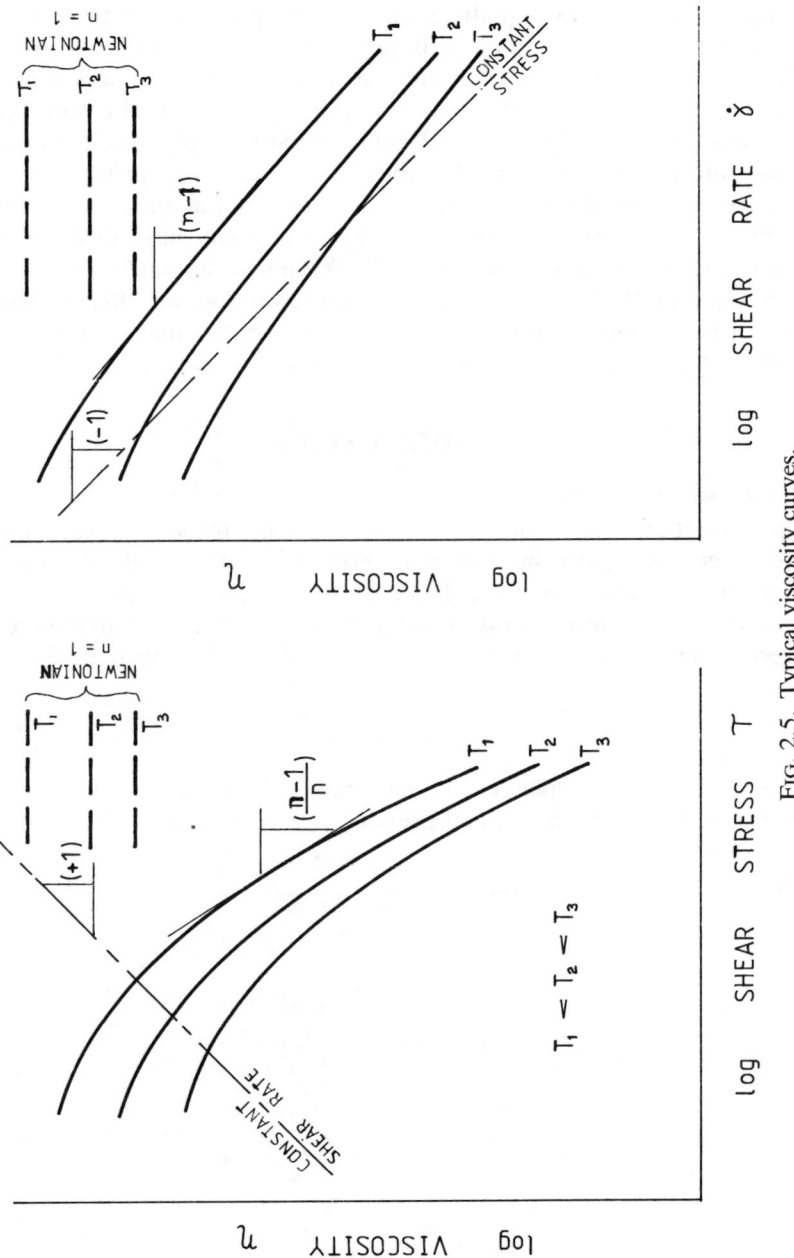

Fig. 2.5. Typical viscosity curves.

complex mathematical equations and for the present purpose, where small changes are concerned, it is sufficient in most cases to use eqns (2.7) and (2.10) with an experimental mean value at an appropriate shear rate and temperature. It is important for any useful calculation that these data are obtained, preferably in the form of viscosity versus shear rate and temperature, for the grade of material in use. These may be available on request from the raw material supplier, but to permit useful process control, the processor should have facilities to obtain his own data (see Section 7.3). A limited amount of data has been published[15,16] on the effect of pressure on viscosity; the increase at high pressures is considerable for polystyrene, but may probably be neglected for polyolefines at normal extrusion pressures.

2.2. SHEAR FLOW

Simple Shear
If a layer of polymer of uniform temperature and thickness h (Fig. 2.6) is sheared by relative motion of its boundaries $W = dx/dt$, the shear stress will be constant (by balance of forces) at all sections, and if viscosity is a material property, it will also be constant at constant stress. Then by eqn (2.5), the shear rate will also be constant:

$$\dot{\gamma} = \frac{W}{h} \qquad (2.11)$$

Note that this is independent of the relation between τ and $\dot{\gamma}$, i.e. it applies to both Newtonian and pseudoplastic behaviour.

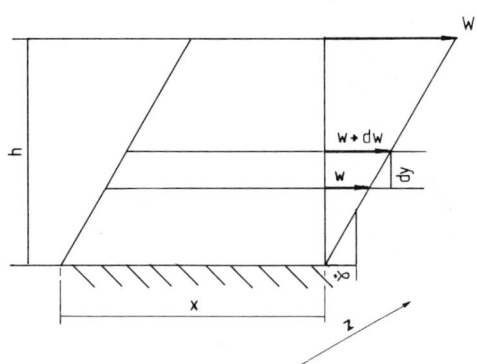

FIG. 2.6. Simple shear.

Circular Capillary (Newtonian)

The flow is considered to be fully developed, i.e. it is purely axial and the velocity profile is independent of axial position. Then the pressure may be considered uniform across a normal cross-section, and will fall in the direction of flow. Using the subscripts y, r and R for values at the corresponding radii, the shear stress and shear rate at radius y are τ_y and $\dot{\gamma}_y = dV_y/dy$, respectively. If the pressure falls from $P+dP$ to P over a length dL, then equating forces on a cylinder radius y and length dL (Fig. 2.7):

$$[(P+dP)-P]\pi y^2 = 2\tau_y \pi y \, dL \qquad (2.12)$$

Therefore

$$\tau_y = \frac{dP}{dL} \cdot \frac{y}{2} \qquad (2.13)$$

Note that:

(i) Shear stress is proportional to pressure gradient
(ii) Shear stress is proportional to radius, since dP/dL is independent of radius
(iii) Shear stress is independent of fluid properties
(iv) Shear stress at the wall $\tau_R = \dfrac{R}{2} \cdot dP/dL$.

Since pressure decreases as L increases, τ takes the opposite sign to $\dot{\gamma}$, and by definition:

$$\tau_y = -\eta \frac{dV_y}{dy} \qquad (2.14)$$

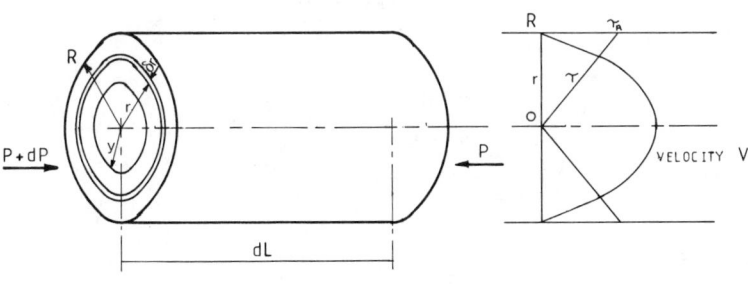

FIG. 2.7. Capillary flow.

Therefore

$$\dot{\gamma}_y \equiv \frac{dV_y}{dy} = -\frac{\tau_y}{\eta} = -\frac{dP}{dL} \cdot \frac{y}{2\eta} \qquad (2.15)$$

so that shear rate is also proportional to radius. Then:

$$-dV_y = \frac{\tau_y}{\eta} dy = \frac{dP}{dL} \cdot \frac{y \, dy}{2\eta} \qquad (2.16)$$

Integrating:

$$\int_0^r dV = -\frac{1}{2\eta} \cdot \frac{dP}{dL} \int_0^r y \, dy \qquad (2.17)$$

giving:

$$V_r - V_0 = -\frac{r^2}{4\eta} \cdot \frac{dP}{dL} \qquad (2.18)$$

Assuming the fluid adheres to the wall (no slip), $V_R = 0$. Then:

$$V_0 = \frac{R^2}{4\eta} \cdot \frac{dP}{dL} \qquad (2.19)$$

and

$$V_r = \frac{1}{4\eta} \cdot \frac{dP}{dL} (R^2 - r^2) = V_0 \left(1 - \frac{r^2}{R^2}\right) \qquad (2.20)$$

giving a paraboloidal velocity profile.

The volume flow in an annulus δr wide at radius r is:

$$dQ_r = V_r \cdot 2\pi r \, dr \qquad (2.21)$$

and total volumetric flow:

$$Q = \frac{2\pi}{4\eta} \cdot \frac{dP}{dL} \int_0^R (R^2 - r^2) r \, dr$$

$$= \frac{\pi R^4}{8\eta} \cdot \frac{dP}{dL} \qquad (2.22)$$

Rearranging:

$$dP = \frac{8\eta Q \, dL}{\pi R^4} \qquad (2.23)$$

Taking eqn (2.15) for shear rate at the capillary wall, and substituting for dP/dL from eqn (2.22):

$$\dot{\gamma}_R = -\frac{dP}{dL} \cdot \frac{R}{2\eta} = \frac{4Q}{\pi R^3} \quad (2.24)$$

which gives the wall shear rate in terms of flow rate and die dimensions.

Infinite Slit (Newtonian)

With the same assumptions as for the capillary, the infinite slit is T wide by H deep, where $T \gg H$, so that drag at the end surfaces can be neglected so that flow is one-dimensional and velocity varies in the H direction only. The subscripts are y, h and X, where $X = H/2$, for the corresponding dimensions measured from the centreline (Fig. 2.8). By force balance on a region T wide by y from the centre:

$$dPyT = \tau_y T \, dL \quad (2.25)$$

Then:

$$\frac{dP}{dL} \cdot y = \tau_y = -\eta\dot{\gamma} = -\eta \frac{dV_y}{dy} \quad (2.26)$$

Again, shear stress is proportional to distance from the centre, to the

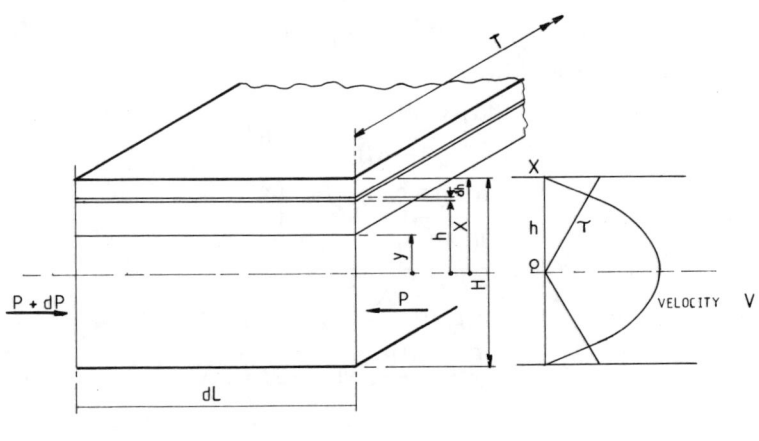

FIG. 2.8. Flow in infinite slit.

pressure gradient, and is independent of the fluid properties. Then:

$$-dV_y = \frac{\tau_y}{\eta} dy = \frac{1}{\eta} \cdot \frac{dP}{dL} y \, dy \qquad (2.27)$$

Integrating:

$$\int_0^h dV = -\frac{1}{\eta} \cdot \frac{dP}{dL} \int_0^h y \, dy \qquad (2.28)$$

giving:

$$V_h - V_0 = -\frac{h^2}{2\eta} \cdot \frac{dP}{dL} \qquad (2.29)$$

But $V_X = 0$. Then:

$$V_0 = \frac{X^2}{2\eta} \cdot \frac{dP}{dL} = \frac{H^2}{8\eta} \cdot \frac{dP}{dL} \qquad (2.30)$$

and

$$V_h = \frac{1}{2\eta} \cdot \frac{dP}{dL} (X^2 - h^2) = \frac{1}{8\eta} \cdot \frac{dP}{dL} (H^2 - 4h^2) \qquad (2.31)$$

giving a parabolic velocity profile (uniform in T direction).

The volume flow in a lamina δh thick at h from the centre is:

$$dQ_h = V_h T \, dh \qquad (2.32)$$

and total volumetric flow:

$$Q = 2 \int_0^X \frac{T}{2\eta} \cdot \frac{dP}{dL} (X^2 - h^2) \, dh$$

$$= \frac{T}{\eta} \cdot \frac{dP}{dL} \left[X^2 h - \frac{h^3}{3} \right]_0^X$$

$$= \frac{2T}{3\eta} \cdot \frac{dP}{dL} X^3$$

$$= \frac{TH^3}{12\eta} \cdot \frac{dP}{dL}, \text{ where } H = 2X \qquad (2.33)$$

Rearranging:

$$dP = \frac{12\eta Q \, dL}{TH^3} \qquad (2.34)$$

Wall shear rate:

$$\dot{\gamma}_x = -\frac{X}{\eta} \cdot \frac{dP}{dL} = -\frac{H}{2\eta} \cdot \frac{dP}{dL}$$

$$= \frac{6Q}{TH^2} \qquad (2.35)$$

The results of these two cases of pressure flow, together with the corresponding results for flow of a power-law fluid, are given in Table 2.1. From eqn (2.23) it is seen that in a capillary of constant radius, and providing viscosity is not dependent on pressure, the pressure gradient dP/dL is constant and equal to P/L. Equation (2.22) may then be rewritten:

$$Q = \frac{\pi R^4}{8\eta} \cdot \frac{P}{L} \qquad (2.36)$$

or putting $K = \pi R^4/8L$ for a die of constant dimensions:

$$Q = \frac{KP}{\eta} \qquad (2.37)$$

Strain Energy in Shear

Referring to Fig. 2.6, consider a lamina δy thick parallel to the direction of shear stress and moving with velocity w. Then the shear force acting over the interface between this and an adjacent lamina having velocity $w + \delta w$ is:

$$F = \tau x z \qquad (2.38)$$

The distance over which this acts in time t is:

$$[(w + \delta w) - w]t \qquad (2.39)$$

and the work done in time t:

$$Ft[(w + \delta w) - w] = \tau x z \, \delta w t \qquad (2.40)$$

But the volume of the lamina is $xz \, \delta y$, so the work done in time t per unit volume is:

$$\frac{\tau x z \, \delta w t}{x z \, \delta y} = \tau \frac{\delta w}{\delta y} t \qquad (2.41)$$

TABLE 2.1
Summary of Flow Equations

	Shear stress at wall, τ_w (N/m^2)	Shear rate at wall, $\dot{\gamma}_w$ (s^{-1})	Ratio, V_{Max}/V_{Mean}	Pressure drop, dP (N/m^2)	Volumetric flow rate, Q (m^3/s)
NEWTONIAN $\tau = \eta\dot{\gamma}$					
Round Tube					
Radius R	$\dfrac{dP}{dL} \cdot \dfrac{R}{2}$	$\dfrac{4Q}{\pi R^3}$	2·0	$\dfrac{8\eta Q\,dL}{\pi R^4}$	$\dfrac{\pi R^4}{8\eta} \cdot \dfrac{dP}{dL}$
'Infinite' slit					
Width T					
Depth H	$\dfrac{dP}{dL} \cdot \dfrac{H}{2}$	$\dfrac{6Q}{TH^2}$	1·5	$\dfrac{12\eta Q\,dL}{TH^3}$	$\dfrac{TH^3}{12\eta} \cdot \dfrac{dP}{dL}$
POWER LAW $\tau \propto (\dot{\gamma})^n$ $\eta = \eta_0\left(\dfrac{\dot{\gamma}}{\dot{\gamma}_0}\right)^{n-1}$					
Round Tube					
Radius R	$\dfrac{dP}{dL} \cdot \dfrac{R}{2}$	$\dfrac{3n+1}{n} \cdot \dfrac{Q}{\pi R^3}$	$\dfrac{3n+1}{n+1}$	$\dfrac{2(3n+1)}{n} \cdot \dfrac{\eta_w Q\,dL}{\pi R^4}$	$\left(\dfrac{n}{3n+1}\right) \cdot \dfrac{\pi R^4}{2\eta_w} \cdot \dfrac{dP}{dL}$
'Infinite' slit					
Width T					
Depth H	$\dfrac{dP}{dL} \cdot \dfrac{H}{2}$	$\dfrac{4n+2}{n} \cdot \dfrac{Q}{TH^2}$	$\dfrac{2n+1}{n+1}$	$\dfrac{4(2n+1)}{n} \cdot \dfrac{\eta_w Q\,dL}{TH^3}$	$\dfrac{n}{4(2n+1)} \cdot \dfrac{TH^3}{\eta_w} \cdot \dfrac{dP}{dL}$

and the strain energy per unit volume:

$$\tau \frac{\delta w}{\delta y} = \tau \dot{\gamma} \text{ N/s m}^2 \tag{2.42}$$

Substituting viscosity from eqn (2.5) gives:

Strain energy per unit volume $= \dfrac{\tau^2}{\eta} = \eta(\dot{\gamma})^2$ N/s m^2 or W/m^3 \hfill (2.43)

In the case of simple shear, substituting from eqn (2.11) gives:

Strain energy per unit volume $= \dfrac{\eta W^2}{h^2}$ W/m^3 \hfill (2.44)

or

Strain energy per unit surface area $= \dfrac{\eta W^2}{h}$ W/m^2 (in simple shear) \hfill (2.45)

For a power-law fluid, substitution of eqn (2.6) in eqn (2.42) gives:

Strain energy per unit volume $= K(\dot{\gamma})^{n+1} = K\left(\dfrac{W}{h}\right)^{n+1}$ \hfill (2.46)

In a circular capillary, from eqn (2.15):

$$\dot{\gamma}_r = -\frac{dP}{dL} \cdot \frac{r}{2\eta} \tag{2.47}$$

For a Newtonian fluid, combining eqns (2.43) and (2.47), the strain energy in an annulus r to $r + \delta r$ is:

$$\left(\frac{dP}{dL}\right)^2 \frac{r^2}{4\eta} \cdot 2\pi r L\, \delta r = \frac{\pi L}{2\eta}\left(\frac{dP}{dL}\right)^2 r^3\, \delta r \tag{2.48}$$

Total strain energy in capillary is:

$$\frac{\pi L}{2\eta}\left(\frac{dP}{dL}\right)^2 \int_0^R r^3\, dr = \frac{\pi L}{8\eta}\left(\frac{dP}{dL}\right)^2 R^4 \tag{2.49}$$

The distribution of local strain energy, given by eqn (2.48), is plotted in Fig. 2.9.

Substituting from the flow eqn (2.22):

Total strain energy $= \dfrac{8\eta L Q^2}{\pi R^4} = Q \cdot dP$ (in circular capillary) \hfill (2.50)

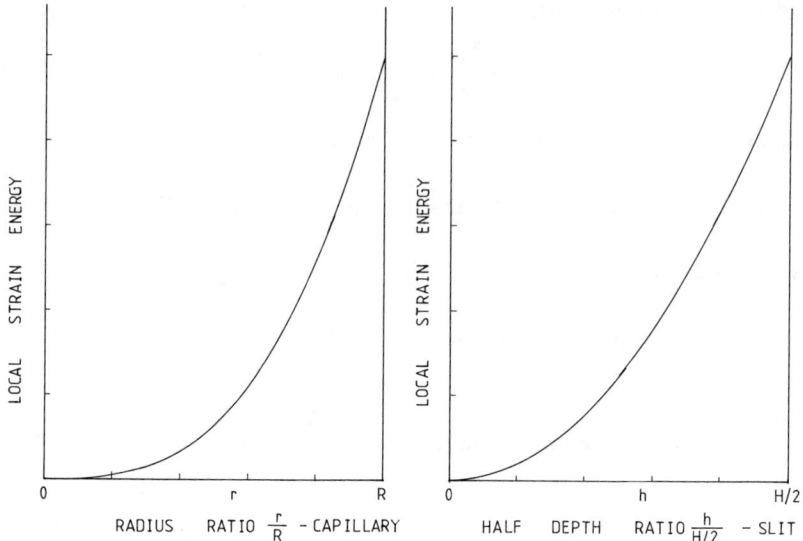

Fig. 2.9. Transverse variation of local strain energy.

where dP is the pressure drop in length L.

Similarly in an infinite slit, from eqn (2.26):

$$\dot{\gamma}_h = -\frac{dP}{dL}\frac{h}{\eta} \qquad (2.51)$$

Strain energy in lamina h to $h + \delta h$:

$$\left(\frac{dP}{dL}\right)^2 \frac{h^2}{\eta} \cdot TL\, \delta h = \frac{TL}{\eta}\left(\frac{dP}{dL}\right)^2 h^2\, \delta h \qquad (2.52)$$

Total strain energy is:

$$\frac{2TL}{\eta}\left(\frac{dP}{dL}\right)^2 \int_0^X h^2\, dh = \frac{2TL}{3\eta}\left(\frac{dP}{dL}\right)^2 X^3 = \frac{TL}{12\eta}\left(\frac{dP}{dL}\right)^2 H^3 \qquad (2.53)$$

since $X = H/2$ as shown in Fig. 2.8. Substituting from flow eqn (2.33):

$$\text{Total strain energy} = \frac{12\eta Q^2 L}{TH^3} = Q \cdot dP \text{ (in infinite slit)} \qquad (2.54)$$

where dP is the pressure drop in length L.

The distribution of local strain energy, given by eqn (2.52), is also plotted in Fig. 2.9.

Strain energy and viscous heating can be derived in more complex cases where an analytical expression is available for shear rate (or shear stress) and (in non-isothermal or non-Newtonian cases) for viscosity. Combined drag and pressure flow is studied in Section 6.5.

2.3. EXTENSIONAL FLOW AND ELASTIC EFFECTS

Extensional Flow

Extensional flow occurs during drawing from the die, under conditions of free boundaries and constant extensional load (i.e. increasing stress), and usually significant cooling. Extensional flow also occurs in adaptors and dies in the transitions from extruder cross-section and in converging to the die lips. In this case the geometric boundaries are defined and flow is approximately isothermal. However, it is accompanied by pressure (capillary) flow due to boundary friction, the energy for both of which coming from the longitudinal pressure gradient. By analogy with eqn (2.43), the strain energy is $\lambda(\dot{\varepsilon})^2$ per unit volume, and unless the transition is very rapid (when non-axial flow may occur—p. 192), the elongational strain rates are modest. The strain energy is stored elastically within the polymer and on release from the constraint of the die much of this energy is released by elastic recovery as the extrudate swells and retracts in length. However, as the melt is only partially elastic, the balance of energy is transformed into heat by viscous dissipation within the die; the longer the residence time in the die, e.g. by increasing die length, the greater the dissipation and the less the die swell. Any extensional flow within the single-screw channel is minimal and is masked by drag and pressure shear flows.

Elastic Effects

These have been discussed above in terms of die swell in relation to elongational flow. Elasticity may be seen when the extruder has just stopped or is extruding slowly, by pulling sharply on the extrudate and then releasing it, when it will (partly) spring back. Polymer melts are also partially elastic in shear flow (e.g. partial recovery in a cone and plate viscometer), in which shear strain energy is stored in the polymer during acceleration (increase of shear rate). This might be seen as a momentary peak in screw torque during start-up, but in practice would be difficult to distinguish from mechanical inertia. As with elongational flow, the elastic strain energy is, with time, converted into viscous dissipation, and so does not appear when the machine is slowed or stopped. The most significant effect is probably when polymer enters the flight clearance from the channel and is suddenly subjected to much greater shear rate; if the melt is non-Newtonian and appreciably elastic, the viscosity will not fall instantaneously to that corresponding to the higher shear rate, giving a transient high stress and accompanying shear heating. Conversely, at the exit from the flight clearance, the viscosity will not instantaneously rise to match the sudden reduction of shear rate; these effects, spaced by an interval of 0·1 s or less, may not be identical and mutually cancelling.

2.4. MEASUREMENT OF VISCOSITY AND ELASTICITY

A detailed survey of rheometers and viscometers would be inappropriate; only the main requirements and limitations will be discussed. Measurement of (shear) viscosity is required at the temperatures and shear rates encountered in the extruder and die, and ready means of checking that these characteristics are maintained (Section 8.4). The processing temperature range for a particular polymer and product is usually known, but data over a range of at least 50°C are desirable so that the temperature coefficient of viscosity (e.g. β in eqn (2.10)) can be reliably determined. Obviously the viscometer must be capable of attaining, and accurately holding, the required temperatures, and be mechanically robust enough to withstand the pressures and torques required by high viscosities. Shear rates in the screw channel are normally in the range $10-100 \, s^{-1}$, which approximates to that of the capillary Melt Flow Indexer (see BS 2782:720A). However, both in the die and over the flight clearance the shear rates may be as high as $2000-3000 \, s^{-1}$, where the Melt Flow Rate gives little indication, owing to non-Newtonian variation of viscosity with shear rate. For similar reasons parallel plate rotational viscometers are of little use, since they represent a range of shear rates from centre to circumference and so for non-Newtonian fluids give only a (weighted) average viscosity over that shear rate range. The cone and plate viscometer gives a uniform shear rate over the radius, but must be capable of operating over a substantial speed range to cover the required shear rates. Although these instruments are capable of producing accurate research data, including elastic recovery in some cases, they are limited to rather low shear rates and stresses, partly due to slippage at the surfaces. The capillary viscometer involves a range of shear rates simultaneously across the capillary radius (eqn (2.15)) but reproduces the shear conditions in simple dies; data are normally produced in terms of the shear rate at the wall (the maximum value) as represented by eqn (2.24) (apparent shear rate) or the pseudoplastic 'true' shear rate (Table 2.1). The corresponding apparent and 'true' viscosities are determined to satisfy eqn (2.5). In capillary flow, either viscosity and corresponding shear rate may be used, as long as the same equation is used in calculating the production die as was used in determining the

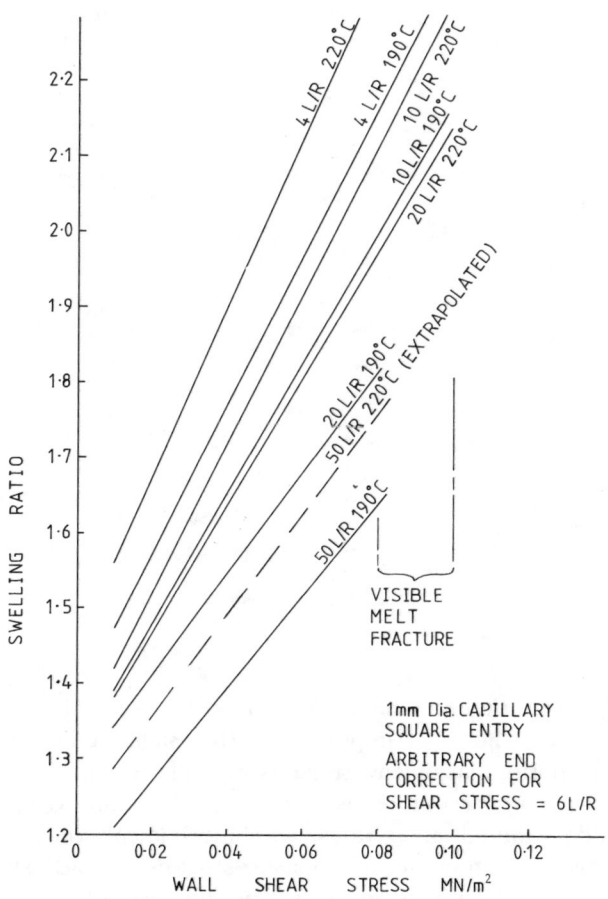

FIG. 2.10. Swelling ratio—GP polystyrene.

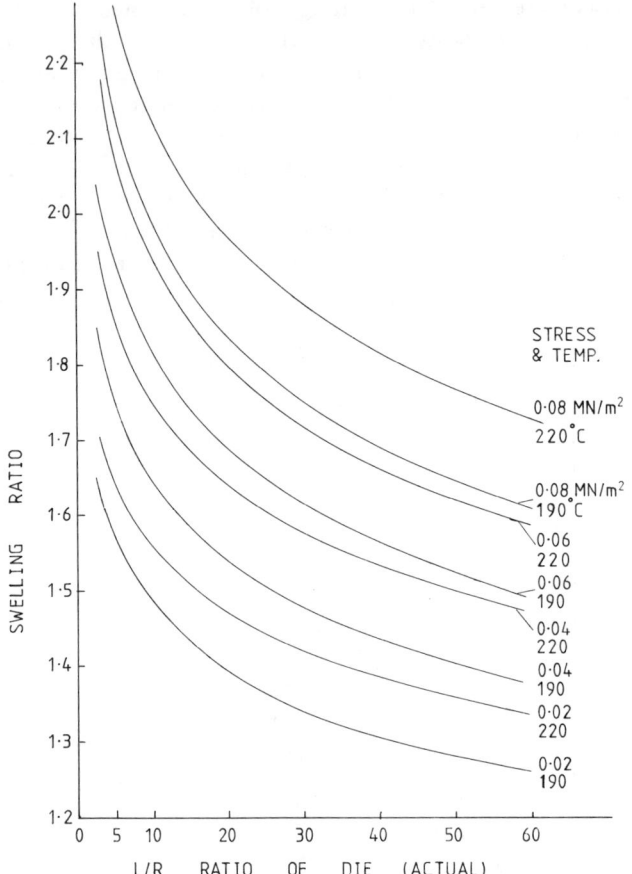

Fig. 2.11. Swelling ratio—polystyrene.

viscosity from experimental data. A number of corrections may be required, of which the most important is the Bagley end-correction, which requires preliminary measurements with dies of at least two L/D ratios. The capillary viscometer is capable of operating over a wide range of predetermined shear rates, with mechanical constant speed drive to the piston, or predetermined shear stresses with adjustable gas pressure. The upper limit is generally set by the onset of melt fracture, which although it nominally occurs at constant shear stress, may cause severe distortion in a square-entry viscometer die which would hardly be apparent in a well-streamlined production die. The viscometer may also be used for determining swell ratios over a range of temperatures and shear stresses, as illustrated by the author's data in Figs 2.10 and 2.11 for a sample of crystal polystyrene.

Chapter 3

THERMAL AND ENERGY PROPERTIES IN PROCESSING

3.1. THERMAL PROPERTIES

The specific volume of polymers in the solid state increases approximately linearly with temperature, so that a coefficient of (cubical) expansion may usefully be used. In the melt state, expansion is also approximately linear with temperature, but usually at a higher rate than in the solid. Semi-crystalline polymers including polyolefines, nylons, polyoxymethylene (POM) and polyethylene terephthalate (PET) also show a fairly sudden increase in specific volume just below the melting point, due to disordering of the crystalline regions to the completely amorphous state in the melt; the magnitude of this increase will depend on the degree of crystallinity in the solid polymer, which, especially with LD polyethylene and polypropylene, is a function, *inter alia*, of the preceding cooling rate. The degree of crystallinity is normally unknown, being only one factor affecting solid density. However, this does not affect density ρ (kg/m^3) and specific volume in the melt, though data on the latter are not always available. The melt density of LD polyethylene at 190°C, for example, is approximately 760 kg/m^3 (0·76 g/cm^3), so it is important that the appropriate value is used in converting mass to volumetric flow and vice versa at processing temperature, Appendix E (Fig. E1).

Specific heat C_p (J/kg K) varies little with temperature in the solid state but is somewhat higher in the melt state. As mentioned in Section 6.1, a constant specific heat is adequate in calculating the effects, e.g. on energy balance, of changes in operating conditions. However, in melting semi-crystalline polymers additional heat is absorbed near the melting point in melting the crystallites, giving rise to a sharp peak in the curve of specific heat; it is thus preferable to use the integrated form which is internal energy (sensible heat) in J/kg versus temperature (Fig. E2), where total energy I at the die is required. This has similar shape to the specific volume curve, and similarly the lower end depends on initial crystallinity in the solid state. At constant pressure, as proposed in Section 6.1, changes in enthalpy may be regarded as equal to corresponding changes in internal energy.

Polymers in both solid and molten states are good thermal insulators, i.e. values of thermal conductivity k (W/m K) are very low in comparison with metals used in construction of extruders and dies. Thermal conductivity of insulators is difficult to measure with accuracy, especially in liquids, and published data on conductivity of polymers at melt temperatures are scarce and conflicting. Some previously unpublished data[17] are presented in Figs E4 and 5.

As discussed below, the low values of conductivity mean that heat conduction within the polymer may be neglected in comparison with conduction in the metal barrel wall, e.g. longitudinally from zone to zone. Its main significance is in limiting heat transfer, e.g. from the barrel heaters in the melting process (Sections 5.9 and 7.1) and dissipating heat generated by internal shear heating. Where such heat flows take place, substantial temperature gradients and temperature differences will be established, especially in thick sections, and these will be slow to disappear.

Thermal diffusivity $\alpha = k/\rho C_p$ (m^2/s) is significant in transient heat conduction and may be more simply measured directly from (transient) cooling experiments than by calculation from the constituent values. However, doubt has been cast on its significance where the latter are varying. It is used in Section 6.5 in relation to dissipation within the metal barrel of viscous heating from the flight clearance. Since much of the existing data are in f.p.s. or c.g.s. units, conversion factors and some typical values are given in Table 3.1.

3.2. THERMAL CONDUCTION

Steady thermal conduction in one dimension is given by the Fourier equation:

$$q = -kA \frac{dT}{dx} \quad (3.1)$$

TABLE 3.1
Thermal Properties—Conversion Factors and Typical Values

Property	SI	c.g.s	f.p.s.
Density	1000 kg/m^3	1 g/cm^3	$\begin{cases} 62\cdot35 \text{ lb/ft}^3 \\ 0\cdot0361 \text{ lb/in}^3 \end{cases}$
Viscosity	1 N s/m^2	10 poise	2420 lb/h ft
Specific heat	1 MJ/kg K	$238\cdot9 \text{ cal/g°C}$	$\begin{cases} 1 \text{ BTU/lb°F} \\ 1 \text{ CHU/lb°C} \end{cases}$
	$4\cdot19\times10^3 \text{ J/kg K}$	1 cal/g°C	—
Thermal conductivity	1 W/m K	$0\cdot00239 \text{ cal/cm s°C}$	$\begin{cases} 6\cdot93 \text{ BTU in/ft}^2\text{h°F} \\ 0\cdot5775 \text{ BTU ft/ft}^2\text{h°F} \end{cases}$
	$0\cdot0419 \text{ W/m K}$	$1\times10^{-4} \text{ cal/cm s°C}$	—
Heat transfer coefficient	$1 \text{ W/m}^2 \text{ K}$	$0\cdot239\times10^{-4} \text{ cal/cm}^2 \text{ s°C}$	$\begin{cases} 0\cdot176 \text{ BTU/ft}^2\text{h°F} \\ 0\cdot176 \text{ CHU/ft}^2\text{h°C} \end{cases}$
Work	1 J	$0\cdot239 \text{ g cal}$	$\begin{cases} 9\cdot48\times10^{-4} \text{ BTU} \\ 5\cdot25\times10^{-4} \text{ CHU} \end{cases}$

Property	Air (at NTP)	Water (at 15°C)	Steel	LD polyethylene
Density (kg/m^3)	1·224	1000	7800	760 (at 180°C.)
Viscosity (N s/m^2)	$1\cdot80\times10^{-5}$	$1\cdot12\times10^{-3}$	—	10^2–10^3
Specific heat (J/kg K)	1005	4190	503	2300–2850
Thermal conductivity (W/m K)	0·0255	0·585	50 (Copper 400)	0·33–0·50
Prandtl No. $(C_p \eta/k)$	0·70	8·0	—	5×10^6
Latent heat (J/kg)	—	$22\cdot6\times10^5$ (steam)	—	$1\cdot3$–$1\cdot7\times10^5$
Thermal diffusivity (m^2/s)			$1\cdot3\times10^{-5}$	$1\cdot5$–3×10^{-7}

where q is the rate of heat flow (W), k is the thermal conductivity (W/m K), A is the cross-section through which heat is conducted (m^2) and $-dT/dx$ is the temperature gradient (K/m) in the direction of heat flow, the negative sign indicating that heat flows in the direction of falling temperature. This equation illustrates that heat flow is necessarily accompanied by, and is proportional to, a temperature gradient. It is also proportional to the conductivity, so that the low thermal conductivity of polymers implies small heat flow and/or large temperature gradients. At a given heat flow, the temperature difference increases with distance (thickness) through which heat is conducted, so that the temperature difference may be appreciable through a thick extruder barrel. An increase of thickness at constant temperature difference implies a smaller heat flow—hence cooling passages are kept close to the mould surface in injection and blow moulds. Where heat is conducted through layers of two or more materials of different conductivity, the heat flow is constant so that:

$$q = -k_1 A_1 \frac{dT_1}{dx_1} = -k_2 A_2 \frac{dT_2}{dx_2} \qquad (3.2)$$

and if the cross-sections are the same, the *heat flux*:

$$\frac{q}{A} = -k_1 \frac{dT_1}{dx_1} = -k_2 \frac{dT_2}{dx_2} \qquad (3.3)$$

Then:

$$\frac{dT_1/dx_1}{dT_2/dx_2} = \frac{k_2}{k_1} \qquad (3.4)$$

that is, the temperature gradients are *inversely* proportional to the conductivities. Taking the temperature of the interface as T', this can be rewritten as:

$$\frac{T_1 - T'}{T' - T_2} = \frac{k_2 \, dx_1}{k_1 \, dx_2} \qquad (3.5)$$

so that the total temperature difference $T_1 - T_2$ is subdivided in proportion to the respective thicknesses and inversely as the conductivities.

If heat is conducted from barrel heaters through a 30 mm thick barrel (k for steel 50 W/m K) and a 5 mm thickness of LD polyethylene ($k = 0.5$ W/m K) with an overall temperature difference of 100°C, the temperature differences across barrel and polymer will be 5.7 and 94.3°C respectively. Heat conduction in two or three dimensions generally requires digital computation, which is rarely justified in steady-state extrusion. However, in the particular case of radial conduction through a cylinder of length L, e.g. an extruder barrel, the heat flow q is constant but the area is proportional to radius and hence the temperature gradient must increase towards the centre. Integration of eqn (3.1) between inner and outer radii R_1 and R_2 gives:

$$q = \frac{2\pi kL(T_2 - T_1)}{\log_e (R_2/R_1)} \qquad (3.6)$$

3.3. NON-ISOTHERMAL FLOW AND HEAT TRANSFER

Velocity Distribution

In Chapter 2 it was shown (eqn (2.13)) that in capillary flow, shear stress increases in direct proportion to radius, from zero at the centre to a maximum at the wall, and independent of the fluid properties. In isothermal flow of a Newtonian fluid, viscosity is constant and, by eqn (2.5), shear rate is similarly proportional to radius. When this is integrated to give velocities, the distribution of velocity is parabolic (strictly, paraboloidal—see mean velocity in Table 2.1). If the fluid is heated from the wall eqn (3.1) indicates a (downward) temperature gradient towards the centre, and thus viscosity will be lower at the wall

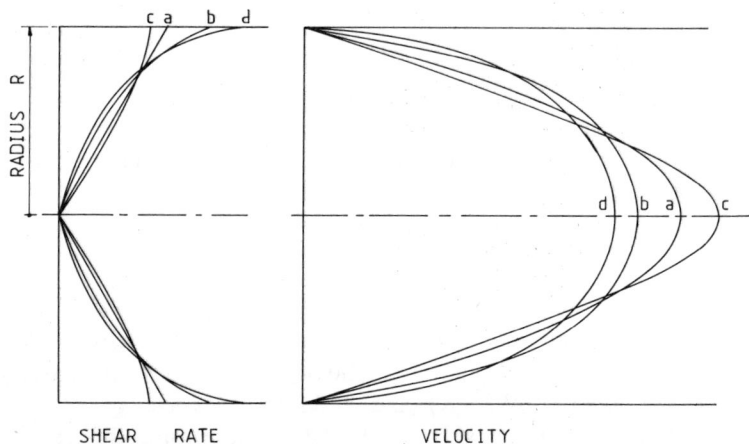

FIG. 3.1. Effects of viscosity variation on velocity distribution. a, isothermal Newtonian; b, heating Newtonian; c, cooling Newtonian; d, heating pseudoplastic.

than in the centre. Thus the shear rate will be increased near the wall and decreased near the centre in comparison with the isothermal case, giving a flattened profile—b in Fig. 3.1 compared with the parabolic curve a. If the fluid is cooled, the temperature and viscosity gradients will be reversed, leading to a lower shear rate at the wall and a more peaked profile (Fig. 3.1 curve c). The curves are drawn in Fig 3.1 to indicate approximately the same mean velocity and volumetric flow rate. In isothermal pseudoplastic flow, the viscosity decreases with increasing shear stress, again giving higher shear rates near the wall and a velocity profile similar to curve b (Fig. 3.1). Heating and cooling of a pseudoplastic fluid again modifies the isothermal velocity profile, e.g. heating will give a profile similar to curve d (Fig. 3.1), which in the extreme tends towards 'plug' or 'rod-like' flow. It is these interactions between temperature and velocity profiles which complicate heat transfer calculations (see below).

Surface Heat Transfer

Where heat is transferred through a surface to or from a fluid, an additional thermal resistance occurs. The effective thickness cannot be determined, so surface heat transfer is usually related to a temperature difference between fluid and solid surface, instead of to a temperature gradient, and the corresponding heat flow is given by:

$$q = hA \, dT \tag{3.7}$$

where dT is the temperature difference and h† is a surface coefficient of heat transfer in W/m² K. Evidently where temperature gradients in the solid are significant, eqns (3.1) and (3.7) may be combined:

$$q = \frac{kA}{x}(T_2 - T') = hA(T' - T_1) \tag{3.8}$$

For heat transfer to cooling water (usually in turbulent flow), the reader is referred to standard works on heat transfer[18,19] for determination of heat transfer coefficients and, where fluid or metal temperatures vary, e.g. from temperature rise between inlet and outlet, for determination of the appropriate mean temperature difference. Reynolds number for flow of polymers in extruders is always much less than 2100, so flow is always laminar. Heat transfer to or from a

---

† The symbol h is conventional, but must be distinguished from screw channel depth in Chapters 5 and 6.

Newtonian fluid is given in terms of the dimensionless Nusselt number hD/k where h is based on arithmetic average temperature difference and D is the diameter of a capillary.

The mass flow rate w† and length of the capillary L are combined in a second dimensionless group, the Graetz number wC_p/kL. For values of Graetz number greater than 10, the theoretical curve is closely approximated by:

$$\frac{hD}{k} = 1 \cdot 75 \left(\frac{wC_p}{kL}\right)^{\frac{1}{3}} \tag{3.9}$$

For a wide slit, this equation may be used, where the thickness H of the slit is equal to $D/2$ and the mass flow rate w is through a width πH, i.e. the same mass velocity (kg/s m^2) as in a capillary of diameter D.

For molten polymers, several corrections are necessary; radial temperature gradients modify the velocity profile from the parabolic form assumed for isothermal Newtonian flow, as does non-Newtonian behaviour (see p. 28). Heat generation due to shear heating and thermal expansion may also be significant; many workers have examined this problem including Forrest and Wilkinson[23] and Abedi.[24] Figure 3.2 which is reproduced from Abedi's work shows both computed and experimental results to give an approximate indication of the range of values to be expected. The reader is referred to the original work for further detail, including the definitions of the variables. The differences in heat transfer between heating (curve 3 of Fig. 3.2) and cooling (curve 2) are primarily due to the effects of coupled temperature and velocity profiles. Curves 4 and 1 show the additional effects of thermal expansion and viscous heating, which according to Abedi tend to oppose each other.

In Fig. 2.9 the variation of strain energy is shown for a Newtonian fluid due to pressure flow in two simple geometries. This strain energy is converted into viscous heating, leading to local temperature increases. If the density is ρ and specific heat C_p, the rate of local temperature rise in the absence of conduction will be (for unit volume):

$$\frac{dT}{dt} = \frac{\eta(\dot{\gamma})^2}{\rho C_p} \tag{3.10}$$

As mentioned above, in non-isothermal and non-Newtonian condi-

† The symbol w is conventional, but must be distinguished from local downstream velocity in Chapters 5 and 6.

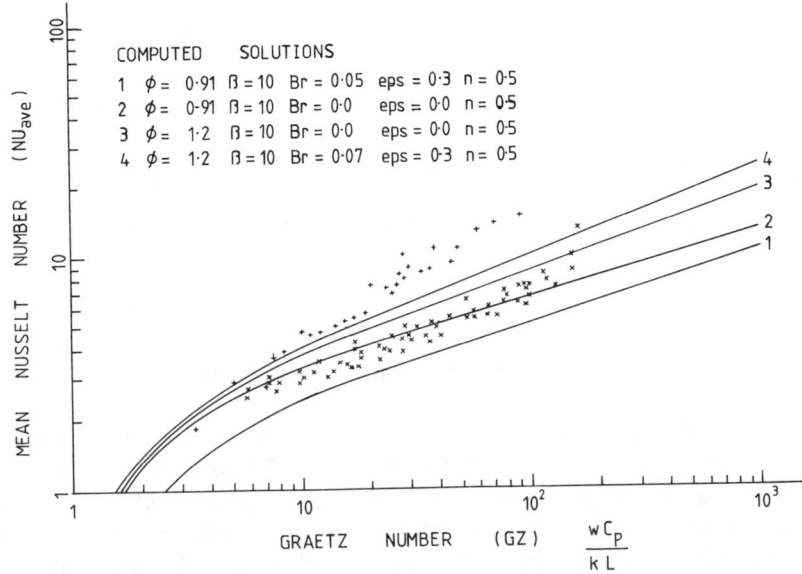

FIG. 3.2. Convective heat transfer to/from polymer melts. +, published experimental data (heating); ×, Abedi's experimental data (cooling).

tions, the shear stress/shear rate relations (eqns (2.15) and (2.26)) will be modified and viscosity is no longer a constant. Thus, aside from changes in the total strain energy, the distributions will change, and consequently so will the local temperature differences—which may persist to the end of the screw. These differences are reduced by transient thermal conduction within the polymer, whose rate is controlled by the thermal diffusivity (p. 25); the low value of thermal conductivity of polymer melts implies that heat diffusion and elimination of temperature differences will be slow.

3.4. MIXING

In flow at high velocity of low viscosity fluids (Reynolds number >2100), turbulence gives rapid and intimate mixing of the stream. In laminar flow, kinetic energy effects are negligible and as seen in Chapter 2, adjacent layers slide relative to each other without interpenetration. Thus mixing between one location and another, which

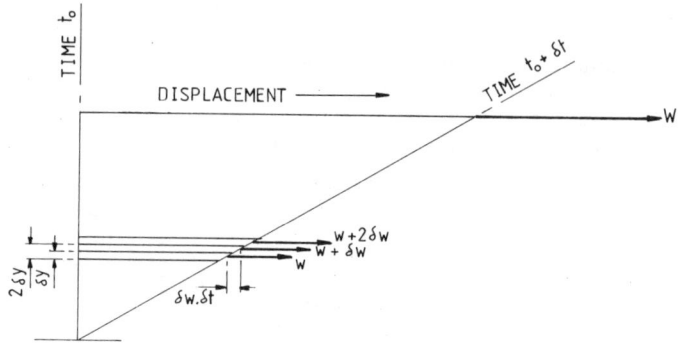

Fig. 3.3. Distributive mixing in shear flow.

may initially have different compositions, is by this relative motion transferring part of the material from one location to another, so tending to mix the material from the two locations and achieve a more uniform composition. This is known as distributive mixing and is characterised by how far two fluid particles, initially adjacent, are separated by the mixing process. Referring to Fig. 3.3, which illustrates a small element of material subjected to uniform simple shear in one direction, points in two laminae (layers) δy thick are adjacent at time t_0. If their velocities in the direction of shear are w and $w + \delta w$, respectively, their *relative* movement at time $t_0 + \delta t$ will be:

$$[(w + \delta w) - w](t_0 + \delta t - t_0) \qquad (3.11)$$

and the total shear strain will be:

$$\frac{\delta w}{\delta y}(t_0 + \delta t - t_0) \qquad (3.12)$$

If the shear stress is uniform, the relative movement of laminae 2 δy apart will be:

$$[(w + 2\,\delta w) - w](t_0 + \delta t - t_0) \qquad (3.13)$$

and the shear strain will be:

$$\frac{2\,\delta w}{2\,\delta y}(t_0 + \delta t - t_0) \qquad (3.14)$$

Thus evidently total shear strain is an appropriate measure of mixing, taking account of both the relative movement and the initial separation of the particles. As shown by eqns (3.12) and (3.14), total shear strain is given by the product of shear rate and the time for which it acts:

$$\frac{\delta w}{\delta y} \cdot \delta t = \dot{\gamma}_y \, \delta t \qquad (3.15)$$

and this is the measure of distributive mixing in laminar flow; if motion occurs in more than one direction, then at any location the shear strain is the time of action multiplied by the vector sum of (local) shear rates. This mechanism of mass transfer not only promotes uniformity of composition but also of temperature, assisting the thermal mechanisms of heat conduction or diffusion. However, in the extruder screw channel the flow velocities are primarily parallel to the cylindrical surface of the barrel, whereas conduction is primarily radial; the former promotes end-to-end and side-to-side uniformity while the latter assists temperature uniformity in a radial direction. The exception is at the sides of the channel (Fig. 5.8) where continuity requires a degree of radial flow.

The foregoing analysis refers to fluids, which deform continuously under a finite (shear) stress. If solid particles, e.g. fillers or pigments are present, they will deform initially and elastically under stress, and recover their original dimensions on removal of stress. To effect mixing of solids or Bingham fluids having a finite yield stress, it is necessary to exceed this stress and cause permanent deformation or fracture, before distributive mixing can occur. In this *dispersive* mixing, it is the stress which must exceed a minimum value for mixing to occur, and this minimum is usually not achieved in the single screw extruder, except possibly over the flight clearance, and then only for a short time interval.

Chapter 4

PRACTICAL EXTRUSION PROCESSES

This chapter makes a brief survey of the common applications of single-screw extruders and how and why these machines differ from other types. It also attempts an explanation of why the requirements of these applications differ, at least in degree, so that when the implications of the general theory in Chapters 5 and 6 are applied to practice in Chapters 7 and 8, it is clear why different materials, processes and products may require different operational and control strategies. These requirements must necessarily be qualitative, and users will no doubt be able to give examples where practice appears to contradict statements made here; the author's hope is that his deductions in later chapters will be helpful in the majority of cases and in the remainder will provide a guide to the right methods and criteria, based on sound analysis.

The single-screw extruder is primarily a drag pump, suitable for working with highly viscous fluids and capable of operation at the high pressures and temperatures consequently required for processing high molecular weight polymers in their thermoplastic state. As explained in Section 5.2, the drag mechanism leads to an output which is more or less influenced by back pressure, and Section 6.6 shows that this also leads to changes in energy balance, so the interdependence of variables is a vital aspect of performance. The single-screw machine is also mechanically simple and robust, permitting high energy inputs at relatively low speeds. The usual screw configuration having 20 or more turns with a pitch similar to diameter gives a long slender machine in which substantial longitudinal temperature gradients can be maintained and controlled, and an appreciable residence time permitting a degree of end-to-end mixing. In addition to this distributive mixing, high shear stresses, especially over the flight tip, give a degree of dispersive mixing for breaking up solid agglomerates, e.g. pigments. As explained in Section 5.2, the complex flow within a relatively narrow channel leads to side-to-side mixing and a fair uniformity of composition and temperature. Coincidentally, the single screw is also an effective conveyor of particulate solids, at approximately the rate required by the melt section, so that gravity feeding is usually sufficient.

The drag mechanism also causes internal shearing of the viscous material being pumped, leading to additional power consumption and temperature rise in the polymer. This inefficiency as a pump may be utilised beneficially by assisting external heating in melting the polymer and in heating it internally rather than by conduction, with consequent reduction of internal temperature gradients.

The single-screw extruder may be used as a continuous stirred reactor for highly viscous materials, e.g. polymerisation, intentional degradation, etc., though differing viscosities of components and product may lead to slip and pumping problems. The end-to-end mixing mentioned previously may also lead to a wide distribution of residence times, undesirable in chemical reaction. The extruder performs a number of other functions, and these are used either principally or incidentally in a wide range of specialised applications. The extruder is essentially suitable for continuous operation; however, it is widely used in many blow moulding machines and most injection moulding machines on an intermittent basis. It performs surprisingly well in this mode, though additional factors to those presented in this book must be taken into account; the former are not readily represented in analytical terms.

The single-screw extruder is thus highly suitable for continuously processing a wide range of synthetic thermoplastic polymers into an equally wide range of finished or semi-finished products. Table 4.1 lists the principal thermoplastics used in extrusion, together with the standard abbreviations.

A large number of 'specialty' polymers are extruded for specific applications; these include some fluoro-polymers, but polytetrafluoroethylene (PTFE) is usually ram extruded because of volume changes and decomposition at high temperatures and extremely high viscosity at lower temperatures. In addition, many raw and unvulcanised rubbers are processed on single-screw extruders and in special cases, e.g. injection and dough moulding, linear prepolymers of phenolic and

TABLE 4.1
Thermoplastics Used in Extrusion

Abbreviation	Base polymer
	AMORPHOUS
ABS	acrylonitrile–butadiene–styrene
PMMA	poly methylmethacrylate
CAB	cellulose acetate butyrate
PC	polycarbonate
PS	polystyrene
PVAC	polyvinyl acetate
PVAL	polyvinyl alcohol
SAN	styrene–acrylonitrile
UPVC	unplasticised (rigid) polyvinyl chloride
PPVC	plasticised (flexible) polyvinyl chloride
HIPS	high impact polystyrene (rubber toughened)
	SEMI-CRYSTALLINE
POM	polyoxymethylene; polyformaldehyde (polyacetal)
EVAC	ethylene vinyl acetate copolymer
PA	polyamide (nylon)
PETP	polyethylene terephthalate (polyester)
PBTP	polybutylene terephthalate
LDPE	low density polyethylene
LLDPE	linear low density polyethylene
HDPE	high density polyethylene
PP	polypropylene (homo- and co-polymers)
TPX[a]	poly 4 methyl pentene 1
PTFE	polytetrafluoroethylene (ram extrusion only)

Based on BS 3502: Pt 1: 1978, with additions.
[a] Trade name of Imperial Chemical Industries PLC

polyester thermosetting polymers are successfully handled. The true thermoplastics are often processed, either precompounded with stabilisers, fillers, plasticisers, pigments, processing lubricants, fire retardants, etc., or these additives are incorporated within the extruder as part of the shaping process. When subsequently crosslinked (e.g. HDPE by radiation) this usually follows the extrusion operation. Rubbers and thermosets are usually precompounded with crosslinking (vulcanising) agents as well as fillers, pigments, etc., so that heating in the extruder commences the crosslinking reaction; it is then a matter of controlling temperature, shear history, residence time, etc., so that the material remains sufficiently thermoplastic in the extruder and die, while crosslinking is completed in subsequent operations.

4.1. SHAPING PROCESSES AND THEIR REQUIREMENTS

Solid Sections

The first of the shaping processes is for solid sections, e.g. rod, strip, profiles and sections. These are used as stock for subsequent machining, beading, gaskets, structural sections (angles, channels, etc.), rainwater guttering, curtain rail, lighting diffusers and many other purposes. Shaping is principally by the melt die, in which uniform flow rate and uniform die-swell (elastic recovery) are pre-eminent. This involves uniform melt and metal temperatures, and control of land length and shear history. Drawdown, if any, is usually no more than to the extent of negativing die swell, and cooling is usually in air or waterbath while supported on a band conveyor or rollers. Sizing is usually only for fine control of shape or dimensions and is frequently omitted. Elastic memory in the polymer and flow patterns in the die may give rise to twisting or lengthwise curling, and attempts to remove these by guides or tension, rather than by correcting the cause, will usually distort the shape and/or dimensions, or give residual stress which may cause distortion in service. Except for small and simple sections, which if flexible may be coiled, these products are usually stored and transported in cut straight lengths.

Hollow Sections

These include circular tube and pipe, square tube for racking and light furniture, and complex hollow sections such as window framing, in some cases incorporating metal sections which are sheathed during extrusion (Fig. 4.1). Applications for tube and pipe include medical,

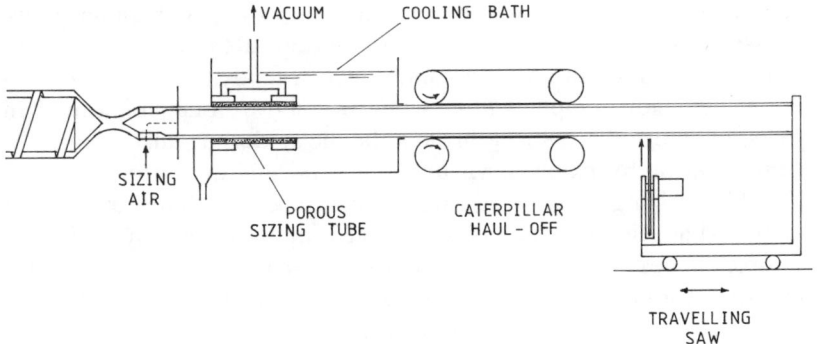

FIG. 4.1. Tube extrusion.

food, chemical, hydraulic, gas and water distribution, effluent and drainage, and conduit and sleeving, e.g. for electrical and telephone cables. Sizes range from medical tubing less than 1 mm diameter to water and drainage pipe 600 mm diameter. The die-cores or mandrels forming the internal shape may be supported on a 'spider' which divides the flow, or at the rear of a crosshead or side-entry die. Note that this also carries the thrust due to the drag of the polymer on the core. Complex sections involve difficulties in achieving uniform flow and avoiding distortion due to non-uniform elastic strain and thermal contraction. In this case, as with open sections, drawdown will be a minimum to control die-swell. However, with symmetrical sections, especially circular tube, drawdown does not distort the section, but reduces both diameter and thickness. It is often used for small diameters, permitting a larger die with lower resistance and higher throughput. Sizing during cooling is almost universal, fixing *either* the internal or the external diameter. The dimensions of the (melt) die and the relation between output and haul-off rate then determine the final thickness and consequently the unsized diameter. Precise internal diameter is required for metering and sealing, e.g. in disposable hypodermics and beer pumps, and precise external diameter is required for use with injection moulded fittings in compression and solvent jointing. In larger pressure pipes, e.g. for gas and water distribution, internal surface finish may also be important in minimising crack initiation and, with chemicals, to reduce environmental stress cracking. In both cases residual stress in the pipe should also be minimised. With large pipes, say over 300 mm diameter, the linear speed during extrusion tends to be low unless a very large extruder is used, and consequently the time for thermal degradation in the die is large. If this is minimised, e.g. with UPVC, by reducing die length, then rapid changes of cross-section occur leading to residual stress, excessive die-swell and transverse orientation of the polymer, instead of the normal longitudinal orientation which tends to occur, especially at low extrusion temperatures. Sizing may be by one or more sizing plates, by internal pressure (through the die spider) against a lubricated sizing tube or by external vacuum on a porous or perforated sizing tube. Since sizing must be accompanied by cooling, the former usually occurs while the product is submerged in a waterbath. Wherever possible, tubes are coiled in lengths of 30 m or greater to reduce subsequent jointing. Large rigid pipes are cut in straight lengths up to 20 m long, depending on transport facilities.

Uniform wall thickness (concentricity) may be achieved either by lateral movement of the spider or by adjustment of the outer die 'lip'. With large tube, the latter may be made sufficiently flexible for local thickness adjustment, using alternate 'pulling' and 'pushing' screws against a rigid ring. Such adjustment should only be used for fine control, since substantial circumferential variations, e.g. in die gap or land length, are likely to lead to non-axial flow and non-uniform swelling, and possibly to lateral pressure differences on the die mandrel, which distorts, tending to negative the adjustment. In such cases, the cause of non-uniform flow or viscosity from the extruder should be corrected. A restrictor ring with a narrow annular gap may also be used well before the die lips to remove uneven flow inherent in the design of the die or adaptor. Low velocities in the die may be used to give good surface finish and low residual strain (and swelling). The consequent low resistance permits use of long die lips which also promotes good surface and low swelling; even so, pressure may be too low for adequate melting and mixing in the extruder, and screen packs are added to increase the back pressure.

Corrugated tube may be produced in line by a shaping operation following or in place of sizing and during cooling. Reinforced tube may be extruded by passing the reinforcement through the die and extruding around it, as in wire covering, or by plaiting or winding the reinforcement onto the cooled lining tube. Further polymer (possibly of different composition) may then be wrapped or extruded over the assembly. The former, integrated process uses die pressure to force the polymer into intimate contact with the reinforcement but places limitations on the type and arrangement of reinforcement, and the use of layers of different polymers.

Wire Covering

In wire covering, which includes both electrical insulation and sheathing of cables for protection, the wire or assembly of insulated cores is pulled by the haul-off through the die, usually of the crosshead type (Fig. 4.2), where the polymer is forced round it by die pressure and into intimate contact before leaving the die lips. The polymer coating is drawn along by the wire and the thickness of the coating depends on the balance between polymer flow rate and wire speed. A typical 'single' may consist of a copper conductor 0·45 mm diameter with an insulating covering of LD polyethylene 0·22 mm thick and extrusion rates in excess of 1500 m/min have been reported. One limitation is

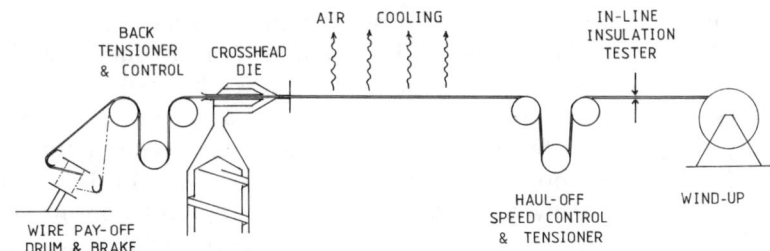

Fig. 4.2. Wire covering.

the tension in the conductor, which may cause cold drawing and loss of accuracy; the other is melt fracture of the polymer causing a rough or distorted surface of poor appearance and liable to give problems in subsequent processes, e.g. coil winding. For fine wires, high speed is desirable to give a reasonable mass flow rate; to offset consequent high die pressures, land lengths are usually minimised and greater die swell is allowed for. Note that the shear conditions in the polymer differ from those in solid rod extrusion in that the inner surface is moving with the conductor causing a drag flow, as distinct from the pressure flow in a simple capillary. Cooling is usually in air, sometimes in a festoon, but waterbaths have been used for the highest speeds. At high speeds, haul-off speed relative to extrusion speed is especially important for insulation thickness, which must be a minimum for economy and uniform for subsequent lay-up or winding. Relative speeds and/or tension between pay-off reel and various components of haul-off and wind-up are important for smooth operation, and must be maintained during acceleration at start-up. Continuity of insulation is obviously vital, but in many cases adhesion to the wire is also important to reduce breakdown due to corona discharge in high voltage AC applications and ingress of moisture in non-electrical applications, where the polymer provides corrosion protection. Thus in submarine telephone cables, where a 15 mm conductor may be covered with a 25 mm thickness of LD polyethylene (to minimise signal loss) extrusion rates will be relatively slow, with water cooling in stages, sometimes with programmed temperatures, to minimise temperature and shrinkage differentials within the insulation and maintain adhesion to the core. With such thick insulation or sheathing, haul-off tension will not be critical, but haul-off speed will determine thickness and a constant speed is essential to avoid surface lumps or ripples. These, apart from appearance and performance aspects, are likely to be detrimental in both handling and subsequent operations such as armouring.

Flat Sheet

Flat sheet extrusion, primarily in UPVC, ABS and polystyrene, is fundamentally similar to that for solid sections; however, as widths between 1 and 2 m are frequently required, uniform distribution of the polymer melt across the die and mechanical rigidity of the latter become major problems. Mechanical design is outside the scope of this book but it should be noted that a deflection of only 0·1 mm in each die lip will cause approximately 20% variation in thickness of a nominal 1 mm sheet. If the simple fishtail die is used, with a small divergence angle to give essentially axial flow, then for wide sheet the die will be massive, residence time of the polymer (re degradation) will be large and the die area subjected to polymer pressure and liable to distortion will be large. Maintenance of uniform die temperature will also be difficult, and uniform flow will not be assured. The coathanger or manifold die, in which the polymer flows transversely in a relatively wide passage before flowing axially through a short thin passage to the die lips, is more compact, has shorter residence time and smaller pressure area. Uniform heating remains difficult. Adjustable die lips and fixed or adjustable restrictor (choker) bars are commonly used to control local flow rates and thickness, but because of die swell and drawdown effects, and the substantial thermal lag in the massive die these adjustments are tedious and interact in a complex way. Thus uniformity and constancy of temperature and viscosity of the polymer at the die inlet are essential. The author recalls a 150 mm extruder feeding polypropylene to a fishtail die approximately 400 mm wide with a multistrand die-plate. When the strands were hauled off at uniform speed, the outer strands were found to be of visibly larger diameter than those in the centre. This was contrary to normal experience, where the longer flow path to the ends causes greater resistance and a lower flow rate. It was discovered that the larger pellets were of higher melt flow rate (lower viscosity) than the smaller pellets from the centre. It was concluded that the temperature of the melt leaving the extruder was higher at the outer diameter than at the centre; this, coupled with the greater residence time, gave greater degradation and lower final viscosity.

The combined effects of temperature and degradation on viscosity outweighed the greater die resistance, leading to a greater flow rate to the outer strands. As described in Section 8.1, any change in output or other operating conditions is likely to alter the spatial temperature distribution from the screw, so that correction by die adjustments would only be valid for one polymer and set of conditions. Die land lengths are usually large to improve surface finish and reduce tendency to swelling. The land length may be varied across the die to compensate for pressure differences due to different path lengths in the manifold; because of the shape of the curve of swelling ratio vs. land length (Fig. 4.3), such variation in land length may cause less variation in die swell than with small land lengths. Sheet is commonly drawn down from the die, leading to a greater reduction in thickness than in width, offsetting to some extent spatial variations in thickness and swelling at the die (due to variations in shear history). However, since the draw-in in width is mainly at the edges, these tend to be of a bulbous form, thicker than the rest of the sheet, and are thus frequently trimmed off. Lateral constraint away from the edges prevents elimination of thickness variation, and if drawdown is too small (slow haul-off) those parts of the sheet which are extruding fast will ripple to conform to the lower linear speed of the remainder. Time-variations of flow rate from the extruder will of course lead to fluctuating thickness of the finished

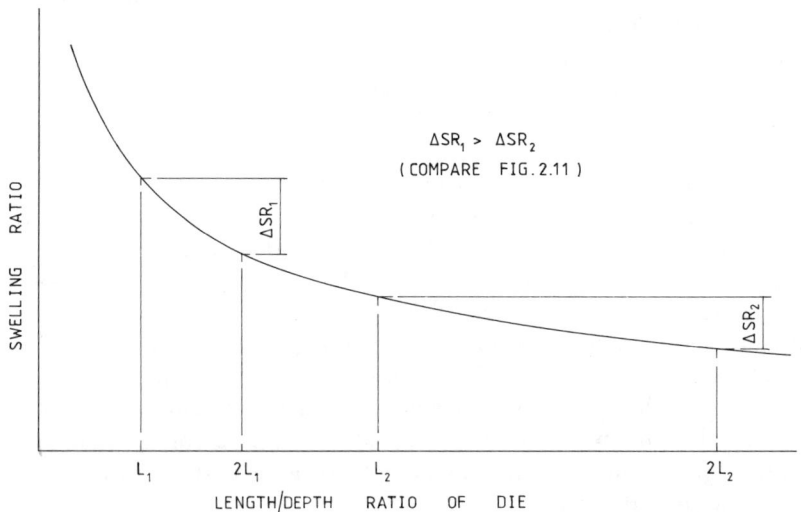

FIG. 4.3. Effect of land length adjustments on swelling ratio variations. (c.f. Fig. 2.11).

sheet despite drawdown. Haul-off speed in relation to extruder output is thus important in determining average thickness and, to some extent, thickness variations. Cooling may be by air, on one side while supported on a band conveyor for thin sheets or on both sides while supported on rollers for thicker sheets. Cooling may also be by waterbath or by passing over/between water-cooled drums. These may have polished, matt or patterned surfaces to give a corresponding surface finish to the plastic sheet, or profiled to give longitudinal corrugations, or lobed to give transverse corrugation. Multilayer sheet of different polymers, grades or colours may be produced by co-extrusion from separate extruders through a common die or through separate dies, the webs then being laminated together between rolls. In the latter case the surface temperatures and freedom from defects and degradation are important in achieving good interlayer adhesion and absence of bubbles, etc. In the former case similar velocities and viscosities are necessary where the streams meet, as well as streamlined flow paths and sufficient pressure before the die exit, to give good adhesion and avoid interpenetration of the layers. Sheet extrusion may also form the first stage of a continuous plant for vacuum-formed products, though the extruded sheet will usually be cooled and re-heated to give accurate temperature control for the vacuum forming operation.

Flat Film
Extrusion of flat film in almost all thermoplastics is in many respects similar to that of sheet, though thickness variations due to mechanical distortion assume even greater importance, since a given absolute die deflection represents a greater percentage on a thinner sheet, and the increased die resistance leads to higher pressures. Higher velocities are also required to give similar mass outputs, and the problem is usually resolved by using high drawdown ratios of 10:1 or more on thickness, giving wider die gaps (say 0·5 mm), lower die velocity and pressure, and improved percentage thickness tolerance. The thinner films are also more flexible, so the unsupported gap between die lips and haul-off/cooling must also be much reduced (20–50 mm). Often this is achieved by shaping the die so that the extruded web is led directly onto the upper surface of a driven and cooled 'chill' roll (Fig. 4.4). Cooling in contact with the polished surface of the roll imparts a good lower surface to the film, while the upper surface relaxes as it cools in air. The roll forms a means of haul-off by surface friction, and its speed

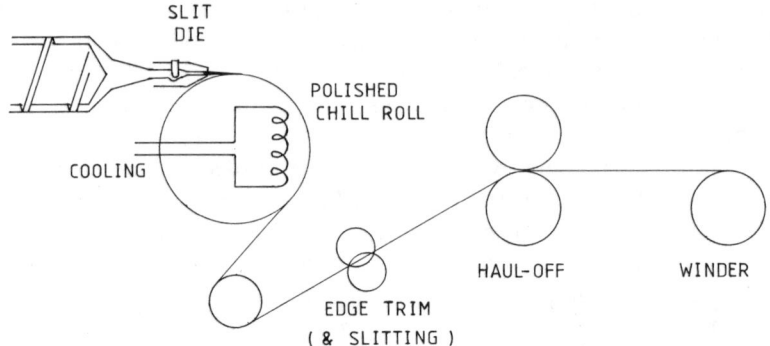

Fig. 4.4. Chill roll extrusion of flat film.

controls drawdown and final film thickness. The high degree and short distance for drawdown requires high elongation to break in the melt, and thus generally high melt temperatures. As many films are manufactured from semi-crystalline polymers, the rapid cooling tends to suppress crystallisation. This is beneficial in giving high elongation of the cooled film for subsequent drawing/orientation processes. In the case of polymers which crystallise fairly slowly, e.g. LD polyethylene, this leads to small spherulites in a substantially amorphous matrix which has high clarity but low stiffness compared with more crystalline films produced by the tubular process. An alternative process is to extrude vertically downward into water with a roller haul-off on the cooled film (Fig. 4.5). By keeping the coolant surface close to the die lips, both drawdown and cooling are rapid, with similar results to the chill roll process. Thus water level and temperature control affect clarity and stiffness, while the haul-off rate controls thickness. Rapid drawing probably assists stability, but uniform extrusion rate and absence of mechanical vibration (affecting the water surface) are also vital. Uniform and constant melt temperature not only assists uniform flow in the die, but also uniform drawing, both of which influence final thickness variation. Since film dies are easily blocked or damaged by solid matter in the polymer, fine filter packs are usually used, further increasing the back pressure on the extruder. It is worth noting that 'gels' of high molecular weight polymer are appreciably elastic and will often elongate to pass through filters and dies, only to recover their shape and cause 'fisheyes', etc., in the finished film. If these cannot be eliminated, it is probably wise to use either a filter mesh

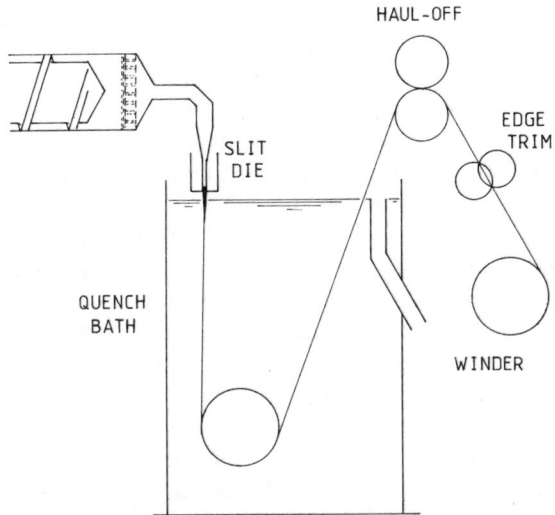

FIG. 4.5. Waterbath flat film extrusion.

appreciably smaller than the die gap to break up any gels passing through so that they are less likely to produce apparent surface defects, or a continuously-changed filter with back flushing to remove the gels before the pressure drop across the filter builds up sufficiently to force the gels through.

Multilayer flat films may be made by similar methods to those for sheet, but the greater importance of optical properties makes the elimination of interfacial defects more vital. Subsequent handling in wrapping machinery makes adhesion more important, especially where one strong or stiff component provides mechanical support for another layer providing mainly chemical resistance or barrier properties. Reduced thicknesses of the layers also make flow- and viscosity-matching more important where two or more components are brought together within the die. The chill-roll process is also extensively used for melt extruding a thin film onto a paper or textile substrate to form a supported coating. The substrate is passed over the roll and the film extruded on top of it, with or without a second roll to improve penetration and surface finish of the film. This surface coating may be only a few micrometres thick so that very high melt temperatures (e.g., >300°C for LD polyethylene) may be required to give low viscosity and uniform thickness to this polymeric film.

Tapes

Tapes for sound and video recording, package strapping, etc., especially in polypropylene and PET, are usually highly oriented in the longitudinal direction. The initial tape, which may be up to 40× 1·5 mm, is made by vertical extrusion into a waterbath, as for flat film, but with limited drawdown. Water temperature is kept low and the distance of the surface from the die is as small as 5 mm to give rapid cooling and low crystallinity. The tape is then reheated to a temperature well below the crystalline melting point and drawn lengthwise by 10:1, or greater, to produce a high degree of uniaxial orientation. It is then 'heat set' by heating to a somewhat higher temperature while held under tension to crystallise in the oriented state and then cooled. The low initial crystallinity gives a high elongation to break which permits sufficient drawing to orient, a low modulus which reduces tension required to draw at high strain rate, and less tendency to reversion (disorientation) before heat setting. Evidently the melt temperature at extrusion must be high enough to eliminate crystallinity present in the feed material, but low enough to permit rapid cooling throughout the relatively thick initial tape. For consistent properties, melt and waterbath temperatures must be closely controlled, while for consistent dimensions, extrusion rate and haul-off speed must be kept in proportion.

Fibres and Monofilaments

Fibres and monofilaments are widely melt extruded in polypropylene, nylon 66 and PET. In the finer gauges this is usually vertically downward into water or a countercurrent of air with a substantial degree of drawdown. Multi-hole spinarets are used as dies, usually immediately preceded by a filter pack of graded sand or porous metal (Fig. 4.6). To give a reasonable mass output, extrusion speeds are high and consequently pressure drops through the filter pack and spinaret are also high. However, as the filter blocks up during a run, its pressure drop will increase and this would normally cause a decrease in extruder output; but to maintain drawdown and properties and minimise tension, which would break the thread-line, the output must be kept closely in step with the haul-off and winding equipment, which would be difficult by adjusting speed of either the extruder screw or the winders. The extruder is therefore usually used as a source of melt which is then metered by gear pumps whose output is insensitive to back pressure. This indicates a requirement for high and constant melt temperature to

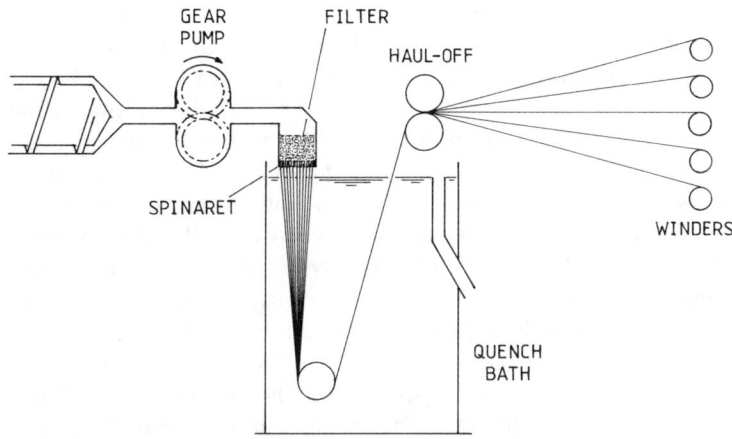

FIG. 4.6. Melt spinning of fibres.

minimise viscosity and pressure drops (cooling is not such a limiting factor with fine threads) and an output which changes rapidly with small changes of pressure (see Section 5.5). On the contrary, where the extruder feeds the spinaret directly the output should change little with pressure, and speed control must be precise but readily adjustable.

The orientation of fibres by reheating and drawing may usually be done at much higher linear speeds than melt spinning and is therefore usually a separate operation out of line.

Netting and Mesh
In addition to adaptations of conventional weaving and knitting processes from filament or tape, netting may be made by several melt processes. Possibly the most common process uses an annular die in which the outer element and mandrel can be rotated or oscillated in relation to each other. Each element contains a number of longitudinal or helical grooves producing a series of strands. As one or both elements are rotated, welds forming a mesh are produced when the grooves in the two halves coincide. Another patented[25] method passes an extruded sheet between grooved rollers, and the indented sheet is then stretched to break the thin portions and create a net—this is capable of economic production of much finer meshes than by an oscillating die.

Tubular Film (Blown Film)

In addition to the processes previously described for flat film, thin films are made by the tubular process, especially from LD and HD polyethylenes, polypropylene and PET. An annular die of the in-line or crosshead type forms a thin tube of 50 mm–2 m diameter and 0·2–1·0 mm thickness. This is flattened and hauled off by a pair of driven 'nip' rolls and inflated by blowing in air, either through the centre of the die or through a flattened tube between the rolls. The blown tube is cooled and solidified by high velocity air from an annular air ring surrounding the tube and some distance from the die; the tube is guided by converging boards or sets of rollers into the nip rollers (Fig. 4.7). It may be wound directly as layflat tube, slit at both sides and wound into two flat reels, or especially for the widest film 2–15 m wide, slit at one side only so that in use it can be opened to the full circumference, though with a visible centreline due to the fold. Extrusion is usually vertically upward, avoiding buoyancy effects on the blown tube and permitting the heavy extruder and wind-ups to be floor mounted and easing handling of raw material and finished product. In small plants the extruder is frequently vertical with an in-line die, both being continuously oscillated to distribute any irregularities in thickness. Thickness variations may be reduced by adjusting a flexible outer die lip; whether this is exaggerated or reduced by blowing depends on the cooling conditions. Where the extruder is horizontal, a crosshead die is convenient, but to improve polymer distribution to the die, an in-line die is sometimes coupled to the extruder by a simple heated pipe bend of small diameter. The ratio of blown tube diameter/die diameter (blow ratio) is usually 1·5:1 to 6:1 and typically 3:1; since the bubble is stretched both transversely and longitudinally, the thickness reduction is approximately 10:1, corresponding to a blow ratio of 3·3:1. The blow ratio is a balance between the hoop stress produced by the internal air pressure and the increase in tensile viscosity due to the air ring cooling; if the latter is very effective, the blown tube diameter will be close to that of the air ring, but otherwise it may continue to expand with less certain control. Since the process is essentially melt shaping, little orientation is normally produced; however, if the haul-off speed is increased beyond that required to avoid sagging of the blown tube, a small degree of longitudinal orientation may be developed. Because of the distance necessary for blowing between die and air ring ('freeze line height'), cooling is relatively slow and the crystallinity of LD polyethylene film will be greater than for

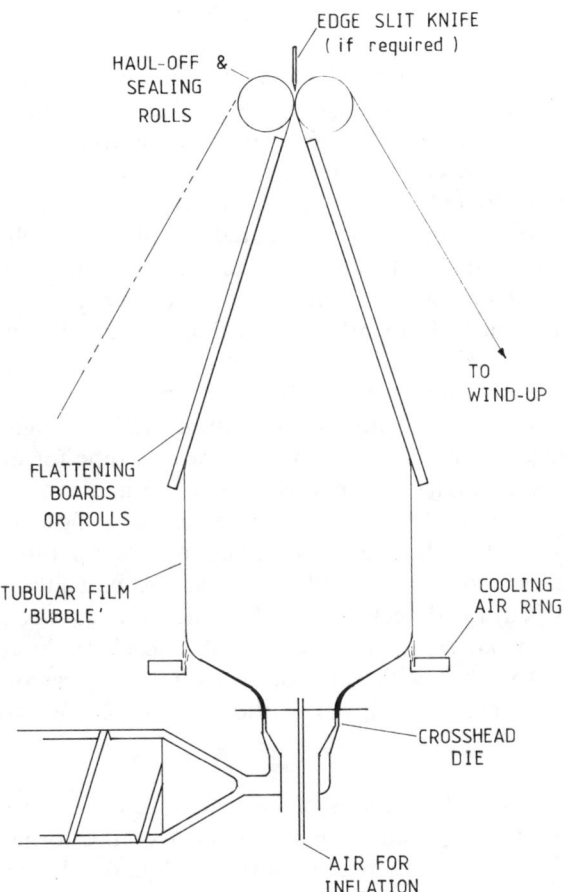

Fig. 4.7. Tubular film extrusion.

chill-roll or waterbath cast film, and will increase with freeze line height. This generally gives a stronger and stiffer film, but of poorer clarity than the chill-roll film. However, slower cooling allows more time for surface defects, e.g. die lines, originating at the die to relax, so optimum optical properties require a compromise in freeze line height. For a fixed freeze line height, increased output rate, increased melt temperature and decreased blow ratio tend to give slower cooling and increased crystallinity—in the last case the resulting greater thickness also delays cooling.

If oriented film is required, this may be done out of line from either chill or blown film, or similarly to tape and fibre, from a 'cast' tube. The latter may be extruded vertically upward, or downward into water, without blowing (increase in diameter) and limited longitudinal drawing, but rapidly cooled to suppress crystallisation. For improved temperature control, the tube is then reheated and blown at well below melt temperature to produce balanced or unbalanced biaxial orientation, depending on haul-off rate, and a further increase in width and decrease in thickness. Evidently tubular film requires complete melting and homogenisation in the extruder, absence of dirt or gels, and a very constant extrusion rate to avoid thickness variations and surface ripples. For blown film, the melt temperature must be high enough to permit the required blow ratio, whereas for cast tube for orientation a compromise is necessary between rapid cooling and avoidance of flow and surface defects. To achieve economic mass flow rates, linear extrusion speeds must be high, and raising melt temperature will raise the threshold speed for onset of melt fracture and give more rapid relaxation of surface defects. Long die lands are usually required to give uniform thickness, good surface finish and elimination of 'spider' lines. These, together with high output rates and narrow die gaps, involve high die pressures, often of the order of 40–50 MN/m^2.

Foams

One of the incidental functions of the single screw extruder is in the processing of thermoplastic foams into most of the forms previously mentioned. This may be by mixing in the feed material a proportion of solid foaming agent which decomposes, typically to nitrogen and carbon dioxide, at the temperatures within the extruder. Owing to the high pressure, these gases remain in solution or as small bubbles until pressure is reduced at the die exit, when foaming occurs. The pore size and distribution is governed by the number and distribution of nucleating points, sometimes achieved by a small percentage of undecomposed foaming agent. Another method is to inject a liquid foaming agent into the melt section of the extruder, which then distributes it throughout the melt. The pressure in the machine maintains the foaming agent as a liquid despite the increase in temperature, until it volatilises and foams the product with the reduction in pressure at the die. Foaming requires fairly constant output for control of product density, especially with liquid injection, which is independently metered, where melt mixing is also required. Close control of temperature, both of the final melt and along the barrel, is required to achieve the correct degree of decomposition of the foaming agent, and at the correct point in the screw. Pressures do not require close control, but must be kept between predetermined limits if output is varied.

4.2. OTHER APPLICATIONS AND THEIR REQUIREMENTS

The single screw extruder fulfils a number of useful functions which do not directly lead to saleable products, although in certain cases these may be integrated with the preceding processes.

Compounding
The commonest of these is compounding of stabilisers, antioxidants, fillers, plasticisers, lubricants, pigments and other additives. This is further discussed under Mixing in Section 7.1 and Quality in Section 8.1; here general considerations will be outlined. As a continuous process, the long-term uniformity of product depends on a consistent composition in the feed material; although the extruder provides a degree of end-to-end mixing, the residence time in the melt is of the order of only 10 s, i.e., the hold-up volume is small in relation to the hourly throughput, thus the extruder can only mix and even out variations within this hold-up volume. As shown in Section 5.2, distributive mixing in the screw channel occurs due to the cross-channel component of flow, and also in a down-channel direction with increasing back pressure, as well as in any devices attached to the screw or preceding the die included specifically to increase mixing. Such devices often impose an increased back pressure on the screw, and part of any improvement in mixing occurs within the screw and could be achieved merely by increasing the back pressure with a screenpack. Shear stresses within the screw channel are usually less than required to fracture agglomerates of particulate solids, including pigments; thus little dispersive mixing occurs in the channel. In the flight clearance, shear rates and stresses are approximately two orders of magnitude greater than in the channel and dispersion readily occurs. However, as shown in Section 5.3, the leakage flow per turn is normally only a few per cent of the total flow and, even if each portion of the melt only passed once through the clearance, it is unlikely that every portion

would experience this in the length of a usual melt section. In fact, the leakage flow is redistributed in the channel flow, from which leakage flow at a later stage is taken randomly, so that some polymer will be sheared two or more times and other portions not at all. In addition, the clearance is typically only 0·1–0·15 mm for a 100 mm diameter screw, so that larger solid particles are unlikely to enter the flight clearance; also the time of shearing is very short—of the order of 0·1 s at 1 rps. In contrast, patented designs of 'barrier' screw and mixers of the smearhead and Dulmage type ensure that all the polymer passes at least once through a close clearance. Thus the single-screw extruder is not usually adequate for dispersive mixing without special screws or mixing attachments, but, especially at high back pressures, it is quite effective as a distributive mixer on a small scale, and forms a useful compounding unit when fed with a blend including masterbatch, or following an 'intensive' high speed mixer for powders. The flow patterns in the screw (see Figs 5.7 and 5.8) give a wide range of residence times, which is undesirable for heat- or shear-sensitive polymers and additives, since some part of the polymer experiences heat or shear for a much greater time than the average.

As output is increased, residence time decreases and the use of distributive mixing attachments becomes desirable. These usually entail substantial pressure drops, giving a high back pressure on the extruder, which also increases the mixing effectiveness of the screw, but at the expense of low output and additional shear heating. Large machines running at low speed are desirable to increase the residence time and increase surface for cooling, providing shear stresses are not unduly reduced. To keep viscosities and shear stresses high, low melt temperatures are desirable, but it may not be possible to apply sufficient cooling to maintain these low temperatures despite shear heating. It is, of course, possible to use the same extruder for compounding and product shaping, but the above requirements for compounding may be incompatible with those for shaping, so that the extruder cannot give its best performance for both functions simultaneously.

Many variants of extruder compounding have been attempted, with varying degrees of success. In one case a dispersion of pigment in a volatile liquid was metered by a piston pump under pressure into the melt pumping section of a single-screw extruder fitted with a patented design of mixing head. Microtoming of the cross-section of the cooled extrudate showed that the circumferential distribution was good, but annular rings of different pigment content were observable, due to lack of a radial mixing component in the mixing head and the point of injection being too near the end of the screw for the latter to achieve thorough mixing. In another case, polyethylene masterbatch containing a high proportion of slip agent was solid fed to a small single-screw extruder, which melted and pumped it through a restrictor into the melt pumping section of a larger single-screw machine handling base polymer. The former gave problems of low and erratic flow rates, until cooling was applied to the screw in the feed section only, but otherwise the process operated steadily with adequate distribution at the end of the large extruder. Proportional control of the speeds of the two machines and monitoring of the pressure at the end of the masterbatch machine gave sufficiently consistent content of slip agent in the final film compound.

Melt Blending

Other mixing operations related to compounding include melt blending, structure modification and intentional degrading. Melt blending includes mixing two polymers of the same type but different molecular weight, mixing two polymers of similar mean molecular weight but different distributions, and more rarely two different polymers, e.g. LD polyethylene with HD polyethylene or polypropylene. In all cases, if the melt viscosities are similar at processing temperature little difficulty need be anticipated, though if the shear sensitivity (non-Newtonian flow index n) is very different for the two components, radial segregation into zones of differing shear rate may occur. If the temperature-coefficient of viscosity is substantially different for the two components, processing temperature ranges may be limited. If viscosities at processing temperature are very different, e.g. due to widely differing molecular weights, the mixture may tend to behave as a suspension of relatively solid, high viscosity component in low viscosity second component. As discussed in Section 5.9 in respect of the melting stage, when a suspension is sheared, the shear rate and viscous heating are higher in the lower viscosity component, causing a temperature rise in the latter which further reduces its viscosity, whereas the high viscosity component is heated mainly by conduction. Thus initial segregation is liable to continue, with poor mixing. A stratagem employed with batch melt mixers in this case is to feed the high molecular weight component first, and add the cold second component only when the former has been sheared and heated to reduce its viscosity; to operate this on the extruder would require special equipment such as an intermediate feed pocket with force feeder.

In all blending operations, as in compounding, large-scale uniformity in the product requires proportional feeding of the components to at least a similar accuracy, and the avoidance of de-mixing. The latter is most likely to occur due to segregation of the feed in the hopper or first part of the screw, e.g. due to different particle shapes or densities causing selective feeding by the screw. Structure modification and degradation may include many chemical reactions, either homogeneous or promoted by additives or catalysts; in addition to its normal functions of melting and pumping, the extruder is usually required to do mechanical (e.g. shear) work on the polymer, to maintain energy balance and to ensure at least a degree of mixing of the product. One example is the selective breakdown by mechanical shearing of a high molecular weight constituent of LD polyethylene to reduce molecular weight spread and improve product properties. This required considerable mechanical drive power and resulted in appreciable temperature rise of the polymer, but the resulting energy input to the polymer was largely absorbed in heating the polymer to a melt temperature suitable for granulation. A second example involved thermal and shear degradation of polypropylene from a high molecular weight as polymerised to a lower value suitable for melt processing. Exclusion of oxygen was important in minimising the oxidative element of degradation, as was accurate temperature control—a change of temperature of a single barrel zone of 1°C produced a measurable change in melt flow rate of the product. The molecular weight distribution of the product was significantly different from that resulting from polymerisation direct to the same average value, giving different mechanical properties, e.g. impact strength as well as different processability. Vacuum extraction was used in the extruder to remove as much as possible of the very low molecular weight products of degradation, as well as catalyst residues, which adversely affected colour and odour of the product.

Degassing, Drying

Vacuum extraction for degassing and/or drying is also frequently used for contaminated or damp feed material, but of itself would be an energy-expensive process of limited effectiveness, because of the restricted residence time and surface area for devolatilisation. It is therefore commonly used in conjunction with an extrusion compounding or fabrication process, as discussed in Section 5.5, where the volatiles do not exceed, say, 5% by mass of the polymer flow rate. Note that extraction in the melt is of limited value, e.g. with damp PET, which will partially hydrolyse at melt temperatures upstream of the vent position.

Filtering
Filtering of solid contamination from the melt, usually between the screw and die, is also commonly associated with fabrication processes, especially those with narrow die gaps such as film and fibre extrusion and thin wire-covering, where contamination would dramatically affect electrical insulation. It is sometimes used as a separate process, e.g. with raw natural rubbers and with scrap recovery; the economic case for the latter is discussed in Section 9.5. Scrap is sometimes re-extruded to obtain a suitable particle shape for re-feeding to the main process (especially sprues and runners in injection moulding). More commonly it is mixed as a proportion of new feed material, either to make a homogeneous granular feed or directly into sheet for vacuum forming or tube for blow moulding; it may also include additives (especially stabilisers) to replace losses between the original feed and the scrap. Thus scrap recovery may include one or more of the steps of blending, compounding, degassing, filtering and granulation. As discussed in Section 5.10, satisfactory feeding of irregular-shaped scrap and avoidance of segregation are probably the most common problems, especially when scrap, alone or blended, is re-fed directly to an extrusion fabrication process.

Dewatering
A special application of extrusion, which has been used commercially with rubbers and often associated with filtering and granulation, makes use of the tendency of the screw to pump high viscosity materials in preference to lower viscosity. This is seen in the rearward displacement of air entrapped in the feed, but in this case is used for dewatering wet rubber. The feed pocket is replaced by a perforated section, for example with a series of narrow longitudinal slits through which the displaced water drains, while the substantially dry rubber is conveyed along the screw to be heated and extruded.

Polymerisation
The single-screw extruder has also been used for continuous polymerisation, using its mixing and heat transfer functions as a high tempera-

ture, high pressure reactor. However, widely differing viscosities of monomer and polymer may cause slip and loss of pumping, and as mentioned under compounding, the spectrum of residence times is broad, tending to an undesirably wide molecular weight distribution in the polymer.

Blow Moulding

The single-screw extruder is widely used in blow moulding of bottles and other hollows, including toys and ducting. In some cases, short lengths of extruded tube are stored and reheated for blowing out-of-line from extrusion. More usually, one or more tubes are extruded, cut off and pinched at one end and blown inside sets of moulds where they cool by contact with the cold mould. The first requirement is for a uniform flow with low elastic memory; the melt temperature should be low to give rapid cooling and reduce 'parison' sag, and for large containers, extrusion should also be rapid to reduce time for sagging. For large containers, where extruder output would form a limitation, the extruder may continuously fill an accumulator, which discharges intermittently and rapidly through the die. In some small machines with a single mould, the extruder screw is run only for the time required to form the parison and then stopped while blowing and cooling occurs; this intermittent operation gives more time for melting, but departs from the steady-state assumptions of the theory given in the following chapters.

Injection Moulding

Modern injection moulding is heavily dependent on the single-screw extruder, although twin screws are also used. Again, the screw usually runs intermittently, though the 'rest' period may only be a few seconds during actual injection. This is necessary where the screw also acts as the ram for injection, a ring non-return valve being fitted at the end of the screw to prevent reverse flow in the screw during injection, which is generally under much higher pressure (up to $200 \, MN/m^2$) than during refilling by the screw (usually less than $5 \, MN/m^2$). An additional complication is that the screw is displaced rearward as it pumps melt into the reservoir formed by the injection cylinder, and since the feed opening is usually fixed, the effective length of the screw progressively decreases during each filling cycle. This does not usually apply to piggy-back machines where the screw preplasticising unit is separate from the injection ram, though a non-return valve and intermittent rotation are still usual. In modern machines under computer control, the screw speed and/or the back pressure against which the screw acts may be programmed to vary during the filling cycle. The first requirement is that the screw should refill the cylinder with the necessary volume of molten polymer within the cooling and 'dead' times of the particular moulding and machine, so that plasticisation does not extend the overall cycle time. The screw must provide a homogeneous melt, frequently including the incorporation of pigments or coloured granular masterbatch, and 'screw-back' pressure is adjustable to assist the mixing in the screw. For thermoplastics, the required melt temperature will tend to be high, giving a low viscosity for thin mouldings and/or long flow paths in the runners and mould, which because of cooling during mould filling must in any case be filled fast, so that high temperature reduces the injection pressure required. However, melt temperature also has important effects on the structure and mechanical properties of the final moulding, while thermally sensitive polymers like UPVC may be subject to degradation, especially of the 'cushion' material left in the cylinder between shots.

Phenolic-type thermosets require very precisely controlled low temperatures (e.g. water/steam heating) to avoid curing of the cushion material before the next shot, while giving sufficient fluidity for the shot to be injected into the hot mould at high velocity so that shear heating raises the temperature of the shot to give rapid curing. Unsaturated polyester thermosets may be injected cold so that the screw does little more than convey the premix to the cylinder.

Thus polymer, machine and product all influence the required temperature; it is important that this is as uniform as possible throughout the shot volume, which varying screw speed or back pressure seeks to achieve, despite the changing effective screw length. It should be noted particularly that during injection the flow within the cylinder is unlikely to be similar to that during refilling,† so that any temperature variation during the latter will not appear in the same sequence in injection. Moisture removal by venting is also included in some plasticising units, with further problems due to axial screw movement. The steady-state theory is only qualitatively applicable to this situation of intermittent

---

† Pearson C. E., The extrusion of metals, Chapman & Hall, 1944, Figs 87, 89, 92 & 94. Tanev N., Brug af blandehoved i snekkesprøjtestøbemaskine, Report Nr. PT, 75.63-A AMT Danmarks tekniske Højskole, 2800 Lyngby Denmark 1975.

TABLE 4.2
Comparison of Process Requirements

Process	Uniform/constant							Maximum(×) or minimum(√)							
	Output, O	Pressure, P	Melt temperature, T_m	Mixing, Mix	Residence time, t_{res}	Specific power, $\frac{E}{O}$	Elastic memory	Output, O	Pressure, P	Melt temperature, T_m	Mixing, Mix	Residence time, t_{res}	Specific power, $\frac{E}{O}$	Shear history	Elastic memory
Rod, section, profile	×	×													
Tube, hollows		×						×[a]	√						
Wire covering, sheathing				×											√√
Sheet, laminating		×	×	×			×	×	×	×					√√
Flat film	×						×		×	×	×				√

Practical Extrusion Processes

Paper coating
Tape
Fibres, filament netting
Tubular film
Foam
Compounding
Blending, degrading
Degassing, drying
Filtering
Scrap recovery
Dewatering
Polymerising
Blow moulding
Injection moulding

[a] Large pipes
[b] Degrading (×), blending (√)

rotation, changing effective length and possibly changing speed and back pressure during the cycle, and predicted effects may be overridden by transients including elasticity of the melt, non-steady-state heat transfer and mechanical/hydraulic inertia in the machine.

Comparison of Requirements

The foregoing brief survey includes many of the commercial applications of the single-screw extruder, sufficient to show its wide importance and the differing requirements of processes of which it forms a part. Thus it is impossible to define an ideal design or method of operation for all purposes, and to make the discussion in Chapters 7 and 8 more specific, Table 4.2 attempts to summarise the important requirements for each process and product type. Of course, high output, uniform temperature, etc., are desirable in all processes; only the more critical items are indicated, and even these will be qualified by the product, e.g. thick tube and sheet require high output, whereas for thin tube and sheet minimum elastic memory may be more critical—the indications should be taken to refer to the more critical product. The table suggests that the non-fabrication processes generally have a wider range of requirements than do those which are primarily shaping, especially in respect of maximum/minimum values. In some cases the requirements of the two groups are contradictory, emphasising that compounding or scrap recovery may be more effective as a separate operation rather than integrated with a shaping process, although requiring more labour, energy and overheads. It appears, not unreasonably, that temperature is the most common criterion, for both uniformity and maximum/minimum value; in several cases not indicated, e.g. film and injection moulding, the temperature must be controlled at a prerequisite value for an individual polymer or operating condition, even though this is not the maximum or minimum achievable.

4.3. COMPARISON OF MACHINE TYPES

The foregoing survey of process requirements raises the question of whether the single-screw extruder is the best machine to meet these requirements. Since a comparison of machines is not the main objective of this book, the following only covers the principal differences and omits the many variants and proprietary designs intended for specific purposes.

The internal melt mixer (e.g., the Farrel–Bridge 'Banbury'[26] and Shaw 'Intermix'[27]) is essentially a batch mixer not intended for direct shaping operations and requiring very different ancillary equipment and method of operation. It is very robustly constructed and capable of very high power inputs, usually controlled by rotor speed and external pressure confining the polymer, leading to high shear stresses and rapid temperature rise. Batch operation permits accurate feeding and proportioning of additives by batch weighing, and automatic weigh/feed systems are eliminating errors due to manual feeding as well as providing a reliable record for quality control. However, the batch output is not often in quantities appropriate to downstream batch processes and storage, on roll mills or heated hoppers, is necessary to give a continuous output in other cases; indeed special hot fed single-screw extruders (slabbers) have been used for continuous granulation or strip extrusion, but not usually for shaping the finished product. The kneading action in the space between the rotors is much more pre-determined, largely by initial rotor design, than the flow in the single-screw channel, and this probably also makes the shear over the rotor tips less random than in the extruder; in any case, the mixing time is independently controllable to give a high probability that the whole batch has been subjected to the higher shear stress. The latter is effective in dispersing pigment agglomerates and takes place at the barrel surface so that energy removal and temperature control are facilitated, where heating/cooling may be programmed throughout the cycle. One of the benefits of batch operation is that severe shear may be imposed without excessive temperatures by a combination of heat build-up from the cold polymer and the heat capacity of the machine scantlings; power may be applied for say 2 min in each cycle, whereas cooling can be maintained for the complete cycle time of say 5 min. The internal melt mixer is thus very effective for dispersing pigments as well as melt blending and intentional degradation, e.g. raw natural rubber mastication; the single-screw extruder requires a preceding high-speed 'intensive' powder mixer or granular masterbatch for the former, while heat and temperature control are problems for the latter. The extruder, however, can give useful distributive mixing and is suitable for continuous and direct shaping of finished products.

Various patented designs of continuous mixer have been developed to give the advantages of the internal mixer coupled with a continuous output. They usually consist of a pair of rotors in a jacketed chamber similar to the internal mixer, with a screw feeder, e.g. on one end of the rotors, which determines the throughput rate. Feeding may then be by gravity, requiring continuous or semi-continuous metering of additives. In this case, rotor speed is no longer available for power control since it is not independent of output, and back pressure by means of a valve is used. However, this would affect flow to a die, making problems for direct shaping. At high outputs, the surface area for cooling is limited, so restricting the level of energy input; the possibility of differing times for energy input and removal no longer obtains. Such machines are no doubt efficient mixers, but in the author's opinion do not constitute a complete compounding–shaping process or compete with the single-screw extruder for those shaping processes which require a modest amount of distributive mixing.

The ram extruder is mechanically simple and robust and is appropriate where very high pressures are required as in hydrostatic extrusion, extrusion of PTFE, and injection moulding. For the first two, the operation can be made continuous at the expense of complication by using two extruders alternately with a changeover valve. Since little flow takes place in the cylinder, heating is by conduction, and therefore is slow and involves temperature differences in the polymer, especially since diameters tend to be large to extend run length. As mentioned on p. 57 in connection with injection moulding, the sequence of flow from the cylinder is complex, so that initial temperature differences will be redistributed in passage through the die. High shear stresses and shear strain are unlikely to occur before the die entrance, so little mixing for temperature or composition will occur. If the usual direct extrusion is employed (force applied to a piston at the rear of the polymer charge) then the pressure at the die will tend to rise as the wall friction decreases due to the diminishing length of the charge, and there will be losses due to mechanical inertia and friction and stored elastic energy in the slightly compressible polymer. The ram extruder is thus an efficient but intermittent pump, but fulfils none of the other functions of the screw extruder, e.g. feeding, melting, mixing, venting and temperature control, which are exemplified in the injection moulder, where the screw carries out all except the final pumping function.

In an unusual design, instead of running as a helix on a thin cylinder, the long narrow channel is wrapped as a spiral on a disc, running against a smooth disc, the flow taking place inwards or outwards. The functions of solids feeding, shear melting and melt pumping with a degree of mixing as in the screw extruder may be expected to take place. Unless the spiral contains as many turns as the conventional screw, the pressure difference across the flight between two turns will be greater, so leakage flow will be greater. The axial thrust between the discs will be large, especially if flow is outward, being the integral of pressure over the whole disc area, as compared with the conventional die pressure times the barrel cross-section; any lack of mechanical rigidity will increase flight clearance and leakage. Note that for a constant drag flow along the length of the channel (eqn (5.10)), the velocity W increases proportionally with radius, so width b or depth h must decrease. In the former case, where shear rate increases with radius, the wide-channel assumption may be invalidated and manufacture will be difficult; in the latter case, shear rate increases as the radius squared and shear heating as the fourth power of the radius. Perhaps the greatest practical disadvantage is the small physical distance (radially) between inlet and outlet, making the maintenance of substantial temperature differences difficult, and reducing operational flexibility.

A variation of the conventional single screw forms the channel on a cylinder of larger diameter and shorter length.[28] This gives a lower rotational speed for the same shear rate, a smaller helix angle (see Ref. 29) and a torsionally stronger screw. However, it involves a greater axial thrust at a given die pressure and the reduced length may give control problems in maintaining axial temperature differences.

A design apparently similar to the spiral disc extruder, but different in principle, is the smooth disc extruder, widely referred to in the literature under various titles including the Weissenberg extruder and the elastic melter. This is claimed to work on the elastic properties of the melt, typified by the 'climbing rod' phenomenon. It consists of a smooth disc rotating close to a similar stationary disc, which pumps material from the circumference towards the centre, where a die may be fixed. The author's experience is that this design did indeed melt polymer very rapidly, but that final melt temperatures were highly non-uniform. Some assistance with solid feeding seemed necessary (possibly a force feeder or a short spiral flight would assist) and the pressure generated was modest (less than 1 MN/m^2). To overcome the latter, a conventional screw extension has been proposed to generate pressure, but it is doubtful whether the temperature inhomogeneity would be sufficiently reduced. Separate control of the disc and screw speeds would further increase mechanical complexity and the limited physical distance between inlet and outlet would again make temperature control difficult.

The most obvious and serious competitor to the single-screw extruder is the twin-screw machine. First let it be said that there is a field for each, and the area of serious competition is perhaps small. Then there are various types and designs of twin, whose merits will not be discussed here. One type has screws which do not intermesh and therefore functions much as the single-screw machine, though there must be additional mixing by the transfer of polymer between the two screws; there must also be additional mechanical complexity. In another case[33] twin screws are provided as far as the vent section, only one of them being continued to raise pressure before the die. This may be regarded as other twin-screw designs with the addition of a venting system, as discussed in Section 5.5. The majority of twin-screw extruders have the two screws intermeshing to a greater or lesser extent. In addition, some have co-rotating screws with the same helix 'hand' while others are contra-rotating with right- and left-hand helices. As a broad generalisation, the performance of the co-rotating designs and those with less intermeshing of the screws falls between that of the single-screw machine and the fully intermeshing contra-rotating types, due to the larger clearances necessary between the screws.[1] The twin-screw machine is thus ideally a positive displacement pump, whose output, like that of the gear pump, is roughly proportional to speed but little affected by back pressure; in fact it may be regarded as a helical gear pump of such length that each 'tooth' describes several helical turns, giving a flow which is predominantly axial rather than tangential. As will be seen in Section 5.2, the single-screw extruder is in contrast essentially a drag pump, whose output is also proportional to speed but reduced by back pressure, and in which the main flow is continuous along the screw channel. In the twin-screw machine, the polymer is trapped in the channel volume between successive meshing points and carried forward as those meshing points together progress forward, giving a much more uniform residence time than in the single-screw machine. Circulation occurs within this trapped volume due to relative motion of screw and barrel, but this is not substantially altered by increasing back pressure. The shear history of the polymer within the channel is thus much less than in the single-screw machine, leading to lower energy input and temperature rise. Departure from the ideal is represented by a number of leakage flows, which must be accurately determined to define the machine performance; these include (a) flow between the volumes trapped in the two screws, (b) flow between successive volumes by leakage at the meshing points, and (c) leakage between the screw flights and the barrel. These flows are to some

TABLE 4.3
Comparison of Single- and Twin-Screw Extruders

	Single screw	Twin screw
Flow type	Drag	Near positive
Residence time and distribution	Medium/wide	Low/narrow (useful for reaction)
Effect of back pressure on output	Reduces output	Slight effect on output
Shear in channel	High (useful for stable polymers)	Low (useful for PVC)
Overall mixing	Poor/medium	Good (useful for compounding)
Power absorption and heat generation	High (may be adiabatic)	Low (mainly conductive heating)
Maximum screw speed	High (output limited by melting, stability, etc.)	Medium (limits output)
Thrust capacity	High	Low (limits pressure)
Mechanical construction	Robust, simple	Complicated
First cost	Moderate	High

extent affected by overall pressure gradients and local pressure differences created by screw motion; they are also affected by the relative directions of motion, velocities and convergence/divergence of leakage paths formed by adjacent screw and/or barrel surfaces. Thus considerable mixing and energy absorption occurs in these leakage paths, but it appears that this is less randomly distributed through the polymer bulk than in the single-screw machine. Change of channel dimensions may be in depth and/or width (a function of pitch) and frequently this is done in several stages, each of constant dimensions, presumably to facilitate manufacture and control of clearances. Frequently mixing sections having little or no forward pumping action are interposed. It appears to the author that 'compression' is largely to take account of changing (bulk) density and that since this is unlikely to be matched exactly, portions of the screws will be running partly filled; it also appears that melting is largely by conducted heat. For reasons which are not altogether clear, twin-screw extruders are usually designed to run at relatively low speed—say, no greater than 2 rps, possibly to restrict heat build-up in the polymers usually processed; this tends to give a relatively low output for the size of machine. Although mechanical design is outside the scope of this book, the complicated screw geometry and critical clearances require precise and costly manufacture, the bi-lobed barrel is more liable to distortion from pressure and thermal stresses than the cylindrical barrel of the single-screw extruder and limited distance between the screw axes causes problems in achieving robust screw drives and thrust bearings. It would be wrong to imply that the twin-screw extruder is unreliable, but the foregoing entail high manufacturing cost and greater care in operation. The comparison between single- and twin-screw extruders is summarised, with possible oversimplification, in Table 4.3. The twin-screw extruder is thus good for compounding, especially of heat- or shear-sensitive polymers including UPVC and polypropylene, for chemical reaction and for shaping processes at minimum temperatures. However, its high first cost and relatively low output are obstacles to its use where the single-screw machine can approach its performance.

Chapter 5

PRINCIPLES OF SINGLE-SCREW EXTRUSION

In this chapter, a simplified analysis leads to equations relating output, pressure and screw speed for the melt pumping section for several typical screw types. The flow patterns within the screw and pressure profiles along it are discussed, together with an examination of leakage flow. Graphical presentations are used to analyse the output/pressure characteristics of the screw/die combination, for simple and some more complex situations, including venting. Corrections to the simplified equations are discussed, leading to those due to pseudoplastic and non-isothermal conditions. The mechanisms of melting and solids conveying are covered qualitatively, since quantitative analysis is not considered helpful at the present for commercial operation.

5.1. FUNCTIONS OF THE EXTRUDER

As discussed in Chapter 2, the mass flow through the die is related to the pressure drop, the die dimensions and the viscosity, the last being mainly a function of temperature. Since the outlet conditions from the die are only slightly affected by the haul-off, the function of the extruder is to deliver to the die the required mass flow at the corresponding pressure and temperature, with minimum fluctuations of these three quantities in space and time (see Section 8.1) and correct composition, e.g. shear history (Section 8.3) and dispersion of additives (Section 3.4). The output of a given extruder screw is mainly dependent on rotational speed, so the relation between mass flow output, screw speed and back pressure will be considered in this chapter, together with the influences of screw dimensions and viscosity, and the interactions with the die pressure.

Note 1: This analysis concerns the screw only, giving $Q = f(N, P)$, i.e. assuming that Q or P can be varied independently of the die, e.g. by altering the die dimensions. But in Section 2.2 it was shown that for the die $Q = f'(P)$; thus in the normal case of an extruder feeding directly to a die, *both* relations must be satisfied and at a given screw speed and temperature only one pair of real values of Q and P will hold (Section 5.5).

Note 2: The equation $Q = f(N, P)$ seems to imply that mass flow (output) can always be raised by increasing screw speed. However, in practice, energy considerations (Chapter 6), product quality (Section 8.1) or stability (Section 8.2) more often limit satisfactory operation. Many texts ignore these factors in analysing performance.

A principal function of the extruder for plastics and rubbers—which it shares with those for clays, bitumen, paste explosives, pastas and even the domestic meat mincer—is thus to pump material in a fluid state against the back pressure created by the resistance of the die. This is the function which is susceptible to hydrodynamic analysis and is the main subject of this chapter. Because this flow is complex a certain amount of both distributive and dispersive mixing occurs (Section 3.4), and energy is transformed into heat (Section 6.1) by shearing, leading to temperature changes and distributions (Section 8.1). Coincidentally, the single-screw extruder, which is an Archimedean screw or 'drag' pump, will also convey solid particles and generate pressure in the solid; this causes shearing and, together with heat conducted from the barrel, generates heat to melt the solid polymer. The extruder may also be used for removing air, volatiles, water, etc., at the feed section and gases or vapours from the melt (Section 5.5). The mixing action may be used for incorporating (compounding) solids or liquids, e.g. pigments or foaming agents, and solids, e.g. dirt may be removed from the melt by mechanical filtering. The mixing action of the screw enables it to be used as a 'scraped' heat exchanger, though cooling is limited by shear heating, and this is also utilised for chemical reaction, e.g. continuous polymerisation when the extruder becomes effectively a stirred reaction vessel. The single-screw extruder thus has many possible functions including melting, conveying, mixing, compounding, pumping, separating and heat exchange, and several of these may have to be optimised in a specific case. However, some are difficult to analyse in theory or control in practice, so mass flow in the melt pumping region, often called the 'metering' section, is usually made the controlled (minimum) factor such that uncertainties in the others are of less significance.

Note: The popular belief that the geometric sections of the screw correspond to the functions of solids conveying, melting and melt pumping, respectively, is not necessarily true (see Sections 5.4 and 5.9) and reference in this book to 'melting region' refers to those parts where this mechanism dominates rather than to a tapered portion as such.

The single-screw extruder for plastics usually consists of a single-start screw of constant pitch and rectangular thread section running in a

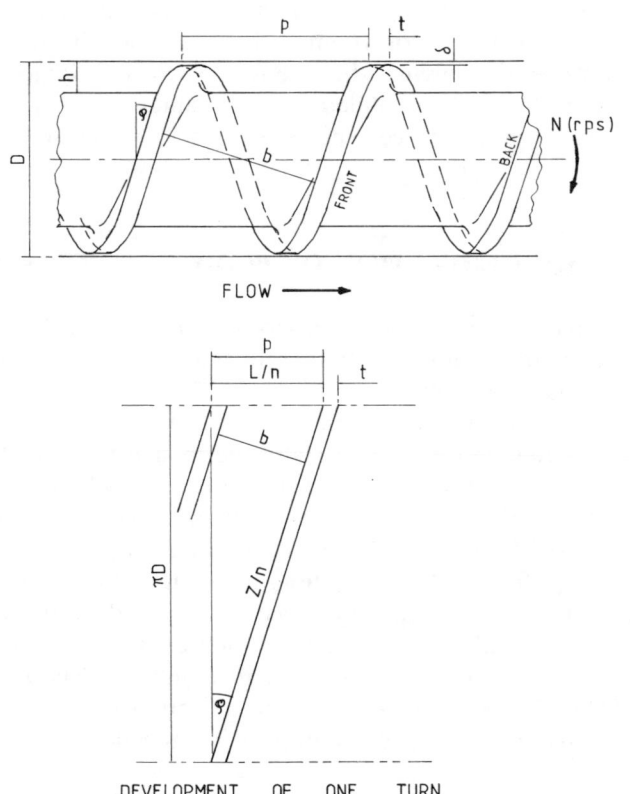

FIG. 5.1. Screw dimensions.

barrel or cylinder of constant diameter. The following analysis can be adapted to multi-start screws by substituting 'lead' for 'pitch' and multiplying the output by 'number of starts'. For rubbers, screws of varying diameter are sometimes used, e.g. in 'slabbers', and frequently of varying depth, with or without constant pitch. In the first case it may be necessary to use mean[30] rather than maximum channel diameter in calculating velocities, and in the second case the 'wide channel' assumption may be invalid, so that shape factors[31] must be included.

In this book the terms 'front' and 'back' refer to the die end and feed end, respectively; applied to the flight they are synonymous with 'leading' or 'pushing' and 'trailing' edges. The terms 'top' and 'bottom' of the channel refer to the outer (barrel surface) and inner (thread root) diameters, respectively. Flight 'clearance' refers to the radial gap necessary for free running, which is assumed constant round the circumference, which is consistent with observation in moderate size machines containing polymer (but see Ref. 34 for very large machines). Wear, especially with glass-filled or other abrasive compounds, may lead to excessive clearance (see Section 5.3) or axial variation in clearance.

5.2. DERIVATION OF FLOW EQUATION

The geometry of a simple screw is shown in Fig. 5.1 with the nomenclature used in this chapter. Note that the pitch p and flight width t† are measured axially, whereas the channel width b is measured normal to the helix. The helix angle ϕ is the true angle, although for convenience the flight is often shown as a straight line in side view. Unless stated otherwise, all diagrams show a left-hand helix with the side nearest to the reader moving down the page, causing a flow from left to right. Evidently a right-hand screw must rotate in the opposite direction to cause the same flow, but the equations are equally valid. The rotational speed is given in revolutions per second (rps) and output in m^3/s or kg/s for dimensional consistency; more theoretical texts may use speed in radians per second where the numerical value will be greater by a factor of 2π. The components of output Q (eqn (5.17)) are derived in *volume* terms and should be multiplied by the density at the mean

---

† Weeks and Allen[35] use t for the helical flight width, i.e. as measured in the workshop, leading to minor differences in eqns (5.12) and (5.13).

Principles of Single-Screw Extrusion

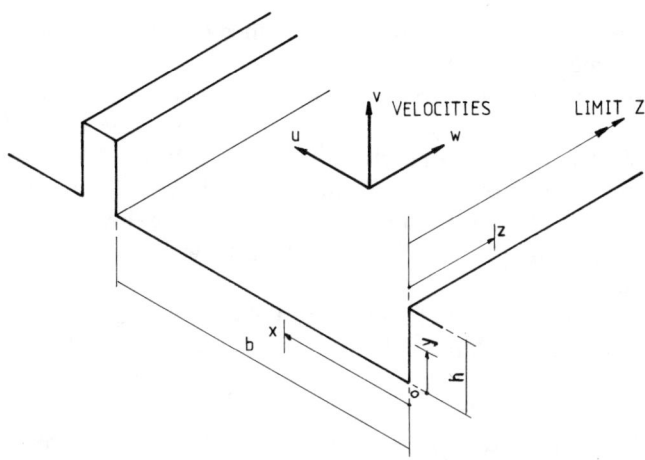

FIG. 5.2. Coordinate system for screw channel.

temperature and pressure in the melt section of the screw to give the theoretical mass output (assumption 8 below). The rectangular coordinate system in the channel is given in Fig. 5.2; the length coordinates are x, y and z in the directions of (normal) channel width, (radial) channel depth and (helical) channel length, respectively. The limiting values are b, h and Z, respectively, where by development it can be seen that:

$$Z = L/\sin \phi \qquad (5.1)$$

$$b = (p - t) \cos \phi \qquad (5.2)$$

$$p = \pi D \tan \phi \qquad (5.3)$$

for a screw having n turns and L is the corresponding axial length. The corresponding velocity components are u, v and w with limiting values U at the barrel surface ($y = h$), V at the channel edge ($x = b$) and W at the barrel surface ($y = h$), respectively.

Assumptions

The assumptions in the simplified flow analysis are:

1. Steady-state operation, i.e. mass flow is constant with time and position. Section 8.2 considers instability.
2. The polymer is a homogeneous fluid, i.e. 'melt pumping' section only. Although partly melted polymer may behave as a fluid and

generate pressure, appropriate values of viscosity and heat content must be used.
3. The polymer adheres to the barrel and screw root surfaces, but not to flight front and back (channel sides)—see Refs 31 and 36 for correction factors where this is significant. These references show that if the screw channel is wide relative to depth (say, $b \geq 10h$) the correction is negligible and the flow may be treated as two-dimensional, i.e. $V = 0$ and $\partial w/\partial x = 0$, with no 'edge effects'.
4. Curvature is neglected, i.e. the helical channel is approximated by a straight channel b wide and h deep. With highly viscous fluids, such as polymer melts, inertia effects giving rise to secondary circulations will be negligible.
5. The channel depth is small compared with the internal barrel diameter ($h \ll D$). This is valid for the melt section of many plastics extruders ($h < 0 \cdot 1D$), making assumption 4 reasonable and minimising the effects of unequal areas and shear stresses at barrel and screw root (top and bottom of channel).
6. Zero leakage flow. This is discussed in Section 5.3. Note that for an unworn screw the flow in the flight clearance is negligible, but as shown in Section 6.3 the power consumption is appreciable and cannot be neglected.
7. The melt is Newtonian, i.e. viscosity is a function of temperature but not of shear rate or time. As shown in Appendix A, this permits a simple integration, and an algebraic solution. In practice viscosity is not an analytical function of shear rate for real polymer melts and if treated as a dependent variable leads to non-linear differential equations requiring digital computation for solution. Although real polymer melts may be time-dependent, in steady-state extrusion such effects are likely to be masked by shear heating and temperature changes. Since the effects of non-Newtonian behaviour may be quite marked, especially on power consumption, this assumption is partially relaxed in Sections 5.7 and 6.4, where the Newtonian solution is modified by treating viscosity as a separate variable to give an approximate solution which accords reasonably with practice.
8. The flow is isothermal. This neglects the effects of heat conduction from or to the barrel and screw and shear (viscous) heating; these are considered in Sections 5.8, 6.5 and 7.2. It also neglects the consequent effects on the longitudinal temperature profile

which usually exists in practice, but is valid for a short section such as might be used in a finite difference calculation. Since viscosity is approximately an inverse function of temperature, use of an average temperature is risky, especially if other factors, e.g. channel depth, are also varying along the length; a weighted average viscosity based on temperature and shear rate would provide a better approximation. This assumption, together with incompressibility, means constant density, i.e. with the first assumption, volumetric flow is also constant.

9. Gravity is neglected. The high viscosity of polymer melts gives low values of Grashof Number[20] so that gravitational flow is negligible and natural convection due to buoyancy[21] is also negligible compared with laminar forced convection.

10. Initially the screw is treated as constant depth, a real condition in many 'metering' sections. The solution is valid for a short section of a gradually tapering screw for finite difference calculation, and for the Newtonian isothermal case (assumptions 7 and 8) may be simply integrated (eqns (5.51), (5.59) and (5.60)). Care must be exercised, however, if it is applied to sharply tapered or stepped screws, not so much because of inertia effects (Reynolds Number is always very small[22]) as possible elastic and thermal effects.

Fenner[10] has published an excellent study of the differences produced by several of these assumptions and has shown that, while they are important for exact design, in most usual cases they can be approximated by correction factors, and thus the effects on *trends* caused by changes of operating conditions are negligible. It is those trends in operation which are the chief concern of the present book, so that useful conclusions may still be drawn from a very simplified mathematical treatment. However, the validity of these assumptions will be reviewed wherever necessary, e.g. Sections 5.6, 6.4 and 7.2.

The author is greatly indebted to Weeks and Allen[35] for the following analysis and that for power consumption in Chapter 6; however he takes full responsibility for the deductions therefrom.

Drag Flow

Assumptions 3, 4, 5 and 6 mean that flow in a straight channel may further be approximated as that between two parallel flat plates moving at an angle ϕ to each other—Fig. 5.3; for convenience the

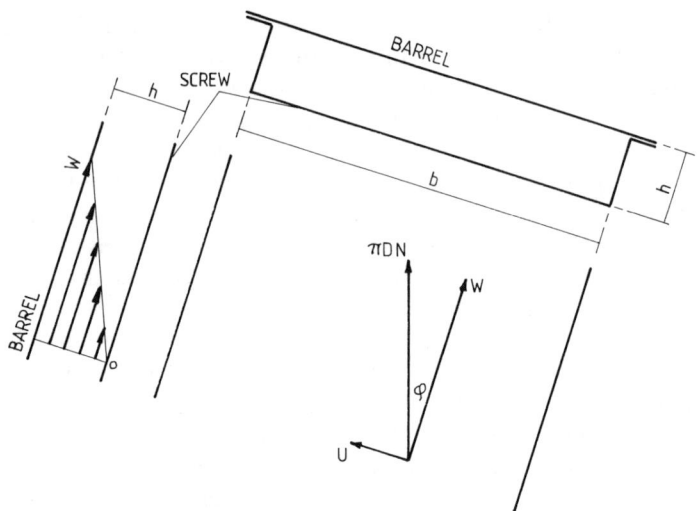

Fig. 5.3. Velocities in drag flow.

lower plate (screw) will be taken as stationary, so that velocities will be relative to the screw, not the barrel—this is obviously immaterial as far as axial components giving gross flow are concerned. Because viscosity is constant (assumptions 7 and 8) superposition of flows is valid and that due to relative motion of the plates is first considered in the absence of a pressure gradient. The velocity of the barrel relative to the screw is its circumference πD times the speed N; barrel velocity equals πDN and is at an angle ϕ to the screw channel. Since the polymer is assumed to adhere to the barrel and screw (assumption 3), its velocity at each limit will be the same as that of the adjacent solid surface, i.e. πDN at the barrel and 0 at the screw. Resolving the barrel velocity along and across the screw channel:

$$W = \pi DN \cos \phi \qquad (5.4)$$

$$U = \pi DN \sin \phi \qquad (5.5)$$

W and U being defined as in Fig. 5.2.

Evidently only the along-channel component W can cause a net flow from the extruder, although U is important in creating a transverse circulation (eqn (5.23)). Now consider an area S of fluid sheared between two parallel plates (Fig. 5.4). The shear force F exerted by one plate and balanced by the equal and opposite reaction of the other

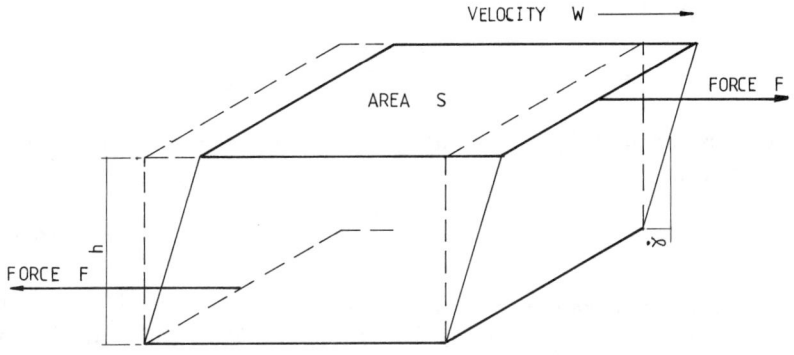

Fig. 5.4. Isothermal simple shear.

is carried by each layer of fluid, so the shear stress τ at all points through the thickness will be this force divided by the (uniform) sheared area:

$$\tau = F/S \qquad (5.6)$$

Since the polymer is homogeneous (assumption 2) and uniform temperature (assumption 8), the viscosity will be uniform. But by definition (Section 2.1):

$$\eta = \tau/\dot{\gamma} \qquad (2.5)$$

and

$$\dot{\gamma} = dw/dh \qquad (2.4)$$

Therefore under uniform shear stress τ, the shear rate $\dot{\gamma}$ will also be uniform, i.e:

$$\dot{\gamma} = W/h \qquad (5.7)$$

giving the velocity distribution shown in Fig. 5.3.

Note that in the real, helical channel, the area is proportional to radius and the shear stress and shear rate therefore decrease towards the barrel, giving a non-linear velocity distribution; assumption 5 minimises this error.

The along-channel flow dQ_D in a layer b wide dh thick is:

$$dQ_D = wb \cdot dh \qquad (5.8)$$

and the total flow:

$$Q_D = \int_0^h wb\,dh = \int_0^h b\frac{W}{h} y\,dy = \left[\frac{Wby^2}{2h}\right]_0^h \qquad (5.9)$$

Since b is constant, $\int_0^h w\,dh$ is the area $Wh/2$ under the plot of velocity vs. depth and:

$$Q_D = \frac{W}{2} bh \qquad (5.10)$$

where Q_D is the *volumetric* flow due to relative motion of barrel and screw, called *Drag flow*, and $W/2$ is the mean down-channel velocity. Substituting for W gives:

$$Q_D = \frac{\pi}{2} DNbh \cos\phi \qquad (5.11)$$

From Fig. 5.1:

$$b = (p - t)\cos\phi \qquad (5.12)$$

and substituting:

$$Q_D = \frac{\pi}{2} DNh(p - t)\cos^2\phi \qquad (5.13)$$

Note 1: As mentioned in the footnote to p. 70, Weeks and Allen[35] define t as helical width, such that eqn (5.12) becomes $b = p\cos\phi - t$

Note 2: Several authors, including Fisher,[2] neglect the flight width t so that $b = \pi D \sin\phi$ leads to a modification of eqn (5.13):

$$Q_D = \frac{\pi^2}{2} D^2 Nh \sin\phi \cos\phi$$

The present book expresses equations for flow and power primarily in terms of channel width b as being more appropriate to performance considerations; any of the above definitions may therefore be used.

Pressure Flow

To separate the effect of pressure, the screw is considered stationary, so that no drag flow takes place. Suppose a pressure is generated at the end of the screw by a separate pump—Fig. 5.5. At the same tempera-

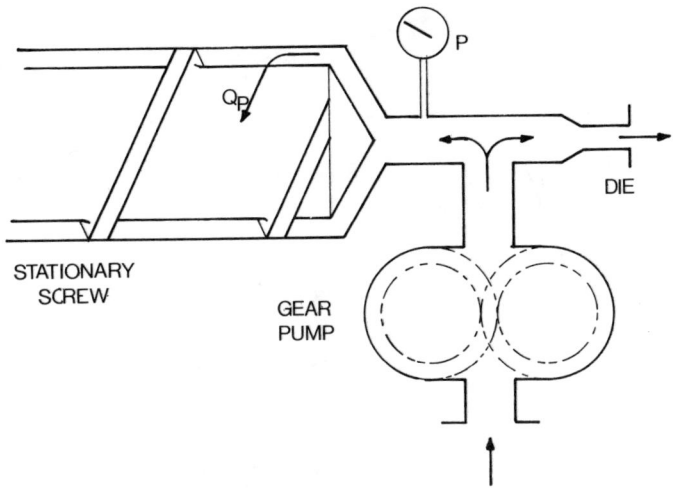

FIG. 5.5. Analogy for pressure (back) flow.

ture (and viscosity) the flow through the die will remain the same as though the pressure were generated by the extruder. However, this pressure will also cause a flow *backwards*, i.e. towards the feed section, in the screw, which behaves as a long die of rectangular cross-section ($b \gg h$, assumption 3). In Chapter 2 the pressure gradient in an 'infinite' slit die T wide by H deep was given by:

$$\frac{dP}{dL} = \frac{12\eta Q}{TH^3} \qquad (2.34)$$

Substituting the screw dimensions and putting $Q = Q_P$ for pressure flow:

$$\frac{dP}{dz} = \frac{12\eta Q_P}{bh^3} \qquad (5.14)$$

or rearranging

$$Q_P = \frac{bh^3}{12\eta} \cdot \frac{dP}{dz} \qquad (5.15)$$

Note that for a positive pressure gradient (pressure rising from feed section to die) the pressure flow is in the reverse direction, i.e. subtracts from the drag flow. The velocity distribution is parabolic (Fig. 2.8) as shown in Fig. 5.6.

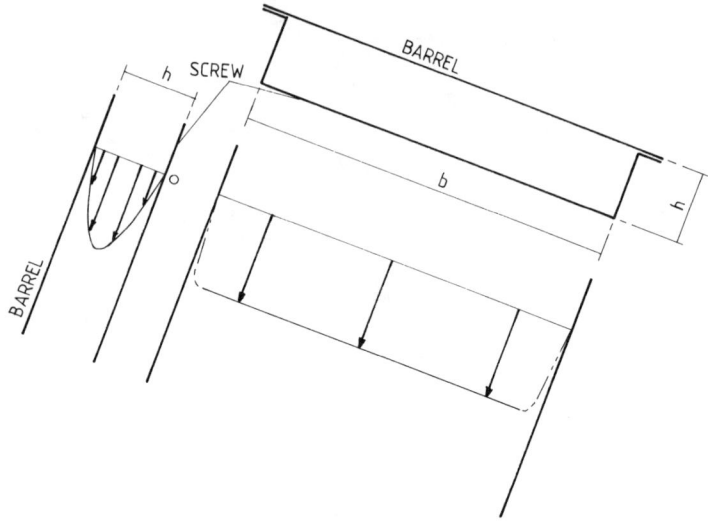

Fig. 5.6. Velocities in pressure flow.

As mentioned on p. 74 the assumption of Newtonian viscosity permits superposition of flows, i.e. the total volumetric flow Q_{Tot} is given by:

$$Q_{\text{Tot}} = Q_D - Q_P - Q_L \qquad (5.16)$$

but by assumption 6, leakage flow Q_L is negligible (see Section 5.3) so substituting eqns (5.10) and (5.15):

$$Q_{\text{Tot}} = Q_D - Q_P = \frac{Wbh}{2} - \frac{bh^3}{12\eta} \cdot \frac{dP}{dz} \qquad (5.17)$$

The same result may be obtained by the more rigorous mathematical method given in Appendix A based on the Navier–Stokes equation of one-dimensional flow. This may also be written in dimensionless form:

$$\frac{Q_{\text{Tot}}}{Wbh} = \frac{1}{2} - \frac{h^2}{12\eta W} \cdot \frac{dP}{dz} \qquad (5.18)$$

This dimensionless output, abbreviated to Q/Wbh, is of great value both theoretically and practically in characterising the operating condition of a single-screw extruder and will be referred to in the following chapters. Some authors have used the ratio (minus) pressure flow/drag flow, which has the following relation to Q/Wbh:

$$\frac{Q_P}{Q_D} = \frac{-(Q_{Tot} - Q_D)}{Q_D} = 1 - \frac{2Q}{Wbh} \qquad (5.19)$$

and

$$\frac{Q}{Wbh} = \frac{1}{2}\left(1 - \frac{Q_P}{Q_D}\right) \qquad (5.20)$$

i.e.

$$0 < Q/Wbh < \tfrac{1}{2} \quad \text{as} \quad 1 > Q_P/Q_D > 0$$

and

$$\tfrac{1}{2} < Q/Wbh < 1 \quad \text{as} \quad 0 > Q_P/Q_D > -1$$

Referring to eqn (5.17), by continuity Q_{Tot} must be approximately constant along the extruder, and therefore for a screw of constant channel width b and depth h,

$$\frac{dP}{dz} = \text{constant} = \frac{P}{Z} \qquad (5.21)$$

i.e. the pressure rises uniformly along a constant depth screw and eqn (5.17) may be rewritten:

$$Q_{Tot} = \frac{Wbh}{2} - \frac{bh^3}{12\eta} \cdot \frac{P}{Z} \qquad (5.22)$$

where Z is the helical length of the melt pumping section and P is the pressure rise over that length. Note that this is not valid for a screw of varying depth whereas eqns (5.17) and (5.18) are valid for an incremental length of such a screw.

The velocity at any point in the channel is given in Appendix A as:

$$w = \frac{Wy}{h} + \frac{y^2 - yh}{2\eta} \cdot \frac{dP}{dz} \qquad (A8)$$

and is the algebraic sum of the velocities due to drag and pressure flows. The net flow will be proportional to the algebraic sum of the areas under the velocity profiles. When $dP/dz = 0$, there is no change of pressure along the screw and no pressure flow (this is quite possible in a parallel 'metering' section if pressure is generated in the preceding tapered portion). Then the velocity profile will be as in Figs 5.3 and 5.7d, all the flow will be positive and $Q/Wbh = 1/2$. If a small back

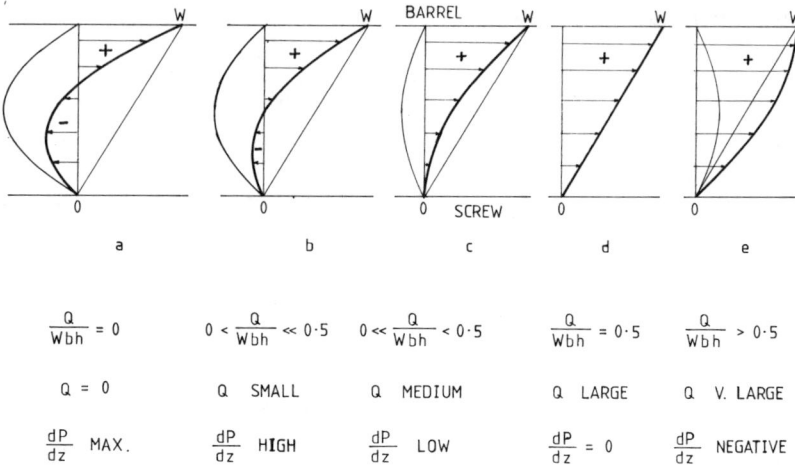

FIG. 5.7. Velocity profiles for several outputs and pressure gradients.

pressure is now introduced, dP/dz will be small and positive, causing a small (virtual) back flow represented by $-bh^3\,dP/12\eta\,dz$, reducing the total flow and hence Q/Wbh is now less than $1/2$; this is shown in Fig. 5.7c. Note that the resultant velocity profile is no longer linear and its slope at the barrel $dw/dy_{(y=h)} = \dot{\gamma}_{\text{wall}}$, the shear rate at the barrel wall is greater than in the corresponding drag flow. If P and dP/dz are further increased, output and Q/Wbh are further reduced and over part of the channel depth the resultant flow will be reversed, as in Fig. 5.7b, partially offsetting the positive flow nearer the barrel wall. At some value of pressure, the positive and negative flows will be equal (Fig. 5.7a) giving zero net flow and $Q/Wbh = 0$. This corresponds to a completely closed die and the pressure is the theoretical maximum achievable with the screw; in practice, shear heating due to 'churning' will modify this and the mechanism of melting will also be disturbed. If the pressure falls through this section of the screw, e.g. due to high pressure generation in an earlier portion of the screw, dP/dz will be negative and the pressure flow will be towards the die, giving a total flow greater than that due to drag flow alone, and $Q/Wbh > 1/2$ as in Fig. 5.7e. Note that the shear rate at the barrel wall is now reduced. The expression Q/Wbh is the ratio of the actual total flow to that theoretically possible if the polymer moved as a plug with the barrel at velocity W and slipped on the screw—like a solid nut prevented from rotating with the screw. It is therefore a kind of 'volumetric efficiency' and compared to the value 1/2 for drag flow gives an impression how much less (or greater) the actual flow is than at drag flow. More importantly, it indicates whether the pressure gradient in the screw is small (Q/Wbh slightly less than 1/2) or large (Q/Wbh tending to zero) or even negative ($Q/Wbh > 1/2$).

In practice it is difficult to envisage what is occurring within the screw, and even if a pressure gauge is fitted before the die, it does not indicate whether the pressure is seriously affecting the performance of the screw. The expression Q/Wbh can be determined in a particular operating condition if the screw dimensions are known. It is only necessary to measure accurately the screw speed (giving W) and the output, e.g. by weighing and dividing by the density *at the melt temperature* to give the volumetric output Q.

Transverse Flow

So far, only the longitudinal or down-channel component of velocity has been considered. The flow transverse to the channel has been analysed by Mohr et al.[42] starting from the Navier–Stokes equation (see Appendix A):

$$\eta \frac{\partial^2 u}{\partial y^2} + \eta \frac{\partial^2 u}{\partial z^2} - \frac{\partial P}{\partial x} = 0 \tag{A10}$$

and ignoring stagnation at the channel sides. In a channel of constant dimensions, the cross channel component:

$$U = \pi DN \sin \phi \tag{5.5}$$

is independent of z, and since the along-channel velocity distribution (Fig. 5.7) is also independent of z, it is reasonable to assume that $\partial u/\partial z = 0$ throughout and hence $\partial^2 u/\partial z^2 = 0$. Eliminating this term and rearranging:

$$\eta \frac{\partial^2 u}{\partial y^2} = \frac{\partial P}{\partial x}$$

Since viscosity η is Newtonian and isothermal, integration gives:

$$u = \frac{1}{2\eta} \cdot \frac{\partial P}{\partial x}(y^2) + Cy + D$$

but $u = 0$ at $y = 0$, and $u = U$ at $y = h$, therefore $D = 0$ and $C = \dfrac{U}{h} - \dfrac{1}{2\eta} \cdot \dfrac{\partial P}{\partial x} \cdot h$ then:

$$u = \frac{Uy}{h} + \frac{1}{2\eta} \cdot \frac{\partial P}{\partial x}(y^2 - yh) \qquad (5.23)(A12)$$

Note that this is of the same form as the along-channel velocity distribution, i.e. a 'drag flow' term due to relative motion of the two surfaces and proportional to distance from the screw, followed by a parabolic term with the pressure gradient. This is shown diagrammatically in Fig. 5.8 which is similar to Fig. 5.7a, the positive and negative areas being equal.

The total transverse flow is given by:

$$Q = \int_0^h uz\, dy = \left[\frac{U}{h}\frac{zy^2}{2} + \frac{z}{2\eta} \cdot \frac{\partial P}{\partial x}\left(\frac{y^3}{3} - \frac{y^2 h}{2}\right)\right]_0^h$$

$$= \frac{Uzh}{2} - \frac{zh^3}{12\eta} \cdot \frac{\partial P}{\partial x} \qquad (A13)$$

But since leakage is assumed negligible, the transverse flow Q must be zero. Then:

$$\frac{Uzh}{2} - \frac{zh^3}{12\eta} \cdot \frac{\partial P}{\partial x} = 0$$

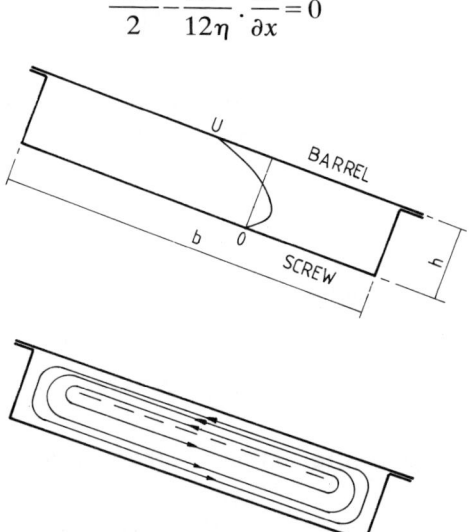

Fig. 5.8. Transverse velocity profile.

rearranging:

$$\frac{\partial P}{\partial x} = \frac{6\eta U}{h^2} \qquad (5.24)$$

or:

$$\frac{1}{2\eta} \cdot \frac{\partial P}{\partial x} = \frac{3U}{h^2}$$

Substituting eqn (5.24) in eqn (5.23):

$$u = \frac{Uy}{h}\left(\frac{3y}{h} - 2\right) \qquad (5.25)$$

giving the velocity at any point independent of the transverse pressure gradient.

In practice the flow near the barrel wall, leftwards in Fig. 5.8, must flow down the front of the flight (inwards) to return, rightwards in Fig. 5.8, nearer the screw. The circulation is completed by a flow upward at the back of the next flight (outwards). The foregoing analysis refers to the central portion of the channel where the radial component is negligible, in accord with assumption 3 that the channel is wide relative to its depth. The implications of this transverse flow for leakage, power and mixing will be discussed under Sections 5.3, 6.3 and 7.1, respectively.

The longitudinal and transverse velocities in the channel combine to form a complex flow pattern which may perhaps best be described as an oblique helix, which is itself centred on the helix described by the channel—a sort of 'coiled coil'. When the die is closed and there is zero net flow $Q = 0$ and substituting in eqn (5.17) gives:

$$\frac{dP}{dz} = \frac{6\eta W}{h^2} \qquad (5.26)$$

When this is substituted in eqn (A8), the latter gives the downstream velocity:

$$w = \frac{Wy}{h}\left(\frac{3y}{h} - 2\right) \qquad (5.27)$$

which is of identical form to eqn (5.25), U being replaced by W. But by eqns (5.4) and (5.5) these are, respectively, $\pi DN \sin \phi$ and

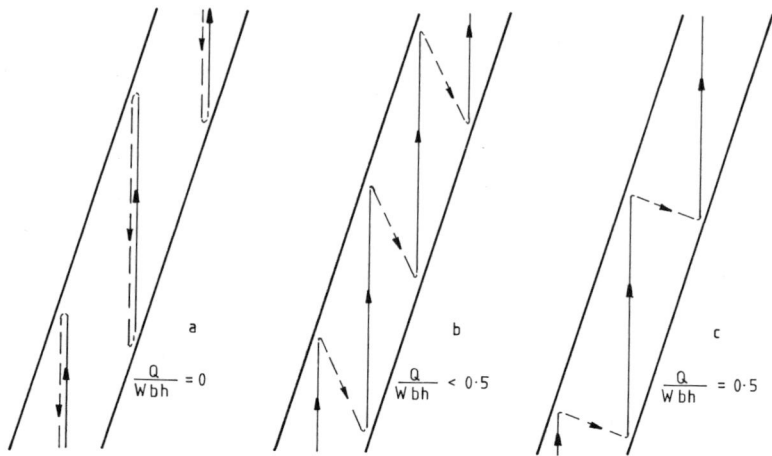

Fig. 5.9. Flow patterns in screw channel (only extreme elements shown).

$\pi DN \cos \phi$, so these vectors may be combined to give:

$$\text{total velocity} = \pi DN \frac{y}{h}\left(\frac{3y}{h} - 2\right) \quad (5.28)$$

in a plane normal to the screw axis, as shown diagrammatically in Fig. 5.9a. When the die is opened and flow commences, the transverse circulation remains but the longitudinal profile changes as shown in Fig. 5.7b, c and d, giving a resultant as in Fig. 5.9b and c. These flow patterns have been experimentally confirmed by Eccher and Valentinotti[43] and Kennaway,[44] the latter producing a ciné film of a transparent model with solid tracer particles in the fluid. Bernhardt[37] gives a more precise representation by a 'flagpole' diagram showing the changing magnitudes and directions of the total velocity vector. For pseudoplastic or other non-Newtonian fluids, where the viscosity is dependent on the total shear in all directions, the longitudinal and transverse flows are interdependent through their effects on viscosity, and superposition of the components is no longer valid. In the same way drag and pressure flows cannot be superimposed, but this will be reconsidered in Section 5.7.

5.3. LEAKAGE FLOW

This treatment is not exhaustive and is included in order to show the order of magnitude of leakage and the factors on which it depends.

Due to differences in the assumptions made, the following expressions differ somewhat from those of other authors (Fisher,[3] McKelvey,[32] Bernhardt[38]); these differences are of less importance than the conclusion that in most practical cases neglect of leakage flow is of less consequence than approximations inherent in the calculation of channel flow. Note that as shown in Section 6.6, it cannot be ignored in power consumption.

Leakage flow in the melt pumping section arises from two causes, (i) the relative motion of screw flight and barrel and (ii) the pressure difference (if any) between the two faces of the flight due to pressure gradients along and across the screw channel. The first is of the form of a drag flow, the polymer being sheared between two parallel faces so that part is carried along with the screw and part retained on the barrel surface. Taking velocities relative to the screw, the velocity of the barrel perpendicular to the channel is $U = \pi DN \sin \phi$ and thus the shear rate in the clearance δ is:

$$\frac{U}{\delta} = \frac{\pi DN \sin \phi}{\delta} \tag{5.29}$$

and the mean polymer velocity is:

$$\frac{U}{2} = \frac{\pi DN \sin \phi}{2} \quad \text{(cf. eqn (5.10))} \tag{5.30}$$

The helical length of one turn (Fig. 5.1) is:

$$\frac{Z}{n} = \frac{\pi D}{\cos \phi} \tag{5.31}$$

Thus the leakage flow due to drag is:

$$Q_{DL} = \frac{U}{2} \cdot \delta \cdot \frac{\pi D}{\cos \phi} = \frac{\pi^2 D^2 N \delta \tan \phi}{2} \text{ per turn} \tag{5.32}$$

Expressed as a fraction of the drag flow in the channel, given by eqn (5.13):

$$\frac{Q_{DL}}{Q_D} = \frac{\frac{1}{2}\pi^2 D^2 N \delta \tan \phi}{\frac{1}{2}\pi Dh(p-t)N \cos^2 \phi} = \frac{\pi D \delta}{(p-t)h} \cdot \frac{\sin \phi}{\cos^3 \phi} \tag{5.33}$$

If the approximation $b = \pi D \sin \phi$ is substituted, eqn (5.33) further simplifies to:

$$\frac{Q_{DL}}{Q_D} = \frac{\delta}{h \cos^2 \phi} \tag{5.34}$$

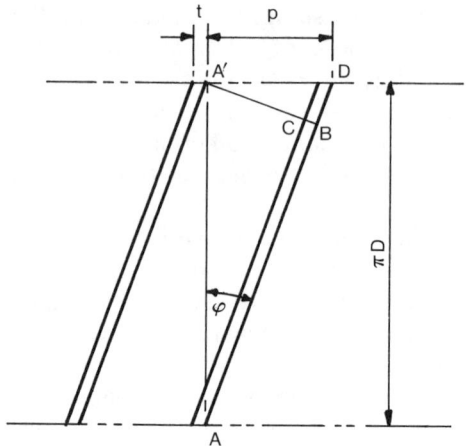

Fig. 5.10. Development of channel

Consider now a longitudinal (down-channel) pressure gradient of dP/dz. Then in Fig. 5.10 the distance $AB = \pi D \cos \phi$ and the pressure difference:

$$P_{BA} = \pi D \cos \phi \cdot \frac{dP}{dz} \qquad (5.35)$$

Point A′ represents the same position on the screw as A, and hence the pressure difference between B and C is also $P_{BC} = \pi D \cos \phi \, dP/dz$ due to longitudinal pressure gradient—positive if dP/dz is positive, i.e. pressure increasing towards the die.

Again applying eqn (2.34) for flow through a slit die, and analogous to eqn (5.15), the leakage flow Q_{PL} due to longitudinal pressure gradient will be:

$$Q_{PL} = \frac{\pi D}{\cos \phi} \cdot \frac{\delta^3}{12\eta} \cdot \frac{P_{BC}}{t \cos \phi} \text{ per turn} \qquad (5.36)$$

i.e. over length A to D in Fig. 5.10. Substituting eqn (5.35):

$$Q_{PL} = \frac{\pi^2 D^2 \delta^3}{12\eta t \cos \phi} \cdot \frac{dP}{dz} \qquad (5.37)$$

However, there is also a transverse pressure gradient between A and C in Fig. 5.10 given by eqn (5.24), i.e. $dP/dx = 6\eta U/h^2$ so that the pressure difference:

$$P_{AC} = b \cdot \frac{dP}{dx} \tag{5.38}$$

Substituting from eqns (5.5), (5.12) and (5.24):

$$P_{AC} = \frac{6\pi\eta DN(p-t)\sin\phi\cos\phi}{h^2} \tag{5.39}$$

In Fig. 5.8 it is seen that the transverse *pressure* flow is left to right, i.e. from leading to trailing sides of the channel, and hence pressure must fall in this direction. Hence P_{AC} is always positive and the total pressure difference across the flight is given by:

$$P_{BC} = P_{BA} + P_{AC} \tag{5.40}$$

This may be substituted from eqns (5.35) and (5.39), but since flow is proportional to pressure difference, the leakage flow due to transverse pressure gradient dP/dx may be calculated separately as:

$$Q'_{PL} = \frac{\pi^2 D^2 \delta^3 N(p-t)\tan\phi}{2th^2} \tag{5.41}$$

The relative importance of drag and pressure leakage flows will evidently depend on the channel pressure gradient, i.e. on whether the screw is operating on a high or low value of Q/Wbh. Thus the leakage flow due to pressure difference is conveniently expressed as a fraction of the channel pressure flow, by analogy with eqn (5.33). Adding eqns (5.37) and (5.41) and dividing by eqn (5.15) (substituting for b from eqn (5.12)):

$$\frac{Q_{PL} + Q'_{PL}}{Q_P} = \frac{\dfrac{\pi^2 D^2 \delta^3}{12\eta t \cos\phi} \cdot \dfrac{dP}{dz}}{\dfrac{(p-t)\cos\phi h^3}{12\eta} \cdot \dfrac{dP}{dz}} + \frac{\dfrac{\pi^2 D^2 \delta^3 N(p-t)\tan\phi}{2th^2}}{\dfrac{(p-t)\cos\phi h^3}{12\eta} \cdot \dfrac{dP}{dz}}$$

$$= \frac{\pi^2 D^2}{(p-t)t}\left(\frac{\delta}{h}\right)^3 \cdot \frac{1}{\cos^2\phi} + \frac{6\eta\pi^2 D^2 \delta^3 N}{th^5} \cdot \frac{\sin\phi}{\cos^2\phi} \cdot \frac{1}{\dfrac{dP}{dz}} \tag{5.42}$$

As an example, consider the following typical values:
Pitch p = Diameter D hence $\phi = 17°40'$
Flight width $t = 0\cdot 1D$ hence $b = 0\cdot 858D$
Channel depth $h = 0\cdot 05D$ (low value taken to maximise proportionate leakage)

Flight clearance = $0.0015D$ (high value taken to maximise leakage)
Substituting in eqn (5.33) gives:

$$Q_{DL}/Q_D = 0.0367 \text{ or } 3\tfrac{1}{2}\% \text{ per turn}$$

and in eqns (5.37) and (5.15) gives:

$$Q_{PL}/Q_P = 0.00326 \text{ or } \tfrac{1}{3}\% \text{ per turn}$$

To estimate the effect of transverse pressure gradient, further assumptions are necessary, e.g.

Diameter $D = 0.1$ m (100 mm or 4 in screw)
Viscosity $\eta = 10^3$ Ns/m^2 (10^4 P)
Pressure gradient $dP/dz = 20$ MN/m^2 per m (73.6 lb/in^2 per inch)
Speed $N = 1$ rev/s

Then from eqns (5.41) and (5.15):

$$Q'_{PL}/Q_P = 0.00107 \text{ or } 0.1\% \text{ per turn}$$

and from eqn (5.39):

$$\frac{Q_{PL} + Q'_{PL}}{Q_P} = 0.00433 \text{ or } 0.43\% \text{ per turn}$$

This pressure term of course only approximates for a real machine, since local increases in shear rate and temperature will increase the leakage flow as a proportion of channel pressure flow, whereas the drag term (eqn (5.33)) is scarcely affected. Note that the leakages are expressed 'per turn' since the leakage in successive turns is primarily in series and only to a limited extent in parallel, i.e. the output of a 10-turn screw would not be reduced by 35% due to drag flow leakage.

Since for a positive output $Q_P \not> Q_D$, the above example shows that in the usual case the drag flow element of leakage is the dominant factor. Analysing the effect of variables on leakage flow, eqn (5.32) shows that the leakage drag flow is proportional to the flight clearance δ, the screw speed N and the square of the screw diameter D. Further, eqn (5.33) shows that for a screw of constant helix angle ϕ (so that $(p-t)/D$ is constant), as a proportion of the channel drag flow, the leakage drag flow is proportional to flight clearance δ, inversely as the channel depth h, and independent of screw speed and diameter. Thus, as a screw wears in service, the leakage increases approximately linearly in proportion to the clearance, but will be proportionately larger for a shallow screw. As with channel drag flow (eqn (5.13) and discussion on p. 89), the leakage drag flow given by this analysis is independent of fluid viscosity, and hence of increased shear rate over the flight and consequent local shear heating. Equations (5.37) and (5.41) show that for a screw of constant helix angle ϕ ($t \propto D$) the (minor) leakage flows due to pressure differences are proportional to the diameter D and the cube of the flight clearance δ. Equation (5.42) shows that, as a proportion of the channel pressure flow, the term which is due to the longitudinal pressure gradient is dependent only on the cube of δ/h, whereas that due to the transverse gradient is more complex as there is no simple relation between longitudinal and transverse pressure gradients—however, it is still proportional to δ^3. The overall implication for operation is that the effect of leakage on output at low back pressures is broadly independent of speed and diameter, is proportional to flight clearance and proportionately greater for shallow screw, eqn (5.34); however, as back pressure is increased, the effect of clearance (wear) on the pressure terms dominates. In other words at high back pressures, a severely worn screw will rapidly lose output, or have to turn faster to generate high pressures at the same output. The implications of leakage for energy consumption and heat generation will be discussed in Section 6.3.

5.4. OUTPUT EQUATIONS AND LONGITUDINAL PRESSURE PROFILES FOR COMMON SCREW TYPES

Analysis of Variables

First the variables in the equation for parallel (constant depth) screws will be analysed. The volumetric flow equation (5.17) consists of two terms representing the drag and pressure flows, respectively. The first is given by eqn (5.10) or (5.13); for geometric similarity, helix angle ϕ is constant and p, t and b (eqn (5.12)) are all proportional to D, so:

$$Q_D \propto D^2 Nh \qquad (5.43)$$

i.e. drag flow is proportional to speed N and channel depth h, and to the square of diameter D. Drag flow can be increased by increasing screw speed and/or using a deeper screw. The effect of changing screw diameter will be considered later under Scale-up in Section 8.5. Volumetric drag flow is also independent of the fluid properties, including viscosity, of the polymer melt; the mass drag flow will therefore be proportional to the polymer melt density and decrease only slightly due to thermal expansion with increase in temperature.

Note that since the dimensionless output Q/Wbh represents the ratio of actual output to (twice) the drag flow, for a constant value of Q/Wbh, the actual output Q_{Tot} will follow the same relationships as drag flow. The second term, representing the (negative) pressure flow is given by eqn (5.15). For geometric similarity ($b \propto D$), the pressure flow is proportional to screw diameter and pressure gradient and the cube of channel depth, but inversely proportional to melt viscosity. Thus the loss of output is proportional to the pressure gradient, but increases very rapidly with increase in channel depth, resulting in a relatively low output from deep screws at high back pressures—see Fig. 5.18. The effect of viscosity is that the loss of output with increasing back pressure will be greater for low viscosity melts, e.g. due to high melt temperature or low intrinsic viscosity such as with nylons or low molecular weight polymers. The effect of shear rate, e.g. due to high screw speeds, must also be considered for pseudoplastic melts—see Section 5.7.

Longitudinal pressure gradients for various screw geometries will be considered in the next section, but the complicated effects of variables can be simplified by use of the dimensionless output Q/Wbh. It is therefore worth examining the relationship between Q/Wbh and the other variables in eqn (5.15). It has already been seen (p. 80) that an increase in (positive) pressure gradient leads to a decrease in Q/Wbh from 1/2 towards zero while a negative pressure gradient corresponds to $Q/Wbh > 1/2$. Equation (5.18) shows that if other factors are constant, a constant value of Q/Wbh corresponds to a constant pressure gradient. However, substituting eqn (5.4) for W gives:

$$\frac{Q_{Tot}}{Wbh} = \frac{1}{2} - \frac{h^2}{12\eta\pi DN \cos \phi} \cdot \frac{dP}{dz} \qquad (5.44)$$

or for constant Q/Wbh:

$$\frac{dP}{dz} \propto \frac{\eta DN \cos \phi}{h^2} \qquad (5.45)$$

Thus for constant Q/Wbh and helix angle ϕ, dP/dz will increase linearly with viscosity η, screw diameter D and speed N but decrease as the square of channel depth h. In other words a constant value of Q/Wbh implies an increase of pressure gradient for a high viscosity polymer and a large or fast-running screw, but a substantial reduction for a deep screw. Thus an increase of temperature, reducing viscosity, implies a reduction of pressure gradient, while an increase of shear rate (W/h in eqn (5.18)) implies an increase in pressure gradient. Alternatively, for a constant pressure gradient, Q/Wbh can only be held constant by adjusting the factors on the right side of eqn (5.45), e.g. if it is required to compare the performance of the same screw at the same pressure gradient, this can be done at constant Q/Wbh if a reduction in viscosity (due to temperature increase) is compensated by a proportional increase in screw speed. This, and other, operating characteristics of a melt pumping section will be discussed in Section 5.5 and Chapter 7. The foregoing analysis is in terms of pressure gradient; the relationships to pressure at the end of the screw are covered in the next section.

Constant Depth Screw

The longitudinal pressure profiles for a constant depth screw will now be derived. In the usual case, mass flow must be constant at all points along the length of the screw. Exceptions occur with additive injection, extraction and some multi-screw arrangements. Thus, except for variations in density due to pressure, temperature or chemical reaction, the volumetric flow will also be constant along the melt pumping section. Equation (5.17) shows that for an isothermal Newtonian fluid in a screw of constant speed N and dimensions D and b, the channel depth h and pressure gradient dP/dz are uniquely related. Thus in a parallel screw (constant h) dP/dz will be constant along the length of the melt pumping section and equal to P/Z (eqn (5.21)) where P is the total pressure increase and Z the helical length of the melt pumping section and $Z = L/\sin \phi$ (eqn (5.1)) where L is the axial length. Equation (5.22) applies and may be rearranged in the form:

$$P = \frac{12\eta WZ}{h^2} \left(\frac{1}{2} - \frac{Q}{Wbh} \right) \qquad (5.46)$$

or:

$$P = K_1 \left(\frac{1}{2} - \frac{Q}{Wbh} \right) \qquad (5.47)$$

where:

$$K_1 = \frac{12\eta WZ}{h^2} \qquad (5.48)$$

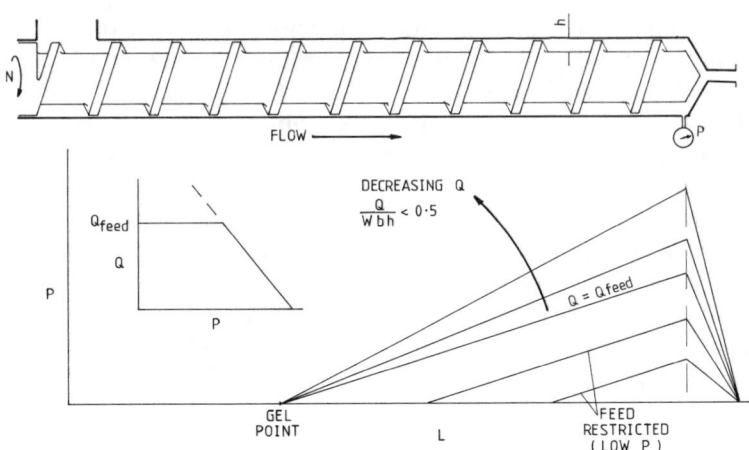

FIG. 5.11. Pressure profiles—parallel screw.

The pressure profile is therefore a straight line as shown in terms of axial length in Fig. 5.11. As the die pressure is increased, dP/dz must increase and eqns (5.17) and (5.18) show Q and Q/Wbh must decrease. In the limit, the output Q and Q/Wbh are zero, and rearranging eqn (5.18) gives:

$$\frac{dP}{dz} = \frac{6\eta W}{h^2} \tag{5.49}$$

or:

$$P = \frac{6\eta WL}{h^2 \sin \phi} \tag{5.50}$$

which represents the maximum gradient and pressure this screw can deliver—but see p. 80 for practical limitations. If the pressure is reduced, e.g. by opening the die, dP/dz will decrease and output Q will increase. In practice, before the die pressure falls to zero, other factors are likely to limit output. The most obvious is the solids feeding and conveying capacity of the feed hopper and screw, which is largely independent of die pressure, and will limit further increase in output rate. Then, as die pressure is further decreased, output remains constant and eqn (5.22) can only be satisfied by decreasing Z; thus dP/dz remains constant but pressure commences to rise further along the screw. At low back pressures, melting will be mainly by conduction and the screw is likely to deliver largely unmolten polymer. The condition of zero pressure gradient, when $Q/Wbh = 1/2$ is thus unlikely to be achieved by a constant depth solid-fed screw. It is, however, perfectly possible for a melt-fed screw (in which effective length is from the front face of the feed opening, where pressure can commence to rise) to operate with either a uniformly increasing pressure along its length or a zero change of pressure. If fed under pressure, a melt-fed screw may also suffer a uniform decrease of pressure, when $Q/Wbh > 1/2$. The same possibilities exist with a constant depth section within a compound screw; the result that there is no change of pressure along that section when $Q/Wbh = 1/2$ is extremely important, especially for a compound screw with a parallel (constant depth) 'metering' section. Limitations on practical output will be discussed under Output Corrections in Section 5.6.

Double-parallel (Stepped) Screw

The merits of a double-parallel or stepped screw (often called a 'nylon' screw) are discussed in Section 7.4. If the point at which pressure commences to rise is at, or forward of, the step, then the pressure profile will be as for the parallel screw and eqn (5.22) will apply, substituting h_2 for h. If, however, pressure commences to rise L_1 axially or Z_1 helically before the step, eqn (5.22) must be modified to the equation due to Weeks and Allen:[35]

$$\frac{Q}{Wbh_2}\left(1+\frac{Z_1 h_2^3}{Z_2 h_1^3}\right) = \frac{1}{2}\left(1+\frac{Z_1 h_2^2}{Z_2 h_1^2}\right) - \frac{h_2^2 P}{12\eta W Z_2} \quad (5.51)$$

This may be rearranged to give:

$$P = K_1\left(K_2 - \frac{Q}{Wbh_2}\right) \quad (5.52)$$

where:

$$K_1 = \frac{12\eta W Z_2}{h_2^2}\left(1+\frac{Z_1 h_2^3}{Z_2 h_1^3}\right) \text{ (similar to eqn (5.48))} \quad (5.53)$$

and

$$K_2 \simeq \tfrac{1}{2} \quad (5.54)$$

Note that this is still a linear relation between P and Q.

Alternatively (assuming D, p, t, ϕ, b, etc., are similar in the two sections), eqn (5.22) may be applied separately to the two sections:

$$Q_1 = \frac{Wbh_1}{2} - \frac{bh_1^3}{12\eta} \cdot \frac{P_1}{Z_1} \quad (5.55)$$

and

$$Q_2 = \frac{Wbh_2}{2} - \frac{bh_2^3}{12\eta} \cdot \frac{P_2}{Z_2} \quad (5.56)$$

But by continuity,

$$Q_1 = Q_2 \quad (5.57)$$

and

$$P_1 + P_2 = P \text{ (total)} \quad (5.58)$$

As for the parallel screw, the pressure gradient within each section will be constant, but generally different in the two sections. Starting from the die pressure at which $Q/Wbh_2 = 1/2$, P_2 will be zero and $P_1 = P$, i.e. the total pressure is raised in the first section. If, as is usual, $h_1 > h_2$, then $Q/Wbh_1 < 1/2$, indicating a positive pressure gradient in the first section; this is illustrated in Fig. 5.12. If the die pressure is now increased, output Q will decrease (eqn (5.51)) and both Q/Wbh_2 and Q/Wbh_1 will be less than 1/2, however, always $Q/Wbh_1 < Q/Wbh_2$ ($Q \neq 0$). Thus pressure will rise throughout the melt pumping section,

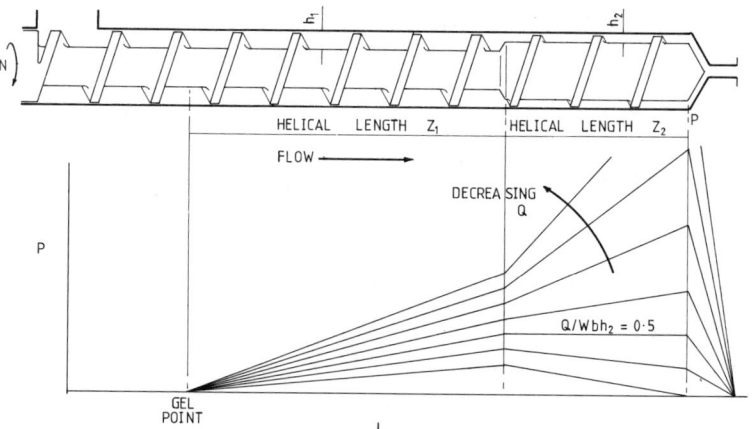

Fig. 5.12. Pressure profiles—stepped screw.

with a step change in gradient (but not pressure) at the change of channel depth. Ultimately, a die pressure will be reached at which output is zero, so that Q/Wbh_2 and Q/Wbh_1 are also zero, representing the maximum pressure achievable with the screw. It will be noted that for the usual ratios of depths h_1/h_2, Q/Wbh_1 will be much less than Q/Wbh_2; however the cubic term in eqns (5.55) and (5.56) dominates, and except at values of Q/Wbh_2 only slightly less than 1/2, dP_2/dz_2 is greater than dP_1/dz_1. The relative magnitudes of the pressure gradients are further examined in Appendix B, which shows, e.g. for $h_1/h_2 = 3$, $dP_2/dz_2 > dP_1/dz_1$, for all values of $Q/Wbh_2 < 0\cdot46$, when $Q/Wbh_1 \simeq 0\cdot15$. At this condition dP_1/dz_1 is only 4% greater than at $Q/Wbh_2 = 0\cdot5$. This means that unless the second section L_2 is very short, most of the pressure is raised in the second section of the screw; however, this is not the case when $Q/Wbh_2 > 1/2$. When die pressure is reduced below that at which $Q/Wbh_2 = 1/2$, output increases so that $Q/Wbh_2 > 1/2$ and dP_2/dz_2 is negative, i.e. pressure falls along the second section. However, Q/Wbh_1 remains less than 1/2 and pressure rises in the first section. The maximum output is achieved when pressure at the end of the screw is zero, i.e. the rise in pressure in the first section is exactly balanced by the fall in the second section. Since in this condition there is resistance to the forward movement through the melting section and output is to some˙extent limited by the melt pumping section, melting by mechanical shearing can take place. It is thus possible for a stepped screw to operate satisfactorily with low or zero die pressure, especially with a high compression ratio (large h_1/h_2) or large L_2/L_1; this can be of use where there is little or no die resistance as in thick tube and section extrusion, or in flow moulding. In effect the second screw section becomes a resistance equivalent to a die, in which melt is forced through it at a greater rate than its own natural pumping rate (when $Q/Wbh_2 = 1/2$). A further increase in output would imply a negative die pressure, where the pressure rise in the first section was insufficient to overcome the 'resistance' of the second section, and evidently Q/Wbh_1 cannot reach a value of 0·5. Screws having the second section deeper than the first ($h_1/h_2 < 1$) have been proposed; in this case the pressure profiles are transposed; at high die pressures, pressure will rise rapidly in the first section and more gradually in the second, whereas at low die pressures, pressure in the first section may be constant zero or even become negative, and rise in the second section to a zero or positive value.

Taper and Taper-parallel Screws

Similar features are to be found with screws having a gradual taper (continuous decrease of h), with or without a parallel section following without a step. In this case the channel depth is assumed to decrease uniformly from h_1 at the commencement of pressure rise over a helical length Z_1 to h_2 at the die or start of the parallel section of depth h_2 and helical length Z_2. Note that in this case, if the point at which pressure commences to rise changes in a real screw, both Z_1 and h_1 will also change, though the taper angle θ remains constant. Weeks and Allen[35] have derived relations between volumetric output and die pressure for the taper screw:

$$\frac{Q}{Wbh_2}\left(1-\frac{h_2^2}{h_1^2}\right) = 1 - \frac{h_2}{h_1} - \frac{h_2 P \tan\theta}{6\eta W} \qquad (5.59)$$

and for the taper-parallel screw:

$$\frac{Q}{Wbh_2}\left(1+\frac{2Z_2\tan\theta}{h_2}-\frac{h_2^2}{h_1^2}\right) = 1 + \frac{Z_2\tan\theta}{h_2} - \frac{h_2}{h_1} - \frac{h_2 P \tan\theta}{6\eta W} \qquad (5.60)$$

where θ is the half angle measured *helically*, i.e.

$$\tan\theta = \frac{h_1 - h_2}{Z_1} \qquad (5.61)$$

Alternatively, eqn (5.17) may be applied to any small element of length dz of the first (tapered) section and rearranged as:

$$\frac{dP}{dz} = \frac{12\eta W}{h^2}\left(\frac{1}{2} - \frac{Q}{Wbh}\right) \qquad (5.62)$$

and integrated between the limits $h = h_1$ at $z = 0$, and $h = h_2$ at $z = Z_1$, remembering that Q is constant and for a constant pitch screw W and b are also constant. By similar triangles:

$$h = h_2 + (h_1 - h_2)\left(\frac{Z_1 - z}{Z_1}\right) = h_1 - \frac{z}{Z_1}(h_1 - h_2) \qquad (5.63)$$

Thus

$$\frac{dh}{dz} = -\left(\frac{h_1 - h_2}{Z_1}\right) \qquad (5.64)$$

Then since $P = 0$ at $z = 0$ by definition of z, the pressure rise in the first section P_1 is given by:

$$P_1 = \int_0^{Z_1} \frac{dP}{dz} \cdot dz = -\frac{6\eta W Z_1}{h_1 - h_2} \int_{h_1}^{h_2} \left(\frac{1}{h^2} - \frac{2Q}{Wbh^3}\right) dh$$

$$= -\frac{6\eta W}{\tan \theta} \left[-\frac{1}{h} + \frac{Q}{Wbh^2}\right]_{h_1}^{h_2}$$

$$= \frac{6\eta W Z_1}{h_1 h_2} \left(1 - \frac{Q}{Wbh_2} \cdot \frac{h_1 + h_2}{h_1}\right) \tag{5.65}$$

which is an alternative form of eqn (5.59).

Equation (5.22) may then be rearranged and applied to the parallel section (if any):

$$P_2 = \frac{6\eta W Z_2}{h_2^2} \left(1 - \frac{2Q}{Wbh_2}\right) \tag{5.66}$$

and combined with eqn (5.65), remembering that $Q_1 = Q_2$ and P (Total) $= P_1 + P_2$:

$$P = \frac{6\eta W Z_1}{h_1 h_2} \left(1 - \frac{Q}{Wbh_2} \cdot \frac{h_1 + h_2}{h_1}\right) + \frac{6\eta W Z_2}{h_2^2} \left(1 - \frac{2Q}{Wbh_2}\right) \tag{5.67}$$

If $Q/Wbh_2 = 1/2$, then the second term in eqn (5.67) is zero (no pressure rise in the parallel section) and since $h_2/h_1 < 1$ both eqns (5.65) and (5.67) give P_1 positive. Since $h_1 > h > h_2$, the differential form of eqn (5.62) shows that dP/dz in the first section is always positive, though tending to zero as $h \to h_2$. To complete the pressure profile it is necessary to examine the trend of the pressure gradient. Differentiating eqn (5.62) gives:

$$\frac{d^2P}{dz^2} = \frac{-12\eta W}{h^3} + \frac{36\eta Q}{bh^4} \tag{5.68}$$

A point of inflexion occurs when $d^2P/dz^2 = 0$ or:

$$\frac{Q}{Wbh} = \frac{1}{3} \tag{5.69}$$

Thus the pressure profile for $Q/Wbh_2 = 1/2$ is as shown in Fig. 5.13, but note that the point of inflexion is independent of Q/Wbh_2 and occurs at all values of the latter greater than 1/3. As die pressure is increased, the pressure gradient dP/dz increases at each point along the length of the screw, and output Q decreases, giving a corresponding decrease in

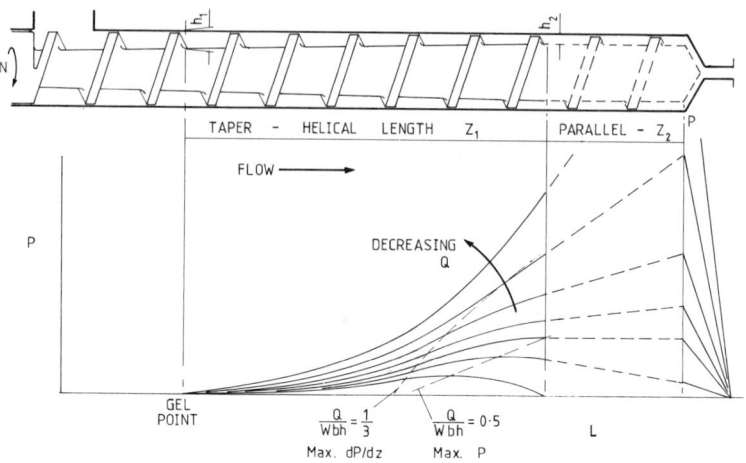

FIG. 5.13. Pressure profiles—taper and taper-parallel screws.

Q/Wbh at each point. At high die pressures giving values of Q/Wbh_2 less than 1/3, the pressure gradient increases continuously along the taper section so that the major portion of the pressure rise is contributed by the later, shallower portion of the screw. Since there is no step change in h at the end of the taper, there is no change in Q/Wbh and dP/dz, so that at all die pressures the pressure profile in the parallel section (if any) is tangential at that point to the curve in the taper section. If there is a parallel section of substantial length then at high die pressures this section will contribute the major portion of the pressure rise, in which case the output/pressure relation will be similar to that of a simple parallel screw of the same depth and slightly greater length than the parallel 'metering' section, i.e. the first term in eqn (5.67) can be neglected in favour of a small increase in Z_2 in the second term. If the die pressure is low so that Q/Wbh_2 is greater than 1/2, pressure will fall at the end of the taper section and in the parallel section, if present. Thus a maximum pressure will occur at some point within the taper at which $Q/Wbh = 1/2$. This point, and that at which the point of inflexion occurs ($Q/Wbh = 1/3$), will move back towards the feed end as die pressure is reduced and output increased; a greater *proportion* of the pressure is raised in the earlier portion of the taper, which becomes more significant in determining the output for a given die pressure. As with the stepped screw, maximum output occurs at zero die pressure, and since positive pressure is generated within the screw, this type also may operate satisfactorily with low or zero die pressure.

An industrial example illustrates the higher pressures possible within a screw than at the die and the value of the ratio Q/Wbh in understanding the conditions within the screw. A 200 mm (8 in.) diameter solid-fed extruder had a taper-parallel screw of pitch = diameter ($\phi = 17°36'$), total length 24 diameters and a metering section depth of 5·59 mm (0·22 in.). At 60 rpm it gave an output of 1091 kg/h (2400 lb/h) of LD polyethylene with a die pressure of only 13·8 MN/m² (2000 lb/in²). Trouble was experienced with granulation due to the high melt temperature and the drive motor tended to be overloaded. With a melt density of approximately 760 kg/m³ (47·4 lb/ft³), Q/Wbh_2 was calculated as 0·672 showing that pressure was falling in the parallel metering section; this only required knowledge of the screw dimensions, speed and output. When a pressure gauge was fitted half way along the metering section it registered 55·2 MN/m² (8000 lb/in²), indicating a pressure at the start of the metering section of 96·6 MN/m² (14 000 lb/in²). As shown by Fig. 5.13 the maximum pressure would be even greater. The immediate remedy was to reduce speed, reducing shear heating and pressures (eqn (5.66) and die equation) and increase barrel temperatures to reduce power input and pressures through reduction in viscosity (eqns (6.27) and (5.66), respectively). A lower output, of course, resulted, but it was judged preferable to continue making good product at low rate, rather than scrap at a high rate, until a screw of different profile could be obtained!

Equation (5.65) is of the form:

$$P = K_3\left(K_4 - \frac{Q}{Wbh_2}\right) \quad (5.70)$$

where:

$$K_3 = \frac{6\eta W Z_1 (h_1 + h_2)}{h_1^2 h_2} \quad (5.71)$$

and:

$$K_4 = \frac{h_1}{h_1 + h_2} \quad (5.72)$$

Also eqn (5.67) may be rearranged to:

$$P = K_5\left(K_6 - \frac{Q}{Wbh_2}\right) \tag{5.73}$$

where:

$$K_5 = \frac{6\eta W}{h_2}\left[\frac{(h_1+h_2)Z_1}{h_1^2} + \frac{2Z_2}{h_2}\right] \tag{5.74}$$

and:

$$K_6 = \left(\frac{Z_1}{h_1} + \frac{Z_2}{h_2}\right) \Big/ \left[\frac{(h_1+h_2)Z_1}{h_1^2} + \frac{2Z_2}{h_2}\right] \tag{5.75}$$

Thus for both the taper and taper-parallel screws P is linear with Q, which will be utilised in the next section.

Some authors, e.g. Bernhardt[39] have identified the value $Q/Wbh = 1/3$ with some form of optimum design. Figure 5.13 shows that for tapered screws it represents neither the maximum output nor maximum pressure, only that point of inflexion in the pressure curve and the maximum value of the pressure gradient. For a constant depth screw, rearranging eqn (5.22) in the form of eqn (5.46) and differentiating P with respect to h, treating Z and Q as constant gives:

$$\frac{dP}{dh} = \frac{-12\eta WZ}{h^3} + \frac{36\eta ZQ}{bh^4} \tag{5.76}$$

and:

$$\frac{d^2P}{dh^2} = \frac{+36\eta WZ}{h^4} - \frac{144\eta ZQ}{bh^5} \tag{5.77}$$

but $dP/dh = 0$ when $Q/Wbh = 1/3$ and then d^2P/dh^2 is negative, indicating a maximum value of P. Thus the corresponding depth h of screw will give the maximum pressure for given values of output and speed; as shown by comparing eqns (5.52) and (5.73) with (5.46) this is approximately true for variable-depth screws at high back pressure. However, to the present author, this seems of doubtful value, since in practice the pressure required is determined by the type of die in use and output is limited by considerations of energy, mixing and stability, as will be discussed later in Chapters 7 and 8.

The foregoing shows the pressure profiles along several common types of screw and their changes with output and back pressure. These can be extended to other screw types, but increasingly complex output/pressure equations can better be solved by use of graphical presentations.

5.5. GRAPHICAL REPRESENTATION OF OUTPUT FOR SCREW/DIE COMBINATIONS, INCLUDING VENTING

Graphical Representation

From eqn (5.10) it is seen that the drag flow is proportional to W and hence to screw speed N (eqn (5.13)). Equation (5.15) shows that the pressure gradient dP/dz, and hence pressure, is independent of screw speed for a Newtonian fluid. Thus at *constant pressure*, output Q is linear with N (eqn (5.22)) being the proportional drag flow minus a constant pressure flow of the form:

$$Q_{Tot} = AN - \frac{BP}{\eta} \quad (5.78)$$

where:

$$A = \frac{\pi}{2} Dbh \cos \phi \quad \text{(from eqn (5.11))} \quad (5.79)$$

and:

$$B = \frac{bh^3}{12Z} \quad \text{(from eqn (5.22))} \quad (5.80)$$

Thus output vs. speed is represented in Fig. 5.14 by a series of parallel straight lines for various back pressures. Equation (5.78) shows that as temperature increases, causing a decrease in viscosity η, the pressure flow Q_P will increase and the total output decrease, as shown by the dashed lines for constant pressure and increasing temperature. Of more practical interest is the variation of output with speed for the combination of extruder and a fixed die. From eqn (2.37), the Q/P relation for a die of fixed geometry is:

$$Q_{Die} = \frac{KP}{\eta} \quad (5.81)$$

where:

$$K = \frac{\pi R^4}{8L} \quad \text{for a capillary radius } R, \text{ length } L \quad (5.82)$$

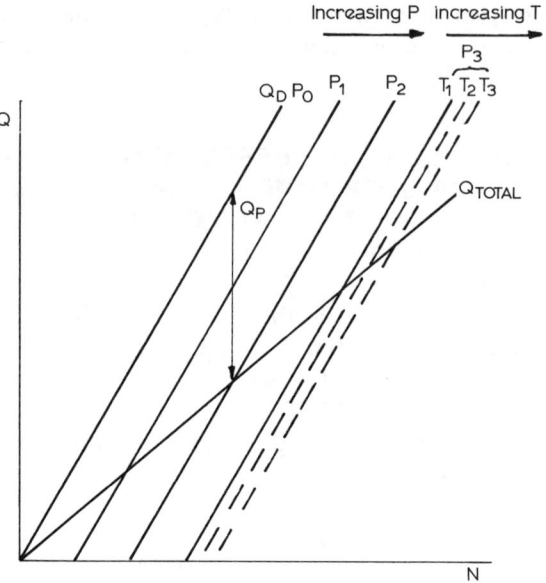

FIG. 5.14. Total output vs. speed.

or:

$$K = \frac{TH^3}{12L} \quad \text{for an 'infinite' slit width } T, \text{ depth } H \quad (5.83)$$

But by continuity:

$$Q_{\text{Tot}} = Q_{\text{Die}} \quad (5.84)$$

Combining eqns (5.81) and (5.84) and rearranging:

$$P = \frac{\eta Q_{\text{Tot}}}{K} \quad (5.85)$$

and substituting in eqn (5.78) and rearranging:

$$Q_{\text{Tot}}\left(1 + \frac{B}{K}\right) = AN \quad (5.86)$$

that is, as shown in Fig. 5.14, the total output and pressure are proportional to speed N:

$$Q_{\text{Tot}} \propto P_{\text{Die}} \propto N \quad (5.87)$$

This is only approximate since in practice temperatures are also likely to change with speed—see Section 7.2. Pseudoplastic fluids are considered in Section 5.7, but for convenience the effect on output vs. speed is included here. Most important is that shear rates and hence viscosities in screw channel and die will be different (eqns (5.78) and (5.81)) and therefore the bracketed term in eqn (5.86) becomes $(1 + B\eta_{Die}/K\eta_{Screw})$. Further, even if the separation of drag and pressure flows is still taken to be valid for a pseudoplastic fluid (see p. 74 and 84), it will be seen in Section 5.7 that wall shear rate is not proportional to pressure or pressure flow. On the other hand, Section 6.4 shows that temperature increase with speed will be less than with a Newtonian fluid. As shown in Fig. 5.35 in Section 5.7, increasing speed leads to increasing shear rate and temperature, causing a decrease in viscosity (η_{Screw}) and increase in the (fictitious) pressure flow. At the same time, increasing flow in the die involves increase of shear rate and decrease in viscosity (η_{Die}) so that pressure does not increase in proportion to output Q; since this affects the total flow whereas screw viscosity affects only the pressure flow, the former might be expected to dominate except at very high pressures. In practice it is often found that output and speed are nearly linear until feed or melting restrictions become apparent (see Section 5.10). At high pressures or with large machines where temperature control is difficult, the output/speed curve may droop, probably due to temperature effects on viscosity.

Equations (5.22), (5.51), (5.59) and (5.60) indicate that for several common types of screw the drag flow is proportional to speed and independent of pressure. Thus in Fig. 5.15 the Q/P diagram shows Q_D as a series of horizontal lines at distances above the origin proportional to speed N. Figure 5.16 shows a similar effect of channel depth h, drag flow at constant speed being proportional to channel depth. According to eqn (5.22), the (negative) pressure flow Q_P is proportional to pressure P and independent of speed N, so Fig. 5.17 consists of Fig. 5.15 with a superimposed series of parallel straight lines for different speeds commencing from drag flow Q_D at zero pressure and decreasing linearly with pressure P; then $Q_{Tot} = Q_D - Q_P$ (eqn (5.17)) is represented by the ordinate. Equations (5.52), (5.70) and (5.73) show that for other types of screw output also varies linearly with pressure. In Fig. 5.18, pressure flow Q_P is similarly superimposed on Fig. 5.16 with Q_{Tot} as the resulting ordinate, but since Q_P is proportional to the cube of channel depth h, the slope of the Q/P line increases rapidly with increase of channel depth. This diagram immediately explains several

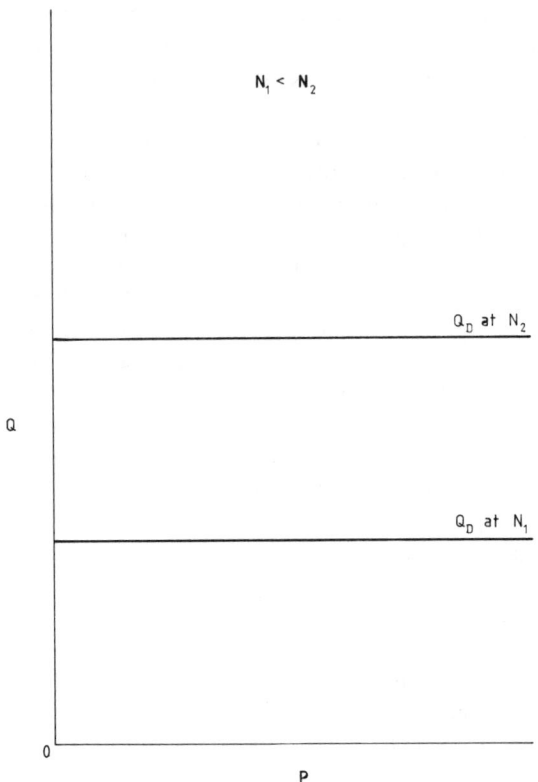

Fig. 5.15. Drag flow vs. speed.

operational features of screws of different depths—as explained on p. 98, final channel depth is of more significance than 'compression ratio' h_1/h_2 and use of the latter term will be avoided as a common but misleading colloquialism. Firstly, for a given increase in pressure, the output of a deep screw will decrease much more than that of a shallow screw—for this reason they are sometimes referred to as 'soft' and 'hard' screws, respectively. Secondly, although the output of a deep screw will be greater than that of a shallow one at low pressures, at high pressures the reverse will be true. The maximum pressure achievable (at zero output) will also be less than that of a shallow screw. The effects of decreasing melt pumping section length Z will be similar to those of increasing depth h, though less dramatic since Q_P varies only

Principles of Single-Screw Extrusion

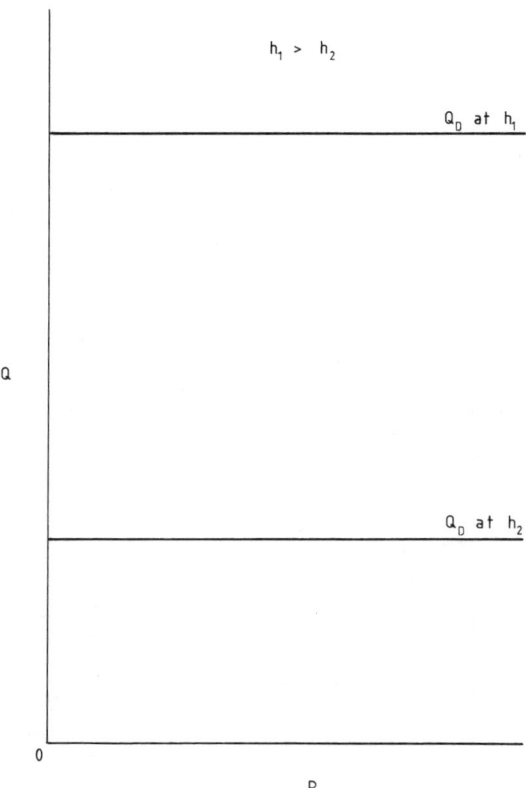

FIG. 5.16. Drag flow vs. channel depth.

linearly with Z (eqn (5.15)). Since Q/Wbh is proportional to Q_{Tot}, the vertical ordinate of Fig. 5.18 may also represent a linear scale of Q/Wbh for a given speed and channel depth, with a value of 1/2 at zero pressure, corresponding to drag flow. Values of $1/2 < Q/Wbh < 1$ represent negative pressure changes by continuation of the output/pressure line to the left of the vertical axis—this will be used in considering two-stage screws etc., in the next section. This output/pressure line is a graphical representation of eqn (5.22) or (5.46) and must be satisfied for the screw in question; however, it gives no information on what values of output, and its corresponding pressure, will be experienced in combination with a die. The output/pressure relation for a die with a Newtonian isothermal fluid is

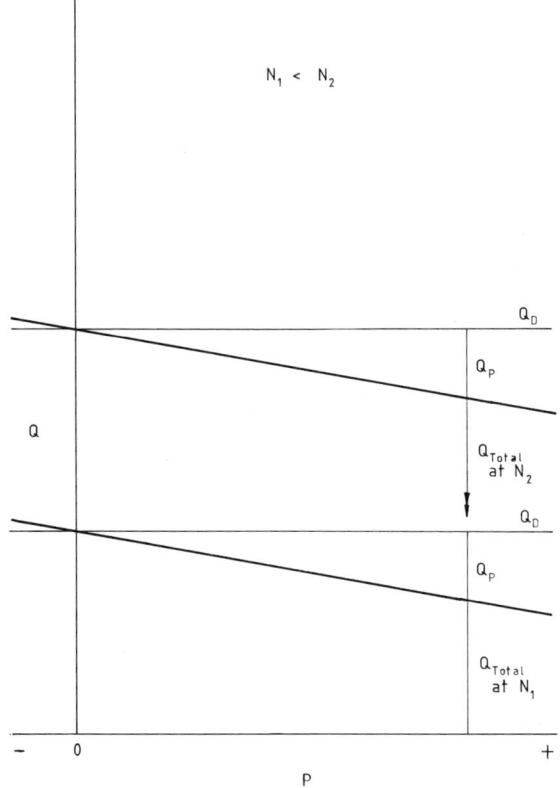

Fig. 5.17. Pressure and total flows vs. speed.

given by eqn (5.81), which is represented by a straight line through the origin. Equations (5.82) and (5.83) show that if the die is narrow and/or long, K will be small and the slope small; if wide and/or short the slope will be large as shown in Fig. 5.19, the slope representing K/η. For a screw and die in steady operation, the relations between output Q and pressure P must both be satisfied, and the intersection of the screw and die lines represents the unique (real) solution to two simultaneous equations. Figure 5.20 shows that for a large die, a deep screw may give a substantially higher output and pressure (point A) than a shallow screw (point B). However, with a small die the shallow screw (point D) may give a higher output than a deep one (point C) at the same speed; this may be overcome by increasing the speed of the

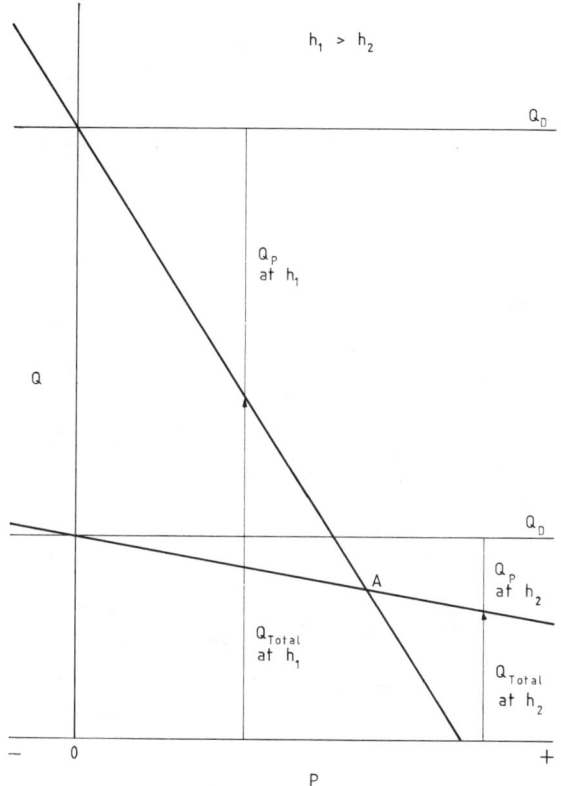

Fig. 5.18. Pressure and total flows vs. depth.

latter, the choice between screw designs, like that of back pressure, then becomes a matter of energy, mixing and stability (see Chapters 7 and 8). Modification to this diagram due to non-isothermal and non-Newtonian operation will be considered in Sections 5.7 and 5.8. The use of this type of diagram to solve more complex problems of operating conditions and especially the effect of changes in variables on the output, will now be exemplified, using the isothermal Newtonian approximation for simplicity.

The output characteristics of several practical screw/die combinations will now be examined. In addition to the output/pressure change or output/pressure gradient relations, e.g. eqns (5.17), (5.22), (5.51), (5.59) and (5.81) which may be deduced for sections of screw or die,

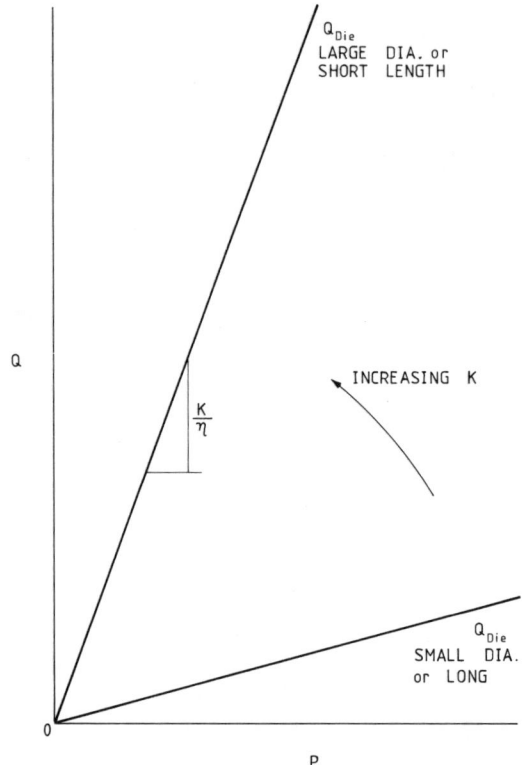

Fig. 5.19. Die characteristics (Newtonian).

continuity requires that for sections in series:

$$Q_1 = Q_2 = Q_3, \text{ etc.} \tag{5.88}$$

and:

$$dP_1 + dP_2 + dP_3 \cdots = 0 \text{ (including pressure drop through the die)} \tag{5.89}$$

In the case of operation in parallel:

$$Q_1 + Q_2 + Q_3 \cdots = Q_{\text{Tot}} \tag{5.90}$$

and:

$$dP_1 = dP_2 = dP_3, \text{ etc.} \tag{5.91}$$

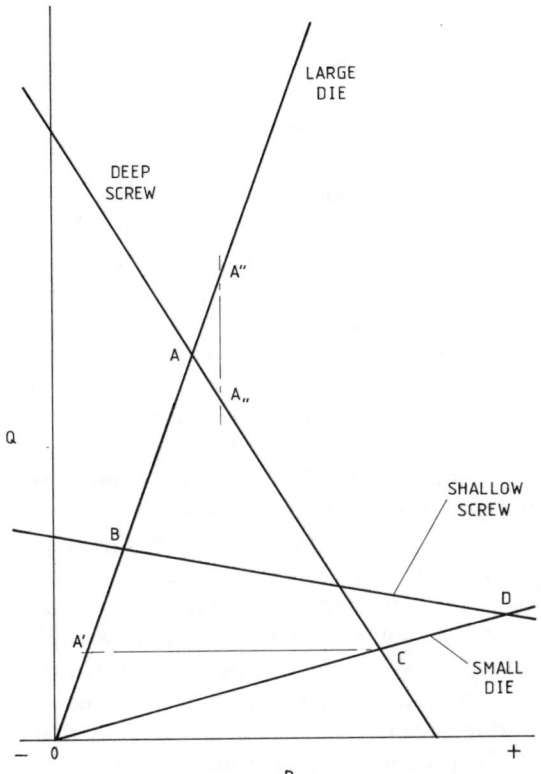

FIG. 5.20. Operating points of screw/die combinations.

Further, where screw sections are on a common shaft, speed N is the same for each section, and where diameter D and helix angle ϕ are also constant, down-channel velocity W and channel width b will be constant.

Venting or Vacuum Extraction

The first example relates to venting or vacuum extraction. The purposes of venting are:

(i)   to remove air trapped in the feed section;
(ii) to remove water vapour on or in solid feed material;
(iii) to remove residual volatiles, e.g. monomer, diluent;
(iv) to remove degradation products produced during heating in the screw.

This may assist in giving non-porous, steady, smooth-surfaced product and in reducing odour, die build-up and 'smoking', e.g. due to high temperatures used in paper-coating with LD polyethylene. Trapped air, and to a partial extent (ii) and (iii), can also be achieved by vacuum feeding—see Section 5.10; (iv) can obviously only be achieved after melting.

Several methods are used of which the commonest is probably a decompression section in the screw with a vent in the barrel over it (Fig. 5.21). All the polymer melt passes over the vent, which is therefore liable to blockage; various shapes, including removable inserts, have been used to reduce blocking of the vent and assist in clearing material forced into the vent, such as that shown in Fig. 5.22 where the vent port is arranged tangentially above the down-going side of the screw, and elongated to increase the area for vapour release, and the trailing edge is well chamfered to assist in re-entrainment—see Section 5.10. Since the principle is that after melting in the first section of the screw, the pressure is reduced to zero under the vent, a second shallow section is usually necessary to raise the requisite pressure before the die. This may be satisfactory for one set of operating conditions, though requiring careful prediction in the design of the screw, but is quite inflexible, since an increase in die pressure is liable to cause vent blockage. Analysing the output/pressure relations, speed is the same for both sections and while the vent is unblocked the first section is operating against zero back pressure so that its output is independent of die pressure. By continuity, if the vent is to remain unblocked, $Q_2 \geqslant Q_1$, so h_2 must be greater than h_1. As the die pressure P_2 is increased, Q_2 will decrease until at $P_2 > P$ (Fig. 5.23) it is less than Q_1, when the vent will block since the second section is not removing material from the vent region as fast as the first section is supplying it. To avoid this, h_2 is often designed considerably greater than h_1, which may allow a slightly higher die pressure P' but, as Fig. 5.18 shows, leads to a steeper gradient of dQ/dP, i.e. output falls more for a given increase in pressure and therefore this design is even more sensitive to changes in die pressure. To permit vapour release, the screw channel at the vent must be only partially filled; if the second section is tapered, it will fill at some point depending on the die pressure. If the die pressure is increased, the same output can be maintained if the channel is filled further back (Fig. 5.24), increasing the effective length, until eventually it fills at the vent, causing blockage. However, as will be seen in Section 5.6, the pumping rate of a

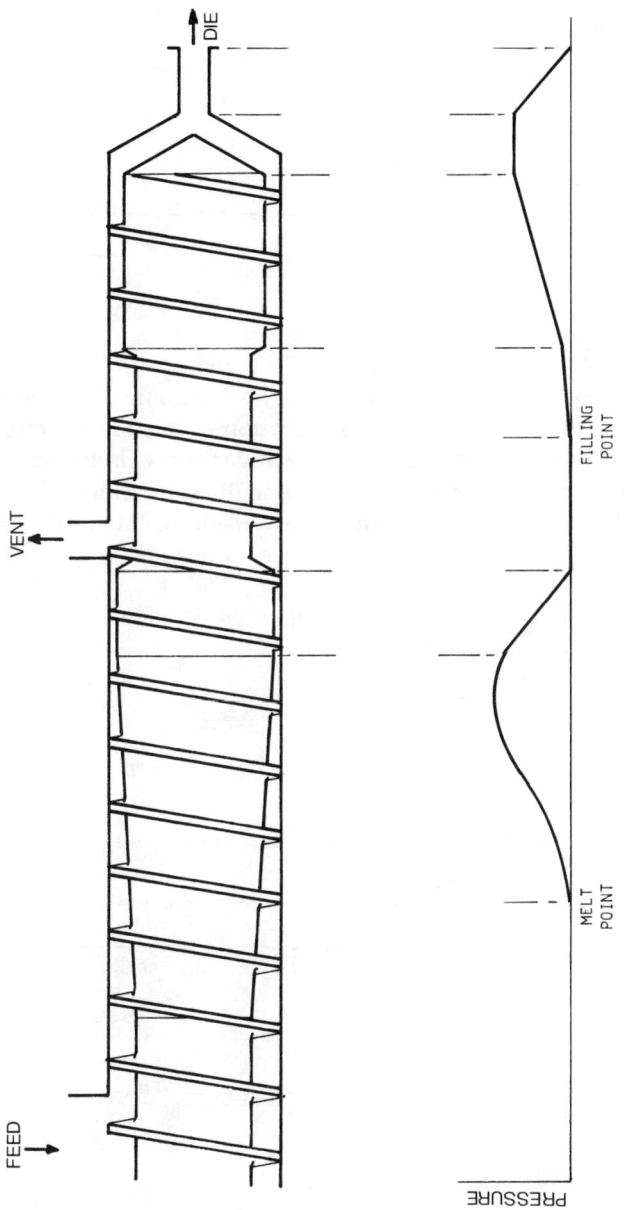

Fig. 5.21. Two-stage vented screw—pressure profile.

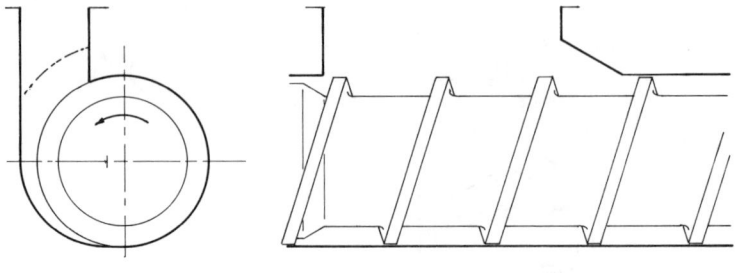

FIG. 5.22. Vent for vacuum extraction.

partially filled screw is very sensitive to the degree of filling and surging is likely to result in otherwise satisfactory conditions; this is aggravated by increasing the depth of the second section h_2 beyond the minimum.

Another method of venting with a similar screw is to make use of the partial filling to avoid the polymer melt passing directly over the vent by placing the latter in the screw itself just behind a flight,[45]

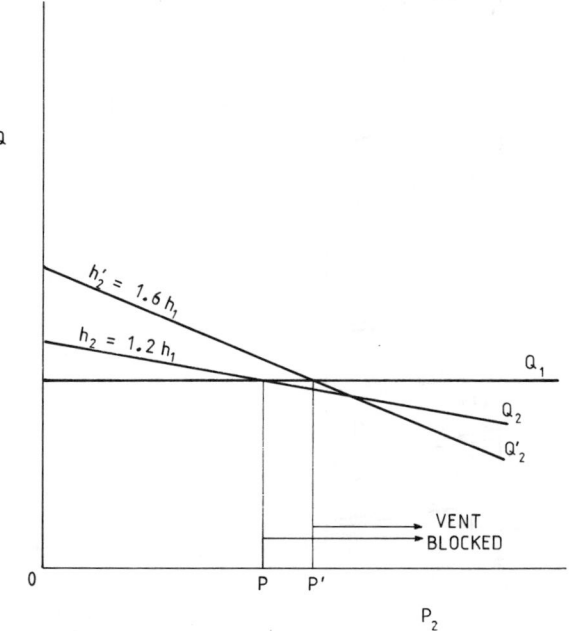

FIG. 5.23. Die pressure effect on venting.

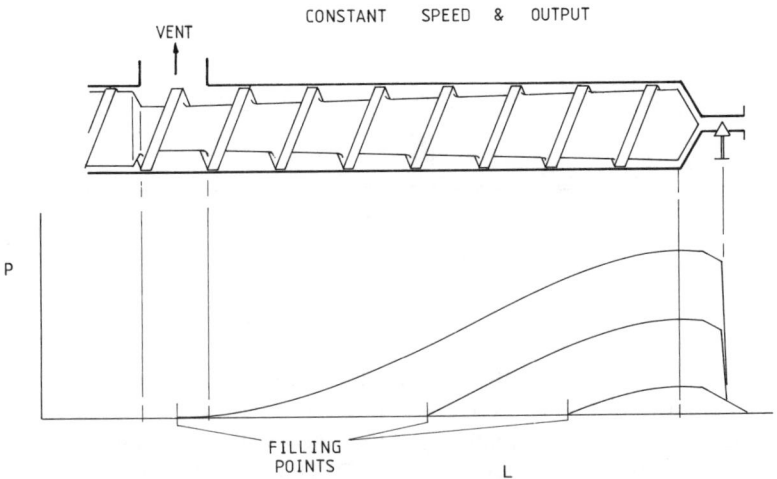

FIG. 5.24. Effect of die pressure on filling position.

where filling is normally last to occur. The vent is then taken out through a hollow screw to a rotary joint at the drive end. The output/pressure characteristics are similar to the preceding method, but the area of the vent is very restricted, it is difficult to incorporate with screw heating or cooling, and if it becomes blocked is inaccessible for manual clearing. Since the first section is operating against zero back pressure, the last part of this section is acting as a resistance to raise pressure and promote melting and mixing in the earlier parts; it has been proposed to replace this with a restriction in the form of a short annulus, e.g. a 'smearhead', permitting a deeper first section with probably a higher output and/or more complete mixing. However, the output of the first section is still independent of die pressure; limitation of the latter and/or surging are still probable. If this restriction is carried to the limit, with a close clearance and reverse-pitch groove for sealing, the flow may be diverted through a bypass channel in the barrel and returned to the screw downstream of the vent position. In this way the melt does not pass over the vent and in the event of accidental blockage, the second section of the screw will tend to pump the vent clear. A refinement of this system is to incorporate an adjustable valve in the bypass[46] so that the back pressure P_1 on the first screw section can be varied to compensate for changes in die pressure P_2—Fig. 5.25. This, in general, will permit higher values of

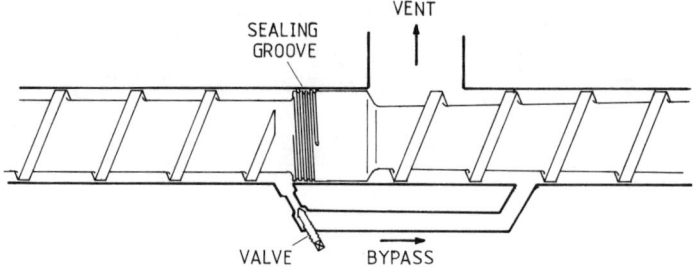

Fig. 5.25. Valved bypass venting system (Bone–Cravens patent).

die pressure without vent blocking, but more importantly enables the depths of the first and second screw sections to be more nearly equal and so reduce the risk of surging. Operation with a given screw may be summarised by a diagram of P_1 vs. P_2 (Fig. 5.26); a pressure gauge upstream of the valve will assist the operator. The author has seen the vent blocked by increasing P_2 and subsequently cleared by increasing P_1 accordingly. This system has an additional advantage of a large melt surface for release of volatiles depending on the filling point of the second section. Actual experience has shown two limitations: (i) thermal degradation due to long residence time of polymer in the seal or vent regions may contaminate the product; (ii) although adjustment of the valve will avoid vent blocking, the increased back pressure on the first screw section will alter the work input and melt temperature. This

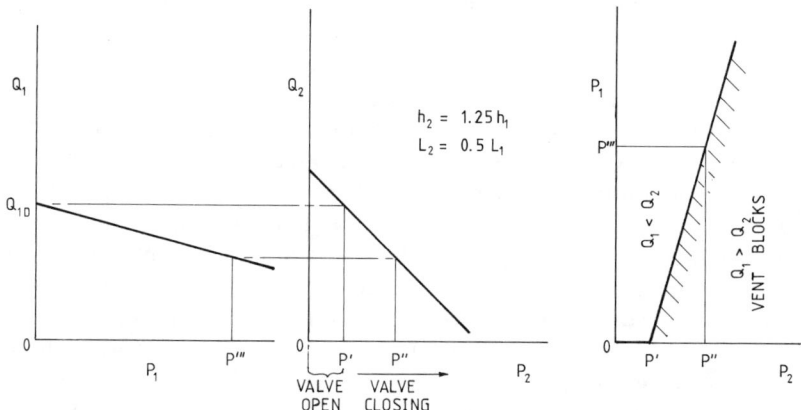

Fig. 5.26. Valve closing vs. die pressure P_2 to avoid vent blocking.

is qualitatively the same as in a simple unvented screw but made more critical by the understandable tendency to try to accommodate venting within the length of a simple screw, thus reducing the effective length for melting and mixing, but expecting similar performance in terms of output, melt temperature, stability, etc. The present author believes it would be more correct to treat the vent and second screw section as an addition to the length of the simple screw and accept a compromise between output and the additional shear history imposed on the polymer by the extra length. Within-the-barrel systems may remove up to about 5% by mass of water vapour—multiple extract points have been used to remove approximately 15% of a volatile monomer down to much less than 1% residual content.

A more radical system[6] is the use of two distinct extruders with a venting vessel (Fig. 5.27) between the first solid-fed machine which melts and pumps the polymer to the vessel, and a second melt-fed machine pumping the devolatilised polymer to the die. The vessel must be heated and may have a stirrer and/or vacuum applied to assist in volatile removal. It can have a large polymer surface and long residence time for separation, and it is claimed that the second machine is surge-free, though this depends on satisfactory feeding of the melt to it. The two extruders may run at different speeds and even be of different diameters; control of relative outputs does not need to be refined because of the large capacity of the vessel. Although the equipment is cumbersome, and the long residence time may give degradation of heat-sensitive polymers, it appears suitable for removal of large proportions of water vapour and other volatiles.

Double-parallel Screw

As mentioned on p. 101, the performance of screws of varying depth can usefully be examined by Q/P diagrams. Consider a double-parallel screw of constant diameter and pitch, with channel depths h_1 and h_2, h_1 being greater than h_2. The corresponding drag and pressure flows will be Q_{1D}, Q_{2D} and Q_{1P}, Q_{2P}, respectively. Figure 5.28 represents these flows and the resultant total flows Q_1 and Q_2 giving in steady state:

$$Q_1 = Q_2 = Q_{Die} \qquad (5.92)$$

The pressure rises P_1 and P_2 are algebraically additive and together equal the die pressure:

$$P_1 + P_2 = P_{Die} \qquad (5.93)$$

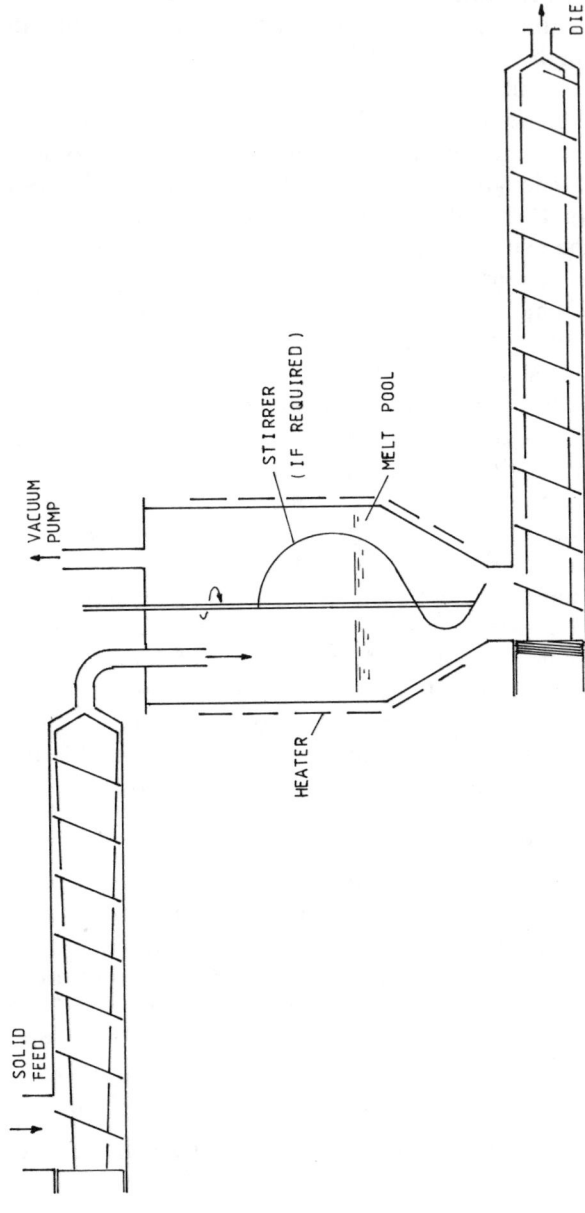

Fig. 5.27. Volatiles removal using two separate extruders.

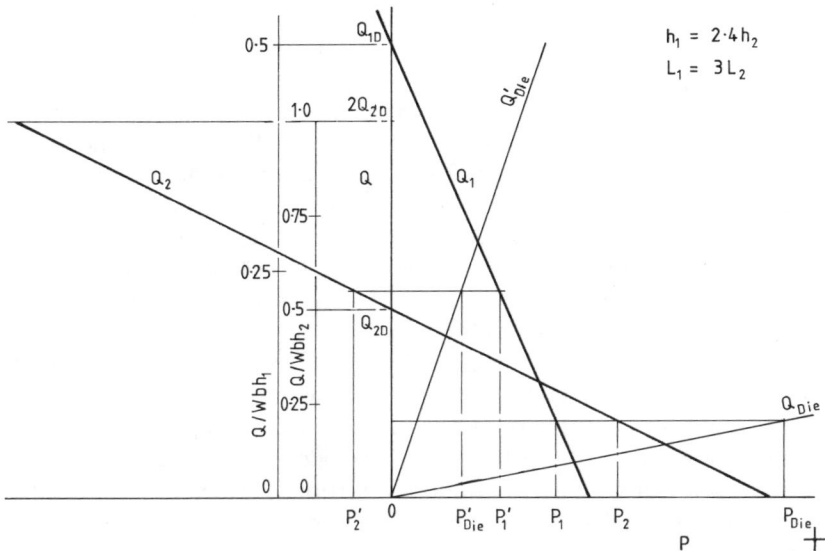

Fig. 5.28. Output vs. pressure for a stepped screw.

Then with a small die requiring high pressure at a low output, Q is determined by a horizontal line, satisfying eqn (5.92) such that the intercepts P_1 and P_2 together equal P_{Die} to satisfy eqn (5.93). If this is replaced by a larger die so that the output becomes Q'_{Die} which is larger than Q_{2D}, $Q/Wbh_2 > 1/2$ and pressure falls in the second section, as shown by a negative intercept of pressure on the Q_2 line. Then Q'_{Die} is given by a horizontal line such that the (arithmetic) difference of P'_1 and P'_2 equals the die pressure P'_{Die}. Note that the vertical scales of Q/Wbh_1 and Q/Wbh_2 give an immediate clue to the pressure and flow conditions in each screw section, for any known value of Q. This method illustrates better than a single Q/P line what is happening in a multi-section screw and can be used for more complicated forms of screw, providing the constraints corresponding to eqns (5.92) and (5.93) are observed; for instance the venting system of Fig. 5.27, paying due regard in calculating values of Q/Wbh to differences in diameter, pitch, speed, etc.

Screw Extruder and Gear Pump

The final example is of a single-screw extruder melting solid polymer and feeding it to a gear pump used for metering the flow. This system has been used in melt spinning of fibres, where the flow through the gear pump is proportional to its speed of rotation, but almost independent of pressure difference, so that as the filter pack becomes partially blocked by dirt and gels, the flow through the spinaret is maintained in balance with the drawdown system, and major adjustment of the latter is not required. In steady conditions the extruder automatically adjusts its output to equal that of the gear pump by raising or lowering the pressure between them. However, if the extruder is set at too low speed, this pressure may fall below, say, $0 \cdot 3\, \text{MN/m}^2$ ($50\, \text{lb/in}^2$) at which the gear pump begins to cavitate, giving erratic output and voids in the product. If speed is too high, the pressure between extruder and pump may rise unacceptably. Consider a system in which a pressure P is desired at the pump inlet at a constant flow rate of Q_{Pump}. A shallow screw, depth h, gives this pressure and output at speed N. If extruder speed is now varied by $\pm 10\%$, the drag flow (the intercept on the vertical axis) will vary by the same proportion. However, as shown by Fig. 5.29, the small slope of the Q/P line for a shallow screw gives a large variation in pressure, including the possibility of insufficient pressure to prevent pump cavitation. Figure 5.30 shows that a deeper screw, giving the same pressure P at a speed N_1, will give a larger

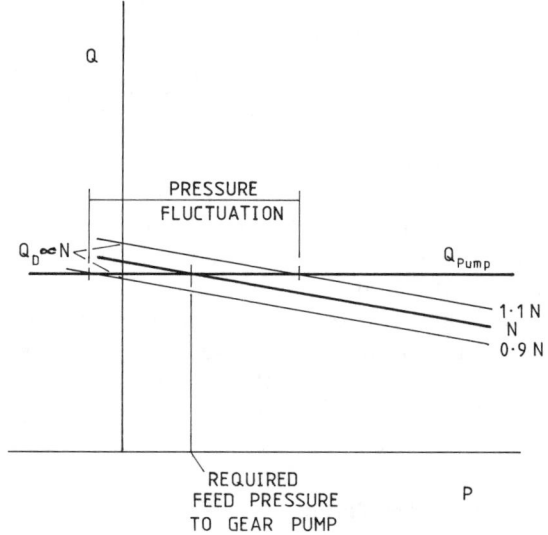

FIG. 5.29. Shallow screw feeding gear pump.

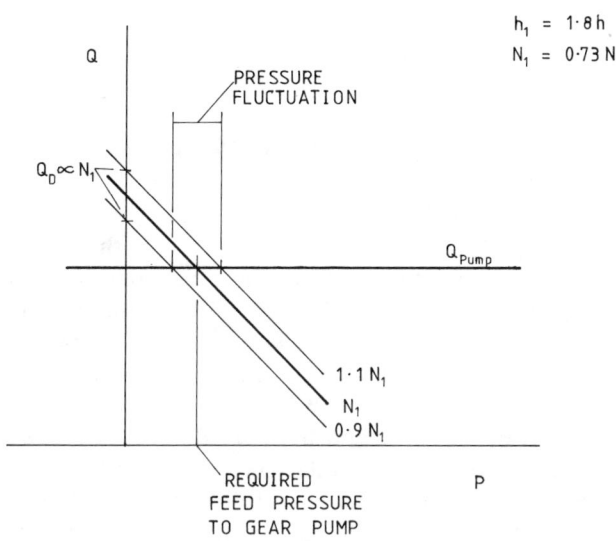

FIG. 5.30. Deep screw feeding gear pump.

absolute variation in drag flow Q_D, but a smaller variation in P due to ±10% speed variation.

Thus the shallow screw, which may be necessary for adequate melting and mixing, requires much closer control of speed, which could only be achieved automatically by a feedback signal from the inlet pressure to the pump. The author recalls an installation in which a 114 mm (4·5 in.) extruder fed twelve gear pumps and spinarets in parallel, involving considerable high pressure heated pipework due to the space required for the drawdown and winding equipment. When one of the pumps was stopped for filter cleaning, a similar situation arose in that Q_{Pump} was effectively reduced by 8·3% while extruder speed N was constant, leading to a temporary increase in pressure. It was then proposed to substitute a screw with vacuum extraction to remove volatiles; the performance of the second screw section would then be similar to Fig. 5.30. However, the output of the first section depends only on screw speed and would not be influenced by the pressure between extruder and gear pump, and the extruder speed must be set to exactly match the flow through the gear pump to avoid pump cavitation or vent blocking. The pressure realised would then depend on the relative proportions of the two screw sections; the diagrams illustrate the replacement of a self-regulating system by one requiring very precise external control.

5.6. OUTPUT CORRECTIONS

Leakage over the flights has been covered in Section 5.3. Assumption 3 on p. 72 neglected the drag due to adherence of the polymer to the sides of the screw flight; the drag flow is not truly one-dimensional and the mean drag velocity is not $W/2$ (eqn (5.10)). Similarly the pressure flow is not completely represented by flow through an 'infinite' slit with no drag at the edges. Fenner[10] has calculated the difference between one- and two-dimensional flow and McKelvey[31] has presented graphs of correction factors for drag and pressure flow for various ratios b/h of channel width to depth. The latter show that for $b/h = 10$, which is typical for extruders for plastics, $K_D = 0\cdot 93$ and $K_P = 0\cdot 94$ and varies only slowly with b/h, so may be neglected in estimating changes due to operating conditions, though not in design. This is not always true in machines handling rubbers, where sometimes variable-pitch screws of constant depth are used giving smaller values of b/h and greater corrections, especially near the die; in the same conditions channel curvature also cannot be neglected, though the effect is small[40] for shallow screws.

A major uncertainty which directly affects the output/pressure characteristic is the effective melt length Z. Experiments by Weeks and Allen[35] have shown that the point of initial pressure rise (sometimes called 'gel point') in a solid-fed extruder can be determined quite reliably by plotting pressure measurements at a series of points along the melt pumping section and extrapolating to zero pressure. The same researchers measured melt temperature by allowing polymer to flow out of the barrel at various points and over thermocouples. This of course gave time-averaged readings, but in the region of the 'gel point' may have given high readings, since any solid particles in the flow could not pass through the relatively narrow passage surrounding the thermocouple. A recalculation by the present author from the measured final temperature and total output, together with the calculated power input in the (known) melt length, giving the enthalpy and mean temperature at the 'gel point', indicated the latter to be between 105 and 110°C for LD polyethylene in the conditions of the experiment, compared with the nominal crystalline melting point of approximately

117°C. This was considered reasonable since at this point the inhomogeneous melt would be in effect a suspension of partly-solid particles in a molten matrix; the suspension might have an average temperature below the nominal melting temperature but would behave as a fluid and be capable of generating hydrodynamic pressure. Experiments made with thermocouples attached to the screw and immersed in the channel flow have been reported by Marshall et al.[47] and Barnett et al.[48] Diagrams of temperature vs. screw length reproduced from these authors by Tadmor and Klein[49] show (i) the existence and nature of time-fluctuations of temperature at a point, (ii) the variation in temperature across the channel width, (iii) the tendency for the fluctuations and variations to increase with screw speed and (iv) the movement of the region of fluctuations towards the die with increase of speed. The implications for the melting process (Section 5.9) and homogeneity at the die (Section 8.1) will be considered later, but they demonstrate that simple measurements of temperature are unreliable in determining the 'gel point' at which pressure begins to rise, but that this point moves towards the die as speed increases. Practical experience suggests that it also moves back, i.e. towards the feed end, with increasing back pressure. Thus the effective melt length is likely to vary with speed and pressure, though a step or special designs incorporating a barrier may restrict this movement. The implications of this movement for the output/pressure characteristic are (i) that as die pressure increases, dP/dz and hence pressure flow Q_P do not increase in proportion, so that the sloping lines of Fig. 5.17 will have a concave upward curve, and (ii) that as speed increases, dP/dz and hence Q_P at a given pressure will increase so that the two sloping lines of Fig. 5.17 will no longer be parallel but increase in slope with speed, i.e. at higher speed, a screw of a given depth will behave more like a deeper screw at the original speed, as Fig. 5.18.

The position of the 'gel point' can be determined experimentally, as outlined above, but is difficult to predict theoretically, since it represents a balance between the melting and pumping mechanisms of the screw, both of which depend in a complex way on operating conditions. This essential information is too often unavailable.

The pumping rate of partially filled screws has been mentioned in connection with pressure profiles (p. 92) and venting (p. 112); various situations are shown diagrammatically in Fig. 5.31. If the channel is filled to a very small extent as in Fig. 5.31a, the melt will form a thin film over the screw surface and rotate without touching the barrel and

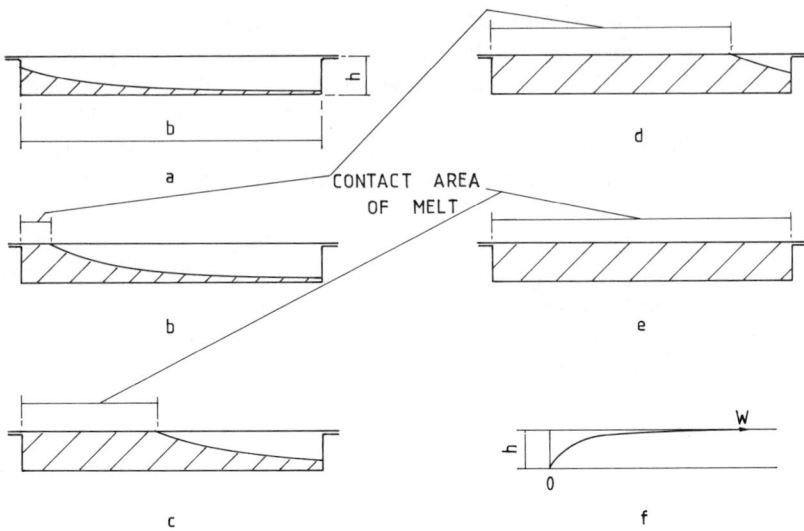

FIG. 5.31. Melt cross-section of partially-filled channels.

thus without any forces tending to move it along the screw. Any material which adheres to the barrel or is trapped in the flight clearance will tend to be collected on the leading face of the screw flight (Fig. 5.31b) by the transverse movement represented by velocity component U (eqn (5.5)). The material represented by the shaded area will be subjected to a shear force at the barrel, opposed by an equilibrium force on the much larger surface in contact with the screw; the shear stress and hence shear rate at the barrel will thus be much higher than at the screw, resulting in a velocity gradient as in Fig. 5.31f. The drag flow will be given by a modification of eqn (5.9):

$$Q_D = \int_0^h wx \, dh \qquad (5.94)$$

but in this case both w and x are non-linear functions of h, so $Q_D \ll Wbh/2$. As the degree of filling increases through the situation represented by Fig. 5.31c, the area in contact with the barrel will increase faster than both the volume of melt, represented by the shaded area, and the area in contact with the screw, so that the shear stresses and shear rates tend to become more nearly equal and the 'pumping efficiency' increases. When the channel is nearly filled, as in Fig. 5.31d, the shear conditions and pumping efficiency will be similar to that of the filled channel as in Fig. 5.31e and a nearly filled channel may therefore be expected to operate similarly to one completely filled, except that evidently the pressure must remain zero, otherwise the channel would fill. However, the non-linear increase in pumping rate with degree of channel filling brings the possibility that a slight increase in feed rate to the screw section may cause a disproportionate increase in its pumping rate, which then pumps the section empty, reducing the pumping rate until the feed again causes an increase in the degree of filling. This is believed to be the cause of the output instability known as 'surging' which is frequently encountered with the second section of conventional vented screws. It is analogous to the classical surging due to the stalling of the inlet stages of turbocompressors.

5.7. PSEUDOPLASTIC FLOW

It has been shown in Fig. 5.7 that, except in the case of drag flow when $Q/Wbh = 1/2$, the velocity profile over the depth of the screw channel is non-linear, and since $\dot{\gamma} = dw/dy$ (eqn (2.4)) the shear rate varies with position, its value depending on the magnitudes of both drag and pressure flow. In Appendix A, the downstream velocity w at a point y from the root of the screw channel is given as:

$$w = \frac{Wy}{h} + \frac{(y^2 - yh)}{2\eta} \cdot \frac{dP}{dz} \tag{A8}$$

Then the shear rate at this point:

$$\dot{\gamma}_y = \frac{dw}{dy} = \frac{W}{h} + \frac{(2y - h)}{2\eta} \cdot \frac{dP}{dz} \tag{5.95}$$

But by rearrangement of eqn (5.18):

$$\frac{dP}{dz} = \left(\frac{1}{2} - \frac{Q}{Wbh}\right)\frac{12\eta W}{h^2} \tag{5.62}$$

and:

$$w = \frac{Wy}{h} + \frac{6W}{h^2}\left(\frac{1}{2} - \frac{Q}{Wbh}\right)(y^2 - yh) \tag{5.96}$$

and substituting in eqn (5.95) gives:

$$\dot{\gamma}_y = \frac{dw}{dy} = \frac{W}{h} + (2y - h)\left(\frac{1}{2} - \frac{Q}{Wbh}\right)\frac{6W}{h^2}$$

$$= \frac{W}{h} + \frac{W}{h}\left(\frac{12y}{h} - 6\right)\left(\frac{1}{2} - \frac{Q}{Wbh}\right) \quad (5.97)$$

At the barrel surface, when $y = h$:

$$\dot{\gamma}_h = \frac{dW}{dy} = \frac{W}{h}\left[1 + 6\left(\frac{1}{2} - \frac{Q}{Wbh}\right)\right]$$

$$= \frac{W}{h}\left(4 - \frac{6Q}{Wbh}\right) \quad (5.98)$$

This is valid for constant viscosity, i.e. isothermal Newtonian flow (assumptions 7 and 8, p. 72), but as shown in Appendix A, for a non-Newtonian fluid, in which viscosity η is a function of shear rate, viscosity will vary across the channel depth and will be a function of Q/Wbh, i.e. a function of the ratio of pressure flow to drag flow (eqn (5.97)). Thus the integration of the Navier–Stokes equation (A1) becomes complex, precluding an analytical solution such as eqn (5.17) which clearly indicates the effects of operational variables. However, an approximate method which takes account of the major effect on viscosity of shear rate (e.g. due to changes in speed or channel depth) and the change in shear rate with increasing back pressure, uses eqn (5.98) to determine the shear rate at which viscosity is taken for insertion in the flow equation. It does not include the effect of changes in viscosity on the shape of the velocity profile, i.e. the distortion of the curves in Fig. 5.7. The method is first to calculate the value of Q/Wbh for a known condition of either output or pressure, and hence the apparent wall shear rate given by eqn (5.98). The intercept at this shear rate on the curve of viscosity vs. shear rate for the polymer in question at the appropriate melt temperature then gives the viscosity to be used in the flow equation (5.17) or (5.22). This enables dP/dz or P to be calculated for the assumed Q or vice versa, with little more difficulty than for the Newtonian case, except that it must be repeated for each value of Q or P.

The implications of eqn (5.98) for pressure profiles and output/pressure characteristics will now be examined briefly. When $Q/Wbh = 1/2$, eqn (5.98) reduces to $dW/dy = W/h$, consistent with Fig.

5.7d, and the second term of eqn (5.17) disappears, giving an output independent of viscosity, as for the Newtonian case. For a constant-depth screw, consider first the condition at which the viscosity of two fluids, one Newtonian and the other pseudoplastic, are equal. Then evidently the pressure and pressure gradient will be equal, and also the output and Q/Wbh (eqns (5.17) and (5.18)). If now the pressure P and pressure gradient dP/dz are increased, and consequently output Q and Q/Wbh decreased, then by Fig. 5.7 and eqn (5.98), wall shear rate $\dot{\gamma}_h$ will be increased and the viscosity η of the pseudoplastic will be decreased, whereas viscosity of the Newtonian fluid remains constant. Then, by eqn (5.17) and compared with the Newtonian fluid at the same output Q and Q/Wbh, Q_P must be the same, but the pressure gradient dP/dz decreased in proportion to the viscosity. Thus as pressure is increased, the pressure profile remains a straight line, but falls progressively further and further below that of the Newtonian case at corresponding outputs. Conversely, as pressure is decreased and viscosity increases, the pressure profile falls above that of the Newtonian case—Fig. 5.32 (compare Fig. 5.11).

Alternatively to the previous paragraph, if the Newtonian and pseudoplastic cases are compared at equal pressure P and pressure gradient dP/dz, the lower viscosity of the pseudoplastic fluid leads to a greater pressure flow Q_P and hence lower output compared with the Newtonian case, i.e. as pressure is increased, the same line on the

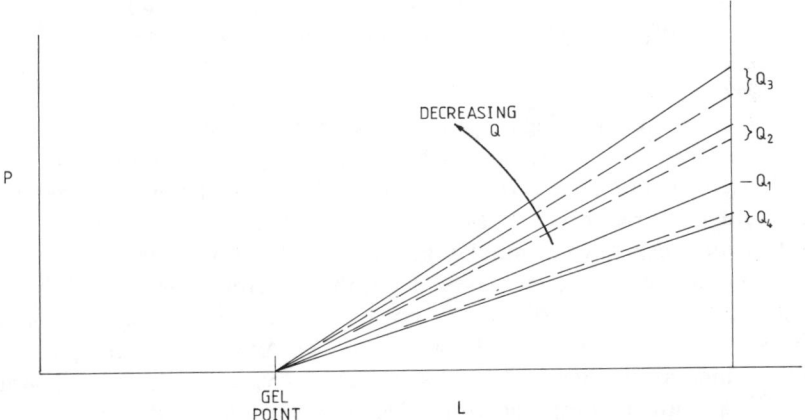

FIG. 5.32. Constant depth screw—effect of pseudoplasticity on pressure. ———, Newtonian; ----, pseudoplastic (viscosities equal at Q_1; c.f. Fig. 5.11).

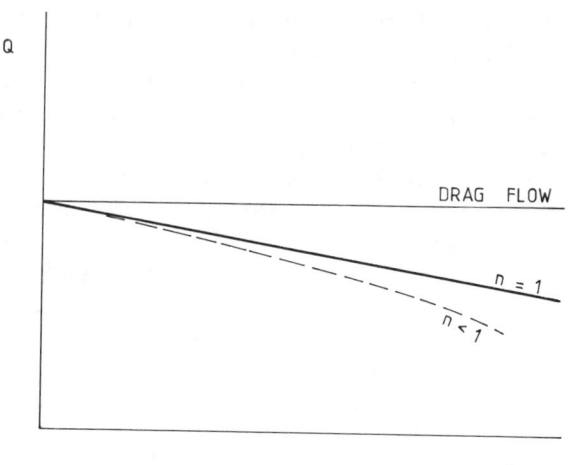

Fig. 5.33. Output/pressure for pseudoplastic. ———, Newtonian; — — —, pseudoplastic.

pressure profile corresponds to a lower output for a pseudoplastic than for a Newtonian fluid. This may be better represented on the output/pressure characteristic (Fig. 5.33, compare Fig. 5.17) where the output line for the pseudoplastic droops below that for a Newtonian fluid. Note that since viscosity is non-linear with pressure, the Q/P characteristic for a pseudoplastic fluid is no longer a straight line; the Newtonian line is tangential to this curve at $P = 0$ if viscosities of the two fluids are equal at shear rates near to that for drag flow, i.e. $\dot{\gamma}_h = W/h$.

For screws of varying depth, e.g. stepped or tapered screws, the above trends will also be observed but, also, the variation of shear rate and viscosity along the screw will cause additional changes in pressure gradient to those shown in Figs 5.12 and 5.13. By differentiating eqn (5.98) with respect to channel depth h, it can be shown that at constant output wall shear rate increases as depth h decreases while $Q/Wbh <$ 1/3, i.e. the same condition as shown in Appendix B for increasing pressure gradient (horizontally on Table 5.1). As shown for the parallel screw, increasing shear rate gives decreasing viscosity and decreasing pressure gradient compared with a Newtonian fluid, i.e. the upper lines in Figs 5.12 and 5.13 will become more nearly straight for a pseudoplastic and a smaller proportion of the total pressure will be generated

TABLE 5.1
Wall Shear Rate from eqn (5.98) vs. Dimensionless Output for Screws of Various Tapers h_1/h_2, Expressed as a Ratio of the Shear Rate at the End of the Screw in Drag Flow W/h_2

$\dfrac{Q}{Wbh_2}$ \ $\dfrac{h_1}{h_2}$	5·0	4·5	4·0	3·5	3·0	2·5	2·0	1·5	1·0
0	0·80	0·89	1·00	1·14	1·33	1·60	2·00	2·67	4·00
0·1	0·78	0·86	0·96	1·09	1·27	1·50	1·85	2·40	3·40
0·2	0·75	0·83	0·92	1·04	1·20	1·41	1·70	2·13	2·80
0·3	0·73	0·80	0·89	1·00	1·13	1·31	1·55	1·87	2·20
0·4	0·70	0·77	0·85	0·95	1·07	1·22	1·40	1·60	1·60
0·5	0·68	0·74	0·81	0·90	1·00	1·12	1·25	1·33	1·00
0·6	0·66	0·71	0·77	0·85	0·93	1·02	1·10	1·07	0·40

Distance along tapered screw →

Decreasing back pressure

in the shallower sections, whereas the portions of lines below the trace of $Q/Wbh = 1/3$ will tend to become more widely spaced compared with a Newtonian fluid, and again will contribute less to the total pressure rise. If in addition $Q/Wbh > 1/2$, then the pressure fall in this section will be greater. These trends are shown qualitatively in Fig. 5.34, remembering that the actual values of pressure P and gradient dP/dz at a point will depend on the value of shear rate at which Newtonian and pseudoplastic viscosities are assumed equal.

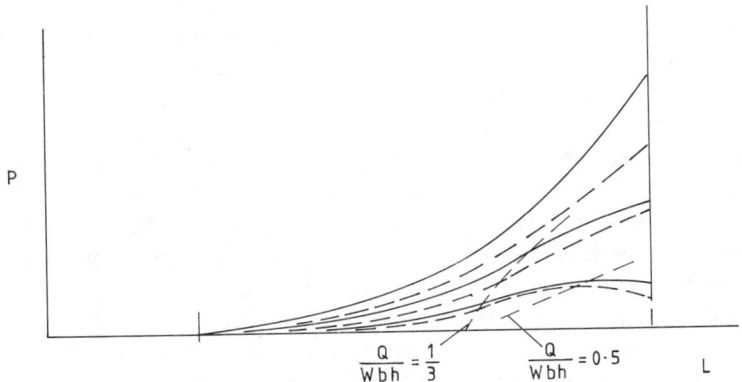

FIG. 5.34. Pressure gradients in pseudoplastic flow. ——, $n = 1$; ---, $n < 1$.

When the screw is considered in combination with a die of fixed geometry, the Q/P relationship for the latter with a pseudoplastic fluid must be included. As shown in Section 2.1 if the rheological equation is approximated by a power-law:

$$\tau \propto (\dot{\gamma})^n \qquad (2.6)$$

$$\eta = \frac{\tau}{\dot{\gamma}} \propto (\dot{\gamma})^{n-1} \qquad (2.7)$$

But in the die:

$$\dot{\gamma}_w \propto Q$$

e.g.

$$\dot{\gamma}_w = \frac{3n+1}{n} \cdot \frac{Q}{\pi R^3}$$

for a circular capillary, Table 2.1. Then substituting eqn (2.7):

$$\eta \propto (Q)^{n-1} \qquad (5.99)$$

Substituting eqn (5.99) in eqn (5.81) gives:

$$Q \propto \frac{KP}{(Q)^{n-1}} \qquad (5.100)$$

or:

$$Q^n \propto KP \qquad (5.101)$$

where $n < 1$. This is a concave upward curve compared with $Q \propto KP$ for a Newtonian fluid ($n = 1$) in Fig. 5.19. Then, the Newtonian screw and die characteristics equations (5.17) and (5.81) must be replaced by Fig. 5.33 and eqn (5.101), respectively, and Fig. 5.20 is modified to Fig. 5.35 and the operating point moves from A to A'.

In short, the effects of a pseudoplastic polymer melt are (i) a varying-depth screw behaves similarly to a somewhat lower compression screw with a Newtonian fluid, and (ii) for a given output, the pressure will be less and increase at a slower rate with output than for a Newtonian fluid. The foregoing assumes isothermal conditions are maintained; the effects of changing temperatures, some of which are implicit in changing shear rates, are considered in the next section. The effects of both pseudoplastic and non-isothermal conditions on power consumption and heat generation are dealt with in Chapter 6.

In estimating the performance of a varying-depth screw at a given

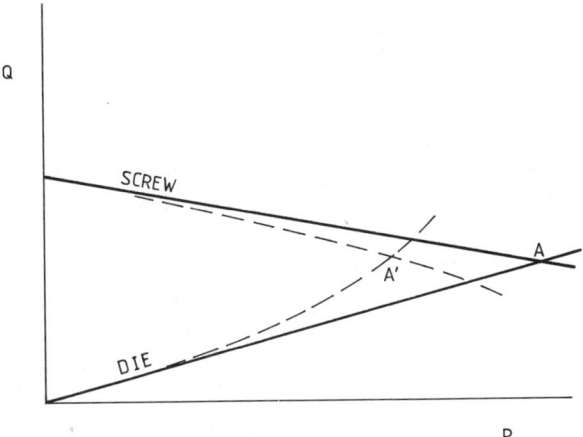

FIG. 5.35. Operating point for pseudoplastic flow. ———, $n = 1$; ---, $n < 1$.

speed and back pressure, a reasonable approximation is to take a viscosity which is an arithmetic average of those equivalent to the range of wall shear rates given by eqn (5.98). Table 5.1 shows the variation of wall shear rate along a screw of uniform taper, for various values of output, represented by Q/Wbh_2 at the end of the screw. This shows that to the left of the heavy line ($Q/Wbh_1 < 1/3$) wall shear rate increases as depth decreases, but that the variation is only large at high back pressures (low Q/Wbh_2). However, at low back pressures, when $Q/Wbh_2 > 1/3$, the maximum shear rate is no longer at the end of the screw, and evidently it would be more correct to take a viscosity which is an average of those corresponding to the maximum and minimum shear rates rather than shear rates at the two ends of the screw. A more accurate method would be to treat the screw as composed of a number of separate sections, each of constant shear rate and viscosity, and combining them by use of eqns (5.88) and (5.89). In the case of a stepped screw, this reduces to using eqns (5.55), (5.56), (5.57) and (5.58) with different values of viscosity in the first two equations. The viscosity to be used in the die equation (5.81) must also be based on the shear rate in the die, e.g. eqn (2.7).

5.8. NON-ISOTHERMAL FLOW

Many of the remarks in the preceding section apply similarly to non-isothermal flow. Firstly it must be remembered that as discussed in

Section 6.2 the temperature of the polymer melt is not the same as the set temperatures on the heater controls, even when steady conditions have been attained; the melt will generally be at a lower temperature except when severe cooling is in operation. As will be seen in Section 6.5 except at drag flow, the variations of shear rate through the channel depth will give rise to non-uniform shear heating. This, coupled with temperature variations 'inherited' from earlier in the screw, and heat transfer from barrel and screw, will lead to transverse temperature variations. That due to shear heating is indicated in Section 3.3 and to heat transfer in Section 3.2. These temperature variations lead to viscosity variations, which again modify the velocity distribution—see Section 3.3 for these effects in a simple capillary. The effects of these on pressure profiles and output/pressure characteristics are extremely complex and cannot readily be quantified in considering operational variables—rather they are to be borne in mind as possible sources of approximation, concentrating more on their direct effect on flow in the die and consequent product variability.

In general, the mean temperature of the polymer melt will tend to rise towards the die, due to a combination of heat transfer to or from the barrel and shear heating within the polymer. The polymer melt temperature can be measured by one or more thermocouples before the die and, with more difficulty, at points along the barrel. The latter may be by thermocouples, etc., fixed to the screw,[47] by intermittently bleeding polymer from the barrel over a thermocouple[35] or, recently, by fibre optics to an external sensor.[53] None of these, however, is generally available to the industrial operator, who must estimate the changes of temperature between the approximate melting point (or softening point) at the start of pressure rise and the final temperature at the end of the screw. An increase in temperature, giving a reduction in viscosity, will increase the pressure flow Q_P and decrease the total output Q at a given speed and back pressure. In addition, shear rate and thus temperature rise will increase with increasing pressure, so that the actual Q/P characteristic droops below the isothermal line, generally with a concave downward curve. As shown by the right-hand column of Table 5.1, this effect becomes severe at low values of Q/Wbh and corresponding high pressure, so the maximum pressure available from the screw is also reduced significantly. Inspection of the horizontal lines of Table 5.1 shows that for a tapered screw, the increased shear rates and shear heating are concentrated at the shallow end, especially at low outputs, where the pressure gradients are greatest. This should be taken into account in estimating the likely temperature profile and the appropriate average temperature to be used in finding a value of viscosity, and may also be of significance in setting barrel temperature profiles in operation. As with pseudoplastic behaviour, a closer approximation can be sought by assuming a series of sections, each at uniform temperature, but the validity of this depends on the accuracy of available data on the actual melt temperatures.

5.9. THE MELTING PROCESS

In Section 5.1 it was mentioned that melting of solid polymers by conduction is usually supplemented in screw extruders by internal shearing. Since polymers generally have low thermal conductivity, this 'shear heating', which does not depend on temperature differences within the polymer, permits more rapid melting than by conduction alone, without risk of burning or degradation. It may even be said that it is the low mechanical efficiency of the single screw as a pump which makes it such an effective means of melting materials such as organic polymers, having high melt viscosity, low thermal conductivity and moderate thermal stability, compared with ram and positive-displacement twin-screw types. For some purposes, including the compounding of rubbers and rigid PVC, this shear heating may become excessive, requiring design compromises to avoid damagingly high temperatures. As mentioned in discussion of 'gel point' variations (p. 121), the position of the latter depends on a balance between the melting and pumping mechanisms, and a common limit on useful output is set by excessive temperature fluctuations or even unmolten polymer at the die, due to deficiencies in the melting mechanism. This has been studied experimentally by a number of workers,[50,54] the usual method being to achieve steady running conditions, stop the screw and apply severe cooling, and either open a split barrel or force out the screw hydraulically. The partially molten material may then be visually inspected and removed for sectioning; this has been especially successful with PVC mixes, which when cold tend not to adhere to metal surfaces.

The present author[55] has reported experiments using a model fluid with similar rheological characteristics in a transparent barrel. Tadmor and Klein[50] have reported, in addition to the measurements of polymer temperature during and after melting, extensive experimental work on the mechanism and progress of melting. Tadmor and Klein also present a series of theoretical models, which in essence assume the solid material rotates almost with the screw and is sheared by contact with the barrel, from which it also receives heat by conduction. The resulting molten film, which is assumed of uniform thickness, is collected by the screw flight and forms a molten stream on the leading face of the flight. They compare theory and experiment, though it is not clear how the former predicts the point at which the solid bed breaks up into softened lumps suspended in the melt. Halmos et al.[56] have improved the model by considering a molten film tapering in thickness from the flight clearance at the trailing edge to a greater thickness, due to melting against the barrel, where it discharges into the melt stream on the leading edge. In the first experiments, the present author[55] found the granules quickly compacted, displacing air rearward, leaving little voidage. With a relatively high barrel temperature the granules first lost their visible identity and further along the barrel became a transparent liquid with some large semi-molten lumps. When barrel temperature was reduced, the torque on the screw increased dramatically and the melting process occupied a considerable length. This mechanism was generally similar to the models of Tadmor and Klein and Halmos et al., i.e. a slow-moving solid bed, a molten layer adjacent the barrel wall and a fast-moving molten stream on the leading face of the flight, which appeared to occupy the full channel depth. The width of this stream gradually increased, no doubt largely by collection of the molten layer by the flight, until the solid bed broke up, giving partially molten particles suspended in the melt. In these experiments the solid bed did not appear to separate readily from the screw surface, as Gale[54] has reported in certain conditions with PVC. This may be because in the author's experiments the 'Perspex' screw had low thermal conductivity and hence would transfer little heat to the solid bed. Different mechanisms might also be expected in the example given by Tadmor and Klein,[49] where the screw temperature was recorded significantly higher than that of the polymer. However, the author's experiments also demonstrated that as screw speed was increased, melting occupied a greater length of the screw, in agreement with the temperature observations of Barnett et al.,[48] but apparently contrary to Tadmor and Klein's[51] predictions.

Another feature, not included in the theoretical models, was that the molten stream moving past the solid bed eroded the latter (Fig. 5.36), probably by a combination of heat transfer and shear. When feeding

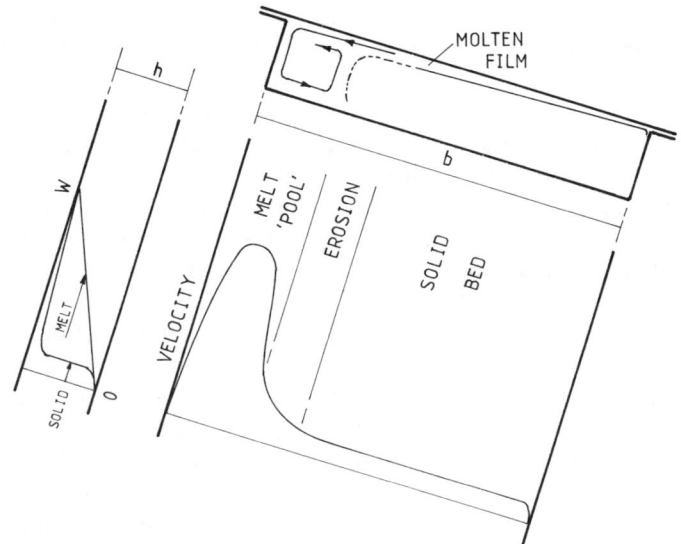

Fig. 5.36. Velocities in melting section.

was temporarily stopped, it was observed that both solid feed and melt pumping sections quickly emptied to the point where only partial contact of the polymer with the barrel surface took place. The melting section, however, remained practically stationary, until it eventually melted by local heat conduction or generation; suggesting that the solid feed tends to force the semi-molten material forward, but that negligible, pumping action is generated in the melting section, justifying the assumption, p. 71, that pressure only commences to rise when the solid bed breaks up to form a suspension. Thus the length of screw required to achieve this condition is important in determining the residual length for pressure build-up, mixing, etc., in the melt pumping section. However, the uniform temperature, viscosity, etc., required at the die for satisfactory extrusion, also depend on the elimination of solid particles *after* break-up of the solid bed. Tadmor and Klein[52] state that experiments confirm that the solid particles follow the same mechanism of melting after break-up of the bed as before break-up; this is difficult to visualise since the suspended particles can only experience the same shear stress as exists in the fluid and, owing to their higher viscosity, the shear rate and shear heating (proportional to stress × shear rate) they experience must be much lower. In addition, external heat can only reach the solid particles by conduction through the fluid augmented by the very small convective heat transfer due to the drag velocity between solid and fluid. It is a matter of practical experience that once the solid bed has broken up it may take a considerable length of screw to complete the melting process. For this reason several mechanical designs, including 'barrier'[57] screws and dual-flight screws[58] have been used commercially, either to control the point of break-up of the solid bed or to hold back solid particles after the break-up. Various operating strategies have also been proposed, mainly with the object of delaying break-up of the solid bed, so that when it does occur, a minimum proportion of material remains to be melted. Screw cooling and variations in barrel temperature profile are obvious stratagems; it is difficult to assess unambiguously the improvements in performance and also to separate the effects from those on other parts of the melting and pumping processes, which also interact in their results.

It is clear that the melting process is complex, depending on polymer properties, screw geometry and operating conditions of speed, back pressure, temperature profile, etc. A simple mathematical solution is unlikely, as is an analysis showing the effects of separate variables. For the present a qualitative understanding of the mechanism of melting together with some general trends seems the best aid to the operator:

(i) Elimination of an unmolten fraction is likely to be more critical for operation than melting the majority to a pumpable suspension.

(ii) At the same screw speed, a shallow screw will melt satisfactorily in conditions in which a deeper screw will not, though generally at a lower output. This may also appear as a scale effect, since large diameter screws are also usually deeper.

(iii) With the same screw, as speed is increased, inhomogeneity of the product tends to increase (Section 8.1), probably as a result of extended melting.

(iv) Combining (ii) and (iii), at the same output, a fast running shallow screw will often give a more homogeneous product than a deep screw running slowly.

(v) With a given screw, increased back pressure will usually improve melting at the same screw speed, but not necessarily at the same output.

(vi) With a given screw and speed, powder feed will take longer to melt than the same polymer fed in granule form.

(vii) A low temperature profile on the barrel will induce shear melting, which at high outputs may dominate over conducted heating, but the mechanical power input will be high. A higher barrel temperature in the melting region will reduce shear and power input, and may result in lower melt temperatures, but at low outputs may be necessary to achieve a reasonable output of molten material.

5.10. SOLIDS CONVEYING

It has already been mentioned (p. 92) that solids conveying may limit the output of an extruder. It must first be recognised that at least three processes are involved, more or less distinct, and any one of these may influence output. These are (i) flow in the hopper or feed pipe, (ii) filling of the screw channel from the feed throat and (iii) conveying by the screw from the open feed section into the closed barrel and compaction in the latter. Flow in the hopper is usually by gravity and is analogous to flow in silos and bins, which is outside the scope of this work. However, some practical points are worth noting, the first being that the hopper should be as wide as possible, especially for feeding coarse or angular material, to reduce the possibility of 'bridging'. It should have fairly steep sides, which should be smooth and free from ledges and bolt and rivet heads, etc., but not so steep as to appreciably lengthen the zone of restricted cross-section. Long pipes are particularly undesirable since they can form a compacted plug of material which is difficult to break up. Internal corners, which may cause a wedging action, should be avoided, e.g. by generous rounding of the corners of a rectangular-section hopper or better by a circular section. The hopper shape must of course be a compromise with the entry into the feed section proper. Tendency to bridging will be increased by factors affecting the friction between particles, including softer materials, angular or irregular shapes, a wide distribution of sizes permitting packing, and superimposed pressure, e.g. by a deep head of material. Fluidising by means of an upward current of air might be expected to reduce bridging, but the author has seen an extruder with an efficient shaft seal and fed with a free-flowing powder, in which air displaced in the screw 'welled up' in the hopper and restricted powder flow. An immediate remedy was to reduce to a minimum the powder level in the hopper and maintain it visually; a more permanent solution was to provide a lateral division near the rear of the hopper, behind which the air could escape freely. Mechanical stirrers or vibrating agitators have also been used to reduce bridging—force-feeding is discussed later. Metal separators and safety grids, though necessary, are potential causes of blockage. Since flow in the hopper is usually not plug flow, but 'funnelling', segregation of masterbatch, colourants, regrind, etc., may also occur (possibly intermittently) due to differences in density, flow properties, etc.

Filling of the screw is usually by gravity, with a relatively close clearance of a hardened feed pocket liner around most of the circumference of the screw to prevent the feed material falling out again and provide some resistance so that it is conveyed into the barrel instead of merely being carried round by the screw. With a purely radial inlet to the screw, the latter can result in material being forced back into the feed throat and hindering the in-flow of fresh material, especially if the particles are elastic or large relative to the dimensions of the screw channel. For this reason the opening is frequently made tangential to the downgoing side of the screw, though some designers argue that at high speeds the screw will 'run away' from the feed, whereas on the upcoming side the screw cuts into the descending column of material. Experience shows that a substantial chamfer is helpful, both on the downgoing side and the end leading to the barrel, in leading material outside the path of the screw flight into the channel rather than ejecting it by hitting a near-radial surface. Especially with large particles or elongated strips it can be seen that a wedging action occurs not only between the screw root and the feed pocket liner, but also between the flight tip and the liner—in absence of a chamfer, the flight tends to cut and/or eject the particles. Experiments[59] carried out on a 37 mm (1·5 in.) extruder with conventional and modified feed pocket liners (Fig. 5.37) showed that with free-flowing solids there was little difference between flow rates with the two designs and in both cases flow rate was approximately proportional to speed up to high values— Fig. 5.38. With materials showing poorer flow in a simple cup flow test, the modified design showed appreciably better flow, in some cases approaching that of the free-flowing samples. Not only was the absolute flow greater, but it was more linear with speed and the limiting values were greater. With strip-cut rubber (Table 5.2), which could only be fed by the conventional design after a separate re-granulation stage, it was found that strips which had once been entrained by the screw then tended to stick together and, if ejected, formed a coarse agglomerate blocking the feed throat. The modified design drew the

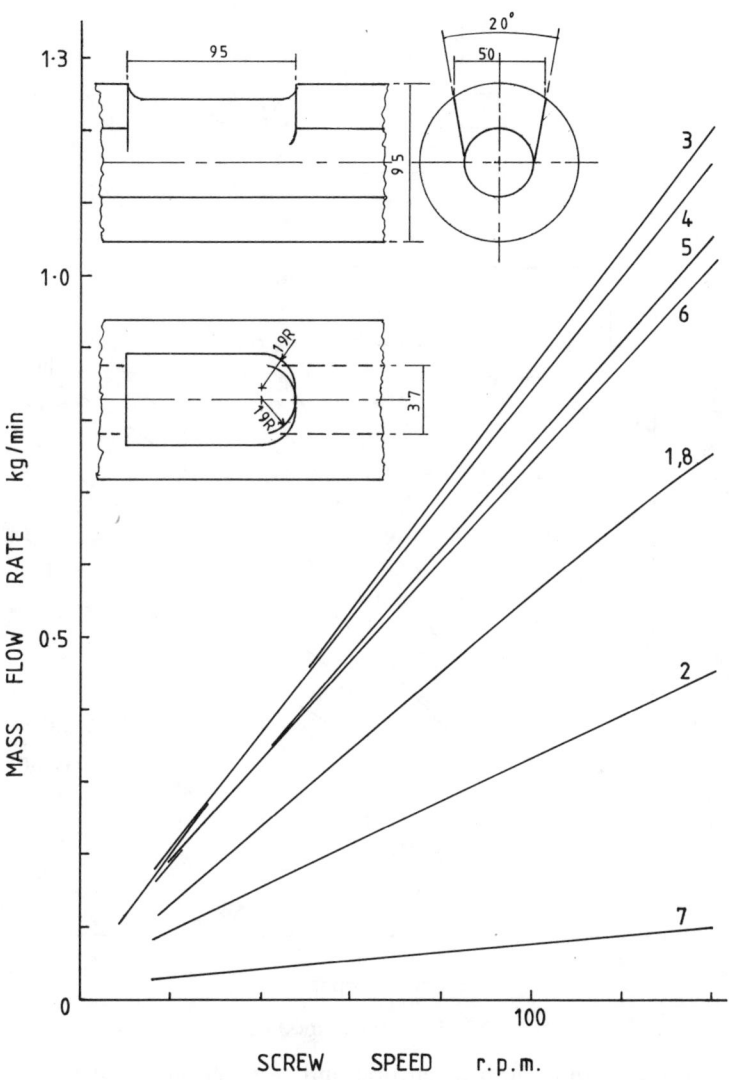

Fig. 5.37. Feed rate—standard liner.

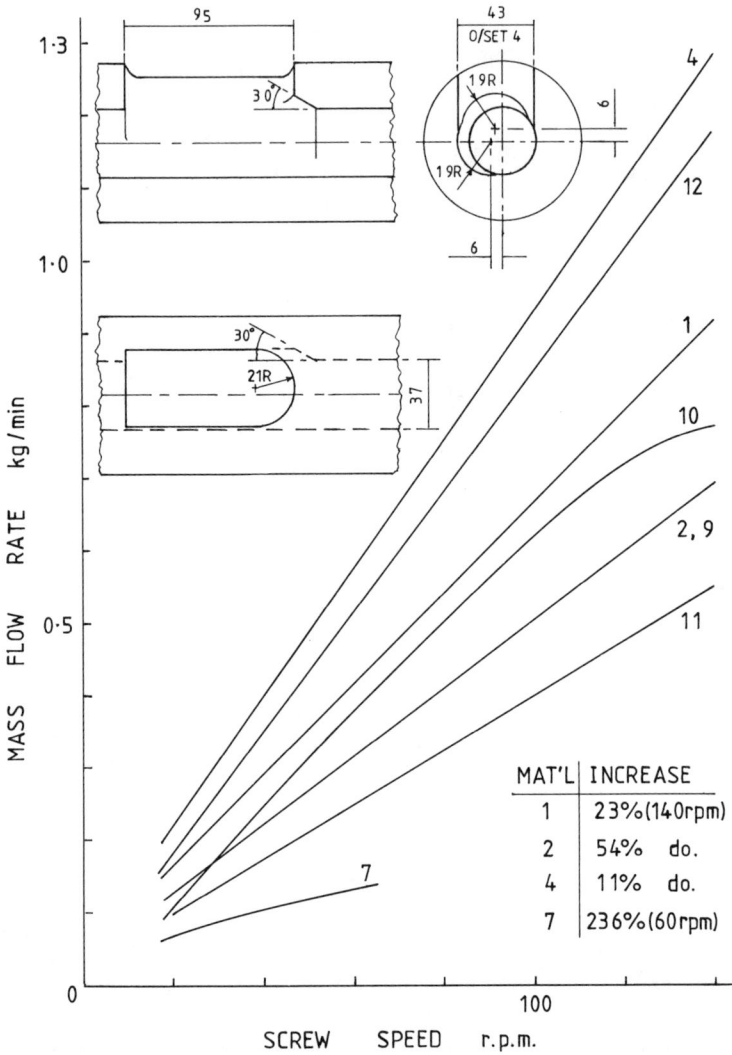

Fig. 5.38. Feed rate—modified liner.

strips in with a minimum of cutting and re-ejection (as in a domestic meat mincer) and permitted satisfactory extrusion though at more limited output than with free-flowing granules. The experiments also covered variations in the length of the opening and some other geometric variations, though the results were less conclusive.

TABLE 5.2
Materials Used in Feeding Trials

No.	Material	Hardness	Shape	Size	Size distribution
1	LDPE	Soft	Cylindrical	2 mm ϕ × 3 mm long	Very even
2	LDPE regrind	Soft	Rough lumps and flakes	4 mm max. dimension	Very uneven
3	HDPE	Hard	Lenticular	4 mm ϕ × 3 mm long	Very even
4	PS	Hard	Rhombic	1·5 mm × 2 mm × 3 mm	Very even
5	LDPE	Soft	Spherical-ended cylinders	5 mm ϕ × 4 mm long	Very even
6	PS	Hard	Cylindrical	2 mm ϕ approx. 4 mm long	Even diameter variable length
7	Synthetic rubber	Soft elastic	Large curled strips	Not specific 20 mm × 8 mm × 4 mm	Very uneven
8	Synthetic rubber	Soft elastic	Roughly tetrahedral chips	Not specific 4 mm height 9 mm × 9 mm base	Uneven
9	Synthetic rubber	Soft elastic	Roughly cubic/ cylindrical lumps	6 mm × 7 mm × 9 mm	Uneven
10	Rubber granules	Very soft elastic	Roughly tetrahedral lumps	5 mm height 9 mm × 9 mm base	Uneven
11	Rubber crumb	Soft	Very small particles almost a powder	—	Variable sized particles
12	PVC powder	—	Powder	—	Even sized particles

The third stage of conveying by the screw has been the subject of theoretical studies, the most notable being by Darnell and Mol,[60] who also published confirmatory experiments. Various authors have made modifications, including the initial assumptions, but all depend on knowledge of the coefficients of friction between polymer and the solid surfaces of screw and barrel; the latter are notoriously difficult to measure reliably but are known to vary widely with temperature. Since it is the difference between friction forces at the screw and the barrel surfaces which causes conveying, the rate achieved is highly dependent on operating conditions. The usual strategy is to maintain the feed pocket liner cold by water jacketing and leave the screw 'neutral'. Especially when extruding polymers with low melting/softening points, this places some restriction on the set temperatures on the rear end of the barrel because of heat conducted through the metal parts, which has even been known to cause solid particles to stick to the front face of the feed opening. However, in one case of a polyethylene masterbatch containing slip agent, cooling of the screw (in the feed section only) promoted satisfactory feeding. The screw surface is usually hardened and highly polished and, in an attempt to improve conveying, some manufacturers have used deliberately roughened feed pocket liners—in at least one case the barrel has been longitudinally fluted to increase rotational resistance without a corresponding longitudinal resistance. Experiments such as Hayward's,[59] where conveying occurs along several turns of screw in an unheated barrel without discharge restriction, generally show solid conveying to be continuous and steady providing the feed is adequate. Feed sections are therefore usually made smooth for ease of cleaning but deep enough to ensure a potential solids flow rate greater than that of the melt pumping section, in which case exact prediction of the former is irrelevant. In the case of large machines of 150 or 200 mm (6 in. or 8 in.) diameter there is little difficulty in providing a great enough depth of feed section which is also several times the size of the feed particles; in small machines it is sometimes necessary to use finer-cut feed material (especially regrind which tends to be irregular), but also the depth is limited by the torsional strength of the screw root.

Some writers have proposed controlling extruder output by 'starve' feeding using some form of continuous weigh feeder; the author's experience is that the solids feeding becomes less regular and possibly affects the melting process, the delivery pressure tends to fluctuate, but the power input decreases only slightly, leading to higher final temperatures. On the other hand Kosel[61] has developed a helically flighted barrel to give vastly increased pressure in the feed section, which he claims also gives considerably quicker melting and increased output. Other workers have employed a robust screw and drive in the feed hopper, referred to as a 'force-feeder', in conjunction with an improved shaft seal, which not only prevents bridging in the hopper but also imposes pressure in the feed section. The author witnessed trials of a 150 mm (6 in.) extruder densifying polypropylene powder for granulation; the output at a given screw speed was increased some 25–30% and melt temperature and degradation of the polymer were decreased significantly. The load on the main drive motor was increased considerably and when feeder speed was increased, the 82 kW (110 hp) main motor stalled before the 2·2 kW (3 hp) feeder motor reached full load, so that careful control was necessary.

Mention has already been made of feeding of strip rubber; the author has experimented with a variety of feedforms including 'cast' (unblown) LD polyethylene tubular film scrap 50–100 mm (2 in.–4 in.)

wide by 2×0·25 mm (2×0·010 in.) thick and PET film edge trim 25 mm (1 in.) wide by 0·025 mm (0·001 in.) thick. In both cases, once the end of the strip was wrapped round the screw and drawn into the barrel it fed naturally and could not readily be held back; eventually this caused overloading of the drive motor if feed rate was allowed to increase, presumably because shearing and melting were taking place at an increased rate in the first section of the screw. In the case of the PET edge-trim, no less than 36 rolls were simultaneously fed into a 50 mm (2 in.) extruder before overloading occurred. With mixed feed materials, segregation was sometimes experienced in conveying as well as in hopper flow. Although outside the scope of this chapter, it is worth briefly recording that the author has fed a melt of high molecular weight LD polyethylene to a 90 mm (3·5 in.) extruder at 400 rpm at rates up to 1090 kg/h (2400 lb/h) using a steam-jacketed hopper of simple V-form with an opening approximately 280 mm (11 in.) square immediately above the screw. Care was necessary to avoid overheating of the extruder bearings and gearbox.

Chapter 6

PRINCIPLES OF ENERGY BALANCE

6.1. ENERGY BALANCE AND EFFICIENCY

The preceding chapter has dealt with the mass output of the extruder and its relation to screw speed and back pressure, but it has been mentioned that energy considerations impose additional restrictions on the useful output in commercial operation. The purpose of the present chapter is to formulate the energy balance of the single-screw extruder, examine the components of this balance and the factors on which they depend, and deduce the overall effects on the energy balance of deliberate changes in operating conditions. These aspects have been largely neglected in the literature, so their inclusion gives greater relevance of theory to practical experience and permits further rational development of operating strategies.

If the extrusion process is considered as a total system, it takes in raw polymer, usually at room temperature, and by energy-consuming processes converts the polymer into semi-finished or saleable products, again usually at room temperature. The specific energy content of the polymer is only slightly changed by virtue of chemical degradation, changes in crystallinity, residual pressure or stresses, changes of shape, etc., and such changes in energy content can rarely be exploited in the final use of the product. Thus the consumption of energy is seen as an inevitable concomitant of the fabrication process and efficiency is defined in terms of specific energy input in J/kg of useful product. However, it reveals little of the consequences for unit output or product quality, for which a closer study of the process and its interactions is necessary. The major energy input is to heat the polymer, either intentionally to the temperature required for shaping or incidentally in random processes such as mixing, and the consequential losses to the surroundings. Rapid production requires that after shaping the temperature and heat content of the product are reduced by large masses of low temperature coolant (see Section 3.3), leading to small temperature rises and low-grade heat in ambient air or cooling water. Proposals have been made for reclaiming this waste heat, which in any case contributes to space heating, thus reducing the overall energy consumption of the process. However, it only affects the process itself if applied, for example, to preheating the raw polymer feed (Section 7.1). Thus the main opportunity for improving the efficiency in energy terms lies in reducing the energy input required by the process; the following analysis therefore concentrates on the energy balance and utilisation within the extruder itself, to give a supply of polymer at the correct temperature and degree of mixing for the shaping process.

The block diagram Fig. 6.1 shows energy input to the extruder from three sources:

(i) enthalpy of feed material I_1;
(ii) mechanical energy to turn the screw E;
(iii) energy from heaters H.

The energy output is in the enthalpy of the product I_2 and losses from the extruder S. Since by continuity of mass, the input and output masses are identical, it is convenient to consider only the increase in enthalpy of the product I where:

$$I = I_2 - I_1 \qquad (6.1)$$

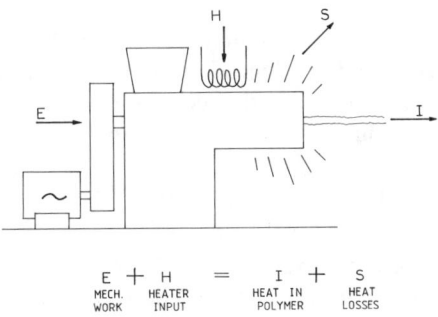

FIG. 6.1. Extruder energy balance.

In steady state, the energy balance is then:

$$E + H = I + S \qquad (6.2)$$

Considering each item in turn, the mechanical power applied to the screw shaft E is available for heating the polymer; if some is converted into frictional heat at the barrel surface or in the feed pocket seal, this is available for heating the polymer or contributes to the losses to feed pocket cooling or in radiation, and so is included in the balancing terms I and S. If E is measured at the drive motor output, then the losses in transmission belts or gears and in the bearings must be included in the total losses S to balance; if E is measured by electrical or hydraulic input to the drive then motor losses must also be included in S, since they cannot contribute to heating the polymer. The energy from the heaters H is conveniently measured as the electrical or fluid heat input to the heaters, since it is difficult to separate heat losses from the heater surfaces and from exposed areas of the extruder itself; if an electric heater were removed, then there would be appreciable heat losses from the additional barrel surface exposed, so it would be unreasonable to attribute this loss separately to the heater. When cooling is applied to barrel, screw or feed pocket, the heat removed must be included either as a negative input in calculating H or as a component of the losses S. Since barrel cooling is normally only applied when the heaters are switched off, the author prefers to count this as a negative input H, whereas cooling to feed pocket (and screw) is applied for other purposes than controlling polymer output temperature, with either barrel heating or cooling, and is thus included in the losses S. The term H thus represents the energy in heating (or cooling) which must be intentionally applied to maintain steady operating conditions.

In the absence of chemical reactions within the extruder, such as polymerisation, crosslinking, degradation, hydrolysis or reaction with additives, the energy input to the constant mass of polymer is the increase in enthalpy plus the pressure energy:

$$I = m \int_{T_1}^{T_2} C_p \, dT + p(V + dV) \qquad (6.3)$$

If gas or vapour is removed in venting, this represents both a decrease of mass and an additional energy loss. The feed material is normally at or near atmospheric temperature and pressure; if preheating or melt-feeding, especially under pressure, is used I_1 becomes significant and must be deducted to give the increase of enthalpy I—eqn (6.1). Otherwise the increase of enthalpy is sufficiently represented by the mass flow in kg/s times the specific enthalpy (heat content—Section 3.1) in J/kg at the end of the machine, reckoned from room temperature. At the end of the screw, the polymer is also at elevated pressure and the corresponding pressure energy (the second right-hand term in eqn (6.3)) must be added to the increase in enthalpy to give the total energy input to the polymer. This may account for up to around 15% of the energy input and many writers[4,41] include it as a separate term. The screw does work on the polymer equivalent to that to 'compress and deliver' in a compressor; however, this is used to force the polymer through the die, where the work is converted to frictional heat, raising the temperature of the polymer and/or die. Since at the die exit the pressure is normally atmospheric, the pressure term may be excluded by taking the energy input I as mass flow times the enthalpy of the polymer at the die exit temperature; in any case this is usually the critical temperature for processing:

$$I = \rho Q \int_{T_1}^{T_2} C_p \, dT \tag{6.4}$$

Since the adaptor and die are rigidly connected to the barrel, heat flow between them is difficult to determine, so it is convenient for the energy balance to treat barrel and die together. Then the heater input H should include heat to the die and the losses S include the heat losses from the die—in the case of large sheet or tube dies, these may be substantial proportions of the totals, but not necessarily equal to each other.

The losses S (Fig. 6.2) include convection and radiation to the surroundings from the barrel, die and heaters, conduction to barrel and die supports and longitudinally to the feed pocket at one end and any die attachments, e.g. sizing plates, at the other. Conduction to the feed pocket reappears in feed pocket cooling and heat to the bearing assembly. Cooling of the screw by air or fluid also removes heat from the system—estimation of these losses is considered in Section 6.2, but the foregoing emphasises the importance of including all factors, for example by measurement under conditions as near as possible to those of actual operation.

The energy balance represented by eqn (6.2) must be satisfied at all steady-state conditions, and the consequences of changes in operating conditions on the separate energy items will be considered later in this

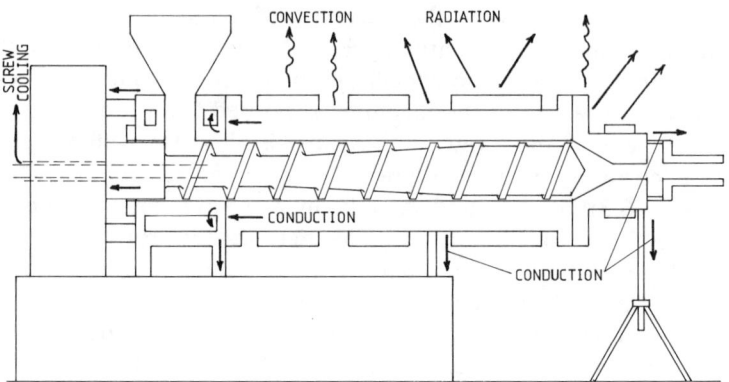

Fig. 6.2. Heat losses.

chapter (Section 6.6) and in Section 7.2. As discussed above, the magnitude of certain terms will depend on the definitions used, so also will the efficiency of the process, whether this is defined as the useful output per unit energy input in kg/MJ, or as the ratio of useful energy to total energy. The former definition depends largely on the maximum temperature and enthalpy reached in the process, which in turn depend on the molecular weight and viscosity of the polymer, the degree of mixing, the yield shear stress of particles to be dispersed, and requirements of downstream stages of the process such as drawing, crystallisation, orientation and crosslinking. A comparative assessment of these is beyond the scope of this book; evidently the achievable efficiencies will differ for different polymers and products. For the present purpose it is more instructive to take the second definition, where process efficiency:

$$\Phi = \frac{I}{E+H} = \frac{I}{I+S} \tag{6.5}$$

which in the absence of preheating reduces to:

$$\Phi = \frac{I_2}{E+H} = \frac{I_2}{I_2+S} \tag{6.6}$$

It is evident that this value of efficiency will depend on the values of E, H and S and hence on how the latter are defined; if power input is measured at drive motor input or output, E and S will be correspondingly greater and the efficiency Φ less than if power is measured at the screw. The effects on efficiency of *small* changes in operating conditions will be similar, but it must be remembered that the efficiency of certain drives such as Shrage and AC commutator motors and magnetic slip couplings varies widely with load and/or speed. Thus changes in efficiency based on electrical input may reflect the efficiency of the drive more than that of the extruder, and comparisons between different machines may be misleading.

6.2. EXPERIMENTAL DETERMINATION OF ENERGY BALANCE

As the energy balance and its effects on the performance of the extrusion process have not been widely discussed in quantitative terms, some comments are given on experimental measurement of the former in respect of semi-technical and production machinery; the special equipment and techniques applicable to research units are outside the scope of this work.

Mechanical Power Input

The most direct methods of measuring mechanical power input to the screw involve concurrent measurements of speed and torque. Speed may be measured by tachometer, but below about 1 rps it is more reliable to integrate over a fixed time period of 1 min or more by automatic or visual counting of revolutions, especially if speed variation with load is appreciable due to slip or regulation in the drive. This may also be valuable for calibrating tachometers at higher speeds. If tachometers are driven from an input shaft, due allowance must be made for slip in magnetic couplings, variable-speed gearboxes, belt drives, etc. A method of torque measurement applied to small machines (especially vertical shaft) which can be supported entirely from the feed end, consists in measuring the torque reaction on the supports, which must be of low friction. Unless the speed reduction exceeds, say 100:1, allowance must be made for the much smaller reaction on the drive motor stator, which must be added or subtracted according to the relative directions of rotation. Care must be taken to minimise the restraint of feed, pipe and cable connections and that due to the product or downstream equipment. Another method is to apply strain gauges to a short section of the drive shaft and conduct the output signal to fixed recorders by telemetering or slip rings. To maximise the signal, a more flexible section (often by reducing diameter) is introduced; this brings the possibility of changing calibration and even drive failure during overload, e.g. at starting. Sharp changes of shaft section usually result in the limited length available in production machines giving non-linear strain patterns and compensation or corrections for temperature and bending stresses are also necessary. Torsional strains have also been measured by optical interference, using mirror systems to amplify the deflection; these have the possibility of incremental output leading to digital signals and long-term stability and can be statically calibrated, but are rather fragile for industrial use. For large machines, systems designed for ship propeller shafts and rolling mill drives have been adapted. One such is the ASEA[62] Torductor in which a section of shaft acts rather like the magnetic core of a transformer; two rings of radial windings surrounding the shaft and fed with alternating current produce a continuous pattern of magnetic fields at ±45° to the shaft axis, the directions of the principal shear stresses due to torsion. Torque on the shaft produces a difference between these principal stresses and, due to consequent changes in magnetic susceptibility of the shaft material, a proportional difference is produced in the mutual inductance between the primary windings and a third ring of secondary windings placed between the two primary rings. The voltage induced in the secondary winding is thus directly proportional to primary voltage and to torque on the shaft. It is claimed to be insensitive to temperature and adjacent changes of cross-section and the effects of bending stresses and varying air-gap are cancelled by integration round 360°. It is claimed to be suitable for most magnetic steels, with appropriate calibration, which may be done statically and is very stable.

Such a torquemeter was used by the author and others on a 90 mm (3·5 in) diameter 20:1 L/D solid-fed extruder driven by a 270 kW (360 hp) electric motor at speeds up to 400 rpm. The author also had considerable experience on several machines up to 26 kW of an indirect system designed by Allen and Hillman.[63] This essentially measured the difference in tension in the two sides of a drive belt, and could be statically calibrated; though allowance must be made for losses in any gearbox and bearings between the drive belt and the screw, it proved accurate and reliable in semi-technical work. With the same proviso, other indirect methods may be used, having the advantages of lower torque at higher speed and more freedom for physical arrangement than measurement directly on the screw shaft. Measurement of torque by reaction on the motor stator requires precautions both in respect of losses in the transmission system and physical restraint from cables. As mentioned on p. 144, if mechanical energy input to the screw is measured by electrical input to the motor, then in addition to the transmission losses, allowance must be made for electrical losses in the motor. These may be obtained from manufacturer's efficiency data or by calibration against a mechanical brake— if the latter is mounted in place of the screw then it will also determine the transmission losses, which vary with both speed and torque. In the case of AC induction motors, speed and torque are related so that a single efficiency curve is sufficient. With DC shunt motors (as often used with thyristor drives), no-load speed is primarily dependent on applied voltage, and the *percentage* reduction in speed with torque is almost independent of voltage, so again a single efficiency curve is approximately correct. However, with variable-speed AC motors, including the Schrage (brush-shifting) and commutator (induction regulator) types, the efficiency is also substantially dependent on the nominal speed setting (i.e. no-load or full-load speed); in some conditions the latter suffers very large losses due to disproportionate circulating currents in the secondary windings. The electrical location of metering instruments must also take account of losses in auxiliaries, e.g. cooling fans.

The measurement of mechanical power input is thus quite complex and for reliable indication, needs to be considered when specifying initial equipment. However, with experience, a simple ammeter will give warning of overload and some appreciation of the *changes* in load with polymer MFR, temperature, speed, pressure, etc.

The necessity discussed on p. 161, for measurement of temperature close to the inner surface of the barrel arises largely from the effects of heat transfer from/to the barrel heating/cooling system. For reasons of fast response, accurate control and reduction of temperature overshoot, the thermocouples providing the control signal for heaters and barrel cooling are situated near to the heating and cooling elements or at least close to the outer surface of the barrel. As shown in Section 3.2, heat flow by conduction is in the direction of, and directly proportional to, the temperature gradient. Thus, for the same 'set' value on the temperature control, the temperature near the polymer will be lower or higher according to whether the heat is being supplied or withdrawn from the polymer (Fig. 6.3). Further, this difference will be greater for severe heating or cooling than for near-autogenous

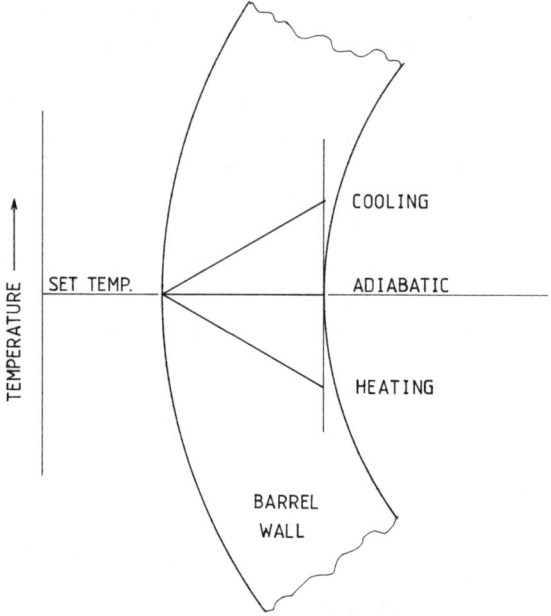

Fig. 6.3. Temperature gradients in barrel.

conditions when radial heat flow will be small. The implications of these temperature changes for operation and control will be considered in Section 7.1, but here it must be noted that they will affect viscosity and power input and correct estimation of the latter must depend on choice of the appropriate temperature.

Heater Energy

The energy added to the system by heaters or removed by cooling is not susceptible to direct theoretical analysis; under the usual system of a closed-loop temperature control, this will automatically adjust the energy input or output to maintain the desired temperatures. This energy is determined as the balancing factor in eqn (6.2), depending on the values of mechanical power E, increase of enthalpy in polymer I and losses S appropriate to these temperatures. It is of course possible to determine the maximum power available, e.g. from the wattage of electric heaters at the corresponding operating temperature or the sensible heat of heating/cooling fluid at the available flow rate and permissible temperature change. This can be used, in conjunction with the energy balance—Figs 6.8 and 6.10—to determine the approximate limiting operating temperature for a given output or vice versa. The reliability of experimental measurement of heat input is highly dependent on the type of control system; if power is applied to resistance heaters by direct connection to the mains supply or under manual control, e.g. through an autotransformer, then the electrical input is simply measured by RMS current and voltage, the power factor being nearly unity. The risk here is that a slight lack of energy balance will cause a slow drift of temperature and related factors including heat losses, the energy discrepancy being mainly absorbed in heating or cooling the barrel and other machine parts. This departure from the steady-state assumption will be significant for small machines or low outputs (as usually used for experimental work) since the heat capacity of the metal parts of the machine is high in comparison with that of the polymer product. With the older time-proportioning controls, a domestic-type integrating wattmeter in each circuit may be sufficiently accurate, though since these are not designed for suddenly changing loads, it is probably more accurate to monitor the total heater load, where different switching times will reduce the proportional load changes. An alternative, described in Appendix C, uses a multichannel event recorder to record the ON and OFF periods of each heater and give an average percentage ON time for each zone—if taken over 10–20 min this allows for time-fluctuations due to interaction between zones. This percentage ON time may then be multiplied either by the rated wattage of each heater and summed or by a separate measurement of the actual wattage at operating temperature. A source of error here is due to the change in heater resistance with temperature, for example:

A given extruder zone is set to 200°C control temperature and during the OFF period the temperature of the heater winding may fall to 180°C, at which the resistance is, say, 58 ohms. When power is applied, the temperature will rise, and may reach 500°C before being switched off. The instantaneous power input at constant voltage is given by V^2/R, i.e. $240 \times 240/58 = 993$ W at 180°C. If the temperature coefficient of resistance over the range 180°C to 500°C is taken as constant at 0·0002 ohms/ohm °C, the resistance and power will be 61·7 ohms and 933 W, respectively, at 500°C, which are the values indicated under continuous operation. If a linear increase of temperature is assumed, the continuous rating of 933 W will underestimate the actual power by some 3%. This error is of course reduced if a steady temperature is reached before switching off.

Phase-angle firing thyristors give a non-sinusoidal waveform while fast-cycling thyristors may only give half-wave output—in either case a simple RMS measurement is inaccurate though in the latter a long-time constant ammeter will give a qualitative indication. The controller manufacturer should be consulted if more accurate measurement is required, but usually it is sufficient to know how the balance between heater and mechanical power input is affected by changes in material, screw or operating conditions. With mains-frequency induction heaters it is necessary to take account of the power factor—usually below $0 \cdot 8$. For fluid heating or cooling the difficulties are mainly in maintaining steady flow rate and accurate measurement of small temperature changes.

It is hardly practicable to estimate heating/cooling energy from temperature gradients within the barrel walls since longitudinal gradients cause conduction from zone to zone within the wall (see Appendix C) and control action causes fluctuations in radial temperature gradient with time. In addition, because of longitudinal conduction and variations in outer temperature between heaters and exposed barrel parts, it is difficult to define the areas over which heat transfer occurs.

Heat Content (Enthalpy) of Polymer

A first requirement for assessing the energy content of the polymer as it leaves the die is accurate knowledge of the mass flow rate. If the downstream process can be interrupted, this is most simply measured by allowing the extrudate to fall into a weighed metal container and cutting the thread by sliding a knife quickly across the die face after a known period—which should be at least 60 s to allow for short-term fluctuations. To avoid disrupting the process, the extrudate at the die may be marked with a wax pencil at the beginning and end of a timed period and then removed for weighing after the cooling process. Measurement of the dimensions near the die is unlikely to be satisfactory, both because of the difficulties of accurate measurement of cross-section and uncertainty of the relevant mean temperature and consequent density. Measurement of dimensions after cooling (and drawing) is likely to be insufficiently accurate; if cut lengths are produced it would be better to time the production of a suitable number of lengths, which can then be weighed. With this, and other indirect methods including weighing the feed material, it is important to ensure that the mass of material 'stored' in the system between the die and the measuring position is the same at start and finish of the timed period.

The specific enthalpy (J/kg) at the die requires knowledge of the mean melt temperature; determination of the appropriate value is discussed in Section 8.1. The corresponding enthalpy may be read from a chart (see Section 3.1) for the appropriate polymer. A difficulty which arises with semi-crystalline polymers is that though the melt state is amorphous and the energy content unambiguous the energy content of the solid feed material processed may differ from that of the sample used for the chart, mainly due to differences of crystallinity, which in turn depend largely on the rate of cooling during granulation or compounding. As explained in Section 3.1, this is not likely to be significant with HD polyethylene, but more so with polypropylene and especially branched LD polyethylene. However, if the feed material is consistent, this will involve a constant error, and thus comparisons between different operating conditions will be unaffected. The same applies to a constant degree of preheating of the feed material. Enthalpies are usually given on a mass rather than a volume basis, so measurement of volume output must be multiplied by the density corresponding to the temperature of measurement. It is important to note that calculation of heat content from specific heat data is generally inaccurate. First, published specific heat figures usually relate to ambient temperature, which may be a sufficient approximation up to the glass transition temperature T_g of amorphous polymers. The specific heat at higher temperatures is usually appreciably greater, so that heat contents based on ambient values of specific heat will be low, both in absolute terms, and for small changes in operating conditions. If the specific heat at operating temperature is available, this may be adequate for the latter purpose, but will overestimate the former. With semi-crystalline polymers, changes in specific heat with temperature may occur, especially between the solid and melt states, but in addition the latent heat of crystallisation involves a further increase in the heat content in the melt state. Even integration of the area under a curve of specific heat vs. temperature is uncertain, since the precise shape of the melting peak is influenced by the speed of response of the measuring instrument (usually in the temperature-scanning mode). The increase of enthalpy is expressed in eqn (6.4) in integral terms purely for convenience.

Energy Losses

If, because of their physical connections, the extruder and die are considered as a single system, then the energy losses from the system are in conduction, convection and radiation to the ground and surrounding atmosphere, conduction to the drive bearings and gearbox, heat loss in feed pocket cooling water and screw cooling (if used). Barrel cooling, which is specific to certain processes and conditions is treated as negative heating and included as such under 'Heater energy'. As discussed in Section 3.2 conduction is proportional to temperature gradient, or temperature difference between two fixed points, and convection is proportional to temperature difference. Radiant heat loss increases at a greater rate, so that heat losses will increase considerably as operating temperatures increase; if the external surface temperature of the extruder was precisely known, these might be calculated according to the equations given in Chapter 3. However, because of the control action, the temperature of the heater element, and hence of the outer surface, may rise considerably above the control temperature while heating, but fall slightly below the latter during OFF periods—see example on p. 151. This control action will cause consequential, but smaller, time-variations in the temperature of other exposed surfaces, which will, since they are heated by conduction, be nearer the control temperature. Conduction along the barrel (and screw) to the feed pocket coolant will also increase with operating temperature, but will also depend on the set temperature profile along the barrel. An example of a calculation of this type is given in Appendix C using experimental measurements of temperature. This Appendix also gives for comparison, experimental measurements of heat loss by measuring the heater input necessary to maintain the pre-set temperature profile when the screw is stationary and there is no polymer output, i.e. when $E = 0$ and $I = 0$, eqn (6.2) reduces to $H = S$. These results show, as expected, that losses rise with temperature and also the effects of altering temperature profile. In this case it is clearly seen how in certain conditions the balance of energy from the various heater zones changes and how zones mutually interact by longitudinal conduction within the barrel wall. For instance the large heater on the die adaptor contributes most of the heat to the rather small die used, while the die heater only modifies the temperature at the die exit. The experimental method is evidently more reliable than calculation and only needs to be done once for any physical configuration, e.g. barrel length and die type. It does require separate readings for different temperatures and temperature profiles, the main limitation being the time necessary to establish steady conditions. Normal variations in ambient temperature will have little effect, though it is important that feed pocket and screw cooling conditions remain as in normal operation. A further limitation of this method is that when the extruder is running, heat is conveyed towards the die by mass transfer of the polymer, and so large longitudinal temperature gradients can be maintained, which in the static test would be reduced by longitudinal conduction.

6.3. POWER CONSUMPTION IN THE SCREW—NEWTONIAN ISOTHERMAL CASE

Mechanical power is absorbed by turning the screw within the polymer mass in the channel and also by shearing between the screw flight and barrel. The former causes internal shearing of the polymer, but only by virtue of a circumferential component of shear stress between the polymer and the barrel wall; it is thus sufficient for total power to consider the screw and polymer as rotating together relative to the barrel, exerting the same component of stress with the relative velocity of the screw to the barrel. Since the shear rate over the flight will be much greater than that over the channel, the shear stresses will differ and the two regions are conveniently considered separately. Components of shear stress which are parallel to the screw axis will cause a thrust on the barrel and equal reaction on the screw and thrust bearing; however, since no motion takes place in this direction, no work is done, and the circumferential or tangential components represent the consumption of power.

Newtonian behaviour implies that viscosity of a given polymer is a function of temperature only, and not of shear rate; this also implies that the effective viscosity (ratio of shear stress to shear rate) in one direction will not be influenced by shear components in other directions and so the components may be treated independently—this is not the case with pseudoplastic fluids. The isothermal restriction implies that the effective temperature and hence viscosity is the same in both directions and also over the screw flight. Further, it implies that viscous heating due to shear both at the surface and within the polymer mass is instantaneously dissipated (e.g. to the barrel and screw) so that temperature and viscosity are not functions of time and do not react on the velocity profile (see Section 3.3). In order to derive expressions for velocity and shear rate, in Chapter 5 the total velocity of barrel relative to the screw πDN was resolved into components along and across the screw channel:

Velocity component along channel $W = \pi DN \cos \phi$ (5.4)

Velocity component across channel $U = \pi DN \sin \phi$ (5.5)

The method followed here, for which grateful acknowledgement is made to Weeks and Allen,[35] is to derive the shear stresses corresponding to these velocity components and recombine them to give the total shear stress in the tangential direction.

From Appendix A the downstream velocity at a distance y from the screw surface is:

$$w = \frac{Wy}{h} + \frac{(y^2 - yh)}{2\eta} \cdot \frac{dP}{dz} \qquad (A8)$$

Then the shear rate at y is:

$$\dot{\gamma}_y = \frac{dw}{dy} = \frac{W}{h} + \frac{(2y - h)}{2\eta} \cdot \frac{dP}{dz} \qquad (5.95)\ (6.7)$$

and the shear rate at the wall ($y = h$) is:

$$\dot{\gamma}_h = \frac{dw}{dy_{y=h}} = \frac{W}{h} + \frac{(2h - h)}{2\eta} \cdot \frac{dP}{dz}$$

$$= \frac{W}{h} + \frac{h}{2\eta} \cdot \frac{dP}{dz} \qquad (6.8)$$

To eliminate dP/dz, eqn (5.18) may be rearranged as:

$$\frac{h}{2\eta} \cdot \frac{dP}{dz} = \frac{6W}{h}\left(\frac{1}{2} - \frac{Q}{Wbh}\right)$$

and substituting in eqn (6.8) gives:

$$\dot{\gamma}_h = \frac{W}{h}\left(4 - \frac{6Q}{Wbh}\right) \qquad (5.98)\ (6.9)$$

The shear stress in the down-channel direction is thus:

$$\tau_h = \eta \dot{\gamma}_h = \frac{\eta W}{h}\left(4 - \frac{6Q}{Wbh}\right) \qquad (6.10)$$

and the tangential component of this stress is:

$$\tau_h \cos \phi = \frac{\eta W}{h}\left(4 - \frac{6Q}{Wbh}\right) \cos \phi \qquad (6.11)$$

As discussed in Chapter 5 under 'Transverse flow', the transverse velocity at a distance y from the screw surface is given by Appendix A as:

$$u = \frac{Uy}{h} + \frac{1}{2\eta}(y^2 - yh)\frac{dP}{dx} \qquad \text{(A12) (5.23)}$$

The shear rate at y is:

$$\frac{du}{dy} = \frac{U}{h} + \frac{1}{2\eta}(2y - h) \cdot \frac{dP}{dx} \qquad (6.12)$$

and the shear rate at the wall ($y = h$) is:

$$\frac{du}{dy_{y=h}} = \frac{U}{h} + \frac{h}{2\eta} \cdot \frac{dP}{dx} \qquad (6.13)$$

Substituting to eliminate dP/dx by:

$$\frac{1}{2\eta} \cdot \frac{dP}{dx} = \frac{3U}{h^2} \qquad (5.24)$$

gives:

$$\frac{du}{dy_{y=h}} = \frac{4U}{h} \qquad (6.14)$$

The shear stress in the cross-channel direction is:

$$\eta \frac{du}{dy_{y=h}} = \frac{4\eta U}{h} \qquad (6.15)$$

and the tangential component of this stress is:

$$\eta \frac{du}{dy} \sin \phi = \frac{4\eta U}{h} \sin \phi \qquad (6.16)$$

The total tangential stress at the barrel surface is thus:

$$\eta \frac{W}{h}\left(4 - \frac{6Q}{Wbh}\right)\cos \phi + 4\eta \frac{U}{h}\sin \phi \qquad (6.17)$$

Combining eqns (5.4) and (5.5) gives:

$$U = W \tan \phi \qquad (6.18)$$

and substituting for U in eqn (6.17) gives:

$$\text{tangential stress} = \eta \frac{W}{h}\left[\left(4 - \frac{6Q}{Wbh}\right)\cos \phi + 4 \sin \phi \tan \phi\right] \qquad (6.19)$$

For an elemental length dz of the channel, this stress acts over an area $b \cdot dz$ giving:

$$\text{tangential shear force} = \frac{\eta W b \, dz}{h}\left[\left(4 - \frac{6Q}{Wbh}\right)\cos \phi + 4 \sin \phi \tan \phi\right] \qquad (6.20)$$

The power absorbed is force × distance/time or force × velocity and by rearrangement of eqn (5.4), the tangential velocity is:

$$\pi DN = \frac{W}{\cos \phi} \qquad (6.21)$$

Thus the power absorbed in a length dz of the channel is:

$$\text{channel power} = \frac{\eta W^2 b \, dz}{h}\left[4(1 + \tan^2 \phi) - \frac{6Q}{Wbh}\right] \qquad (6.22)$$

As shown in Section 5.3 under Leakage, the pressure flow through the flight clearance is usually small compared with the drag flow, and in view of other approximations (see Sections 6.4 and 6.5) it is sufficient to consider only the latter. Because this is simple shear flow, independent of pressure gradient in the channel, it is unnecessary to resolve it into down-channel and cross-channel components. As for channel flow, the tangential velocity is given by:

$$\pi DN = \frac{W}{\cos \phi} \qquad (6.21)$$

and the shear rate in the flight clearance δ is:

$$\dot{\gamma}_{fl} = \frac{W}{\delta \cos \phi} \qquad (6.23)$$

Principles of Energy Balance

The corresponding shear stress is:

$$\tau_{fl} = \frac{\eta W}{\delta \cos \phi} \quad (6.24)$$

and this acts over an area $t \cdot \cos \phi \cdot dz$ giving a shear force:

$$\text{flight shear force} = \frac{\eta W}{\delta \cos \phi} t \cos \phi \, dz \quad (6.25)$$

for a channel length dz.

The velocity is $W/\cos \phi$ as before, giving:

$$\text{power absorbed over flights} = \frac{\eta W}{\delta \cos \phi} \cdot t \cos \phi \, dz \cdot \frac{W}{\cos \phi}$$

$$= \frac{\eta W^2 t \, dz}{\delta \cos \phi} \quad (6.26)$$

The total power absorbed in length dz of the melt pumping section is:

$$E_{dz} = \frac{\eta W^2 b \, dz}{h} \left[4(1 + \tan^2 \phi) - \frac{6Q}{Wbh} \right] + \frac{\eta W^2 t \, dz}{\delta \cos \phi} \quad (6.27)$$

For a screw of constant pitch p and depth h in the melt pumping section the *total* power in a channel of helical length Z is given by eqn (6.27) with dz replaced by Z, where:

$$Z = \frac{L}{\sin \phi} \quad (5.1)$$

and L is the corresponding axial melt length. For a screw of constant diameter D but varying depth h, eqn (6.27) may be integrated, or more simply approximated by summing the power absorbed in a number of elements, each of constant depth and to each of which eqn (6.27) applies with the appropriate value of h. Note that for constant pitch, W, b, ϕ and t are constant, and the flight clearance δ is also normally constant; the volumetric flow Q is constant by continuity, so a changing value of h implies also a changing value of Q/Wbh. The right-hand term is independent of h and so may be calculated directly from the value of Z as for a constant depth screw.

The power absorbed in the melt section is that likely to be affected most by changes in operating conditions, including the consequent change in effective melt length. In the usual single-screw extruder without force feeding, the power absorbed in conveying the solid polymer is likely to be a small proportion of the total, and little affected by operating changes except to screw speed, to which it would be expected to be approximately proportional. The several possible mechanisms of melting are extremely complex, as has been discussed in Section 5.9, and the temperature and shear rate will vary with position in the channel as well as from turn to turn. Although calculation of power consumed in melting is important in design, it is doubtful whether in the present state of the art it will contribute to an understanding of correct operation. As has been suggested previously, more useful information is obtained by experimental determination of the position of initial pressure rise, as an indication of the start of the effective melt length. As stated in Section 5.6, experience indicates that this position tends to move towards the feed end with increase of back pressure and towards the die with increase of speed; if these changes can be measured experimentally, a sound basis will be established for the effect of melt length Z on power in eqn (6.27). If it is necessary, for example, to determine the maximum output with the existing drive motor, it is suggested that to the power calculated from eqn (6.27) should be added the mass output multiplied by the increase in enthalpy (J/kg) from feed temperature to the softening point (say 30°C above T_g) for amorphous polymers or crystalline melting point for semi-crystalline polymers. The total power thus obtained will be an overestimate for small machines and low speeds, where considerable energy for melting is received from the heaters, but with larger machines or higher speeds it will become progressively more accurate. At the adiabatic point, where the heaters are supplying only the losses (see Section 6.6) a close approximation should be achieved.

In the present work it will be assumed that the specific enthalpy (J/kg) of the polymer at the beginning of the melt pumping section is constant, irrespective of the relative contributions from heaters and mechanical power to the screw up to that point. Thus the changes in power input in the melt pumping section, and hence changes in melt temperature, due to changes in operating conditions will be given by eqn (6.27), although it is appreciated that the total power input and the energy balance will also be affected by associated changes in the mechanism of melting. Before analysing eqn (6.27) in terms of the effects of the independent variables (speed, back pressure, screw diameter, length, depth and flight clearance), the appropriate value of temperature for determining viscosity must be discussed. Since eqn (6.27) is based on the shear stress at the barrel wall, the temperature and viscosity at this position are relevant, not the 'mean mixed' temperature for the whole channel; the latter is difficult to measure reliably even by sophisticated sampling or integrated multiple-measuring methods. In practice, the temperature near the barrel will be influenced to some extent by cross-channel circulation (Section 5.2), especially in conditions of severe heating or cooling, and by the distribution of shear heating within the channel (Section 3.3). It will also be strongly affected by shear heating in the flight clearance (Section 6.5) so the best compromise using the isothermal equation is to take the time-averaged temperature of the barrel wall *close* to the inner surface.† The longitudinal variation of temperature experienced in practice is best allowed for by applying eqn (6.27) to successive length elements in a similar way to that suggested for varying-depth screws. The effects of shear rate and temperature variations on viscosity and power will be considered in Sections 6.4 and 6.5; the purpose here is to determine what experimental values are most appropriate for use with the Newtonian isothermal approximation represented by eqn (6.27). It may be noted that the form of the equation keeps separate the power absorbed in channel and flight clearance so that it is possible simply to insert different values of viscosity in the two terms of the equation to give a closer approach to the practical situation without impairing the simplicity of the equation and its implications for operation.

Analysis of Variables—Power Equation

The previous section emphasises experimental measurement because of the problems of theoretical calculation of certain items in the energy balance. In particular, as discussed in Section 5.6 and p. 160, melt length Z changes in a complex way, especially with speed and back pressure. For simplicity, the following discussion will ignore this interaction and deductions in this and the following chapters regarding power input, energy balance, etc., must be modified by a correction to allow for the accompanying changes in melt length Z, especially with large changes of speed or pressure.

The increase in heat content I (enthalpy) of the polymer is directly proportional to the mass flow and change in specific enthalpy, but the latter requires laboratory data and in semi-crystalline polymers is

† See p. 149. Appropriate thermocouple tappings at several points along the barrel should be provided during initial manufacture, as these temperatures provide a simple but important signal for operational control.

dependent on the (usually unknown) degree of crystallinity of the feed material. However, for small changes of operating temperature, a constant (melt) specific heat may be an adequate approximation. The heat losses S are almost impossible to calculate with accuracy, but require a preliminary experiment (see Appendix C) with the appropriate physical arrangement (barrel length, die and adaptor assembly, heater rating and location, feed pocket and screw cooling flows and temperatures) and *external* temperatures. The heater inputs H are difficult to predict (except for the maximum available) and may be uncertain in measurement—this is perhaps best treated as the dependent variable or balancing factor in eqn (6.2).

In contradistinction, the analysis leading to eqn (6.27) gives a basis for estimating the mechanical power input E to the screw, and the effects upon it of changes in the independent variables η, D, N, Z, h, δ and pressure P. Both channel and flight terms in eqn (6.27) are directly proportional to viscosity; thus power input increases with polymer molecular weight (decrease of melt flow rate) and decreases with increase of temperature. By eqn (5.4) for a constant helix angle ϕ:

$$W \propto DN \qquad (6.28)$$

Thus both terms of the power equation increase in proportion to the square of the speed N^2. For constant helix angle ϕ and flight proportions, the channel width b and flight width t are proportional to diameter D, so that both terms are proportional to the cube of diameter D^3 and directly proportional to melt length Z and L (eqn (5.1)). The channel and flight power terms are inversely proportional to depth h and clearance δ, respectively. Although in the first term the helix angle ϕ only appears explicitly within the brackets, eqns (5.1) to (5.4) show it to be implicit in W, b, t and Z (for constant L), so the effect of varying helix angle ϕ on power is complex—for small values of ϕ, when $\tan^2 \phi \ll 1$, and constant flight proportions (t/p constant), $b \propto \sin \phi$ (eqns (5.2) and (5.3)) and power E is approximately proportional to $\cos^2 \phi$ for constant D, N, h and L, predicting a small decrease of power with increase of helix angle.† Since only the drag element of leakage flow is considered, the second term of eqn (6.27) is independent of pressure; the latter is represented in the first term by the dimensionless group Q/Wbh, and as discussed on p. 80 (Fig. 5.7) a

---

† Compare McKelvey,[29] Fig. 10.11, p. 255. Q increases with $\theta(\phi)$ up to $\theta = 30°$. Efficiency Q/Power decreases ∴ Power must *increase*.

decrease in the latter corresponds to an increase in back pressure. Thus eqn (6.27) indicates that the channel power increases as back pressure increases, notwithstanding the decrease in output Q given by eqns (5.17), (5.51), (5.60), etc. Thus for a given screw design and speed, Q/Wbh is a convenient way of including the effect of back pressure on power, and for a constant helix angle ϕ and pressure P, the square bracket in the first term of eqn (6.27) is constant. However, in other cases, eqn (5.45) represents the effect of independent variables on pressure at constant Q/Wbh. This reinforces the assertion following eqn (5.18) that the magnitude of Q/Wbh is valuable in characterising operating conditions. Then for constant helix angle ϕ and constant dimensionless output Q/Wbh, eqn (6.27) may be simplified to:

$$E_{\text{channel}} \propto \frac{\eta D^3 N^2 L}{h} \qquad (6.29)$$

and:

$$E_{\text{flight}} \propto \frac{\eta D^3 N^2 L}{\delta} \qquad (6.30)$$

These will be considered in conjunction with output in Section 6.6 in the effects on the energy balance and in Section 8.5 under Scale-up.

6.4. PSEUDOPLASTIC ISOTHERMAL APPROXIMATION

The numerical example given on p. 88 for leakage flow indicates that the flight clearance δ is usually one to two orders of magnitude smaller than the channel depth h; therefore the shear rate over the flight will be much greater than in the channel (eqns (5.29) and (5.98)). Since the two terms of eqn (6.27) are independent, a crude approximation to pseudoplastic behaviour is to insert different values of viscosity into each term, corresponding to the respective shear rates (see p. 161). As the example in Section 6.6 shows, the strict Newtonian isothermal application of eqn (6.27) indicates a high proportion of total power is absorbed over the flights; use of viscosities appropriate to the shear rates in channel and flight substantially reduces the contribution of the latter. This procedure also allows the inclusion of the effect of screw speed on viscosity, which decreases with increase of speed, offsetting to some extent the effect of the latter on power. For example, if the power-law approximation given in eqn (2.7) is used:

$$\eta \propto (\dot{\gamma})^{n-1} \propto (N)^{n-1} \qquad (6.31)$$

and eqn (6.27) becomes:

$$E \propto (N)^{n-1} \cdot N^2 \propto (N)^{n+1} \quad (6.32)$$

Since $n<1$, the Newtonian relation of eqn (6.27)

$$E \propto N^2 \quad (6.33)$$

is replaced by

$$E \propto N^{n+1} \quad (6.34)$$

where $1<(n+1)<2$, i.e. the power increases less rapidly with speed for the pseudoplastic than the Newtonian case.

Figure 5.7 shows that for a given speed N, as Q/Wbh decreases from 0·5 and pressure increases, the down-channel wall shear rate increases in accordance with eqn (5.98). This expression is a factor of eqn (6.11) and can be identified as part of eqns (6.22) and (6.27), i.e. the effect of changes in shear rate is linear in the power equation. A further approximation allowing for the effect of pressure is thus to take for the channel a viscosity η evaluated at the wall shear rate given by:

$$\dot{\gamma}_h = \frac{W}{h}\left(4 - \frac{6Q}{Wbh}\right) \quad (5.98)$$

This expression is strictly only applicable to Newtonian flow and as shown in Fig. 3.1 underestimates the wall shear rate and overestimates the viscosity in pseudoplastic flow at the same flow rate; however, it is qualitatively correct, partially offsetting the increase in the bracketed term of eqn (6.27) and giving a closer overall assessment of the effect of pressure on power input.

6.5. POWER IN NON-ISOTHERMAL FLOW

On p. 161 the appropriate temperature for evaluation of viscosity was discussed, including the usual longitudinal temperature profile. Radial temperature variations, e.g. due to the effects of shear heating in the channel, will lead to a non-linear velocity distribution for drag flow and a further departure from eqn (5.98) where pressure flow is included (Fig. 3.1). In general, for a given net flow rate, the shear rate near the wall will increase and viscosity decrease with increase of temperature near the wall above the bulk average; this is also the situation where strong external heating is applied. This will lead to a lower power input than predicted from the bulk temperature, suggesting the actual barrel wall temperature be used, as mentioned on p. 161.

The example on p. 185 shows that in the usual case, the power consumption over the screw flight, calculated on isothermal Newtonian assumptions, (eqn (6.27)) is appreciable compared with that in the screw channel, despite the much smaller area over which it acts (proportional to width $t.\cos\phi$ rather than b). In extreme cases, the flight power may actually exceed the channel power. If the much higher shear rate in the flight clearance is taken into account (p. 163), for a pseudoplastic fluid the contribution to total power from the flight is reduced, but still significant (see example on p. 185).

If the polymer in the channel and its surroundings are assumed to be in thermal equilibrium, then shear heating in the channel is absorbed in maintaining the temperature of the melt in the channel and the greater power dissipation over the flight clearance must produce an imbalance and consequent local temperature rise. Any polymer which adheres to the screw flight and rotates with it will be continuously subjected to this heating, whereas the remainder will be sheared and heated while in the clearance and then will return to the lower shear rate in the channel, where sensible heat may be lost again by conduction and convection in the surrounding channel flow. In the extreme, polymer adhering to the barrel will be heated during approximately 10% of each revolution by the passing flight tip and relax during the remaining 90% to a quasi-steady state, which is likely to be above the average temperature in the channel. The metal of the screw flight will be continuously heated and reach a steady-state temperature dependent on heat flows within the screw. As mentioned on p. 199, in the absence of intentional cooling, the screw is largely insulated and heat flows to other parts of the screw (e.g. to the channel root and overall towards the feed end) are likely to be small and steady. Because of its very small volume, the polymer in the clearance has negligible heat capacity, and because of its small thickness cannot maintain substantial temperature differences; it therefore represents a negligible heat sink and its temperature will follow closely that of the surroundings. The barrel is subjected to a cyclic heating, and both during passage of the flight and the channel, the barrel will conduct heat away from points on its inner surface in all directions.

If the barrel (assumed to be initially at the temperature of the channel) constituted negligible thermal resistance, the heat generated over the flight would be conducted away and the isothermal pseudoplastic calculation (see example p. 185) would remain valid. However, a finite thermal resistance would cause a transient temperature increase in the flight clearance and consequently a reduction in viscosity, heat generated and power absorbed. It is believed that this has not previously been studied in the literature and the purpose here is limited to a preliminary study to assess the order of magnitude of the effect and how it may be influenced by operating conditions.

In Appendix F an example is taken which is generally similar to that on p. 185 except that the channel depth h and clearance δ are taken as 10 and 0·15 mm instead of 5 and 0·10 mm, respectively, to represent a less extreme case. The polymer chosen is a typical MFR 2·0 low density polyethylene having viscosities at a shear rate of $30 \, s^{-1}$ (typical of the screw channel) of $2200 \, Ns/m^2$ at 150°C and $730 \, Ns/m^2$ at 200°C, giving a temperature coefficient of $\beta = 0·0134$ in the simplified equation:

$$\eta = \eta_0[1 - \beta(T - T_0)] \qquad (6.35) \, (2.10)$$

At a shear rate of $2000 \, s^{-1}$ (typical of the flight clearance), the viscosities are $124 \, Ns/m^2$ at 150°C and $61 \, Ns/m^2$ at 200°C, giving a value of $\beta = 0·0102$. The corresponding average values of pseudoplasticity index n are 0·3152 at 150°C and 0·4090 at 200°C. From these experimental data, melt viscosities are calculated for various temperatures and flight clearance shear rates corresponding to several rotational speeds. The energy dissipated in shearing is then calculated from eqn (2.43) for unit volume and multiplied by the clearance volume for one turn (pitch) of the screw, and by the time of shearing (e.g. 0·1 s at 1 rps) to give the work done over one pitch during one transit of the flight. As an indication of the totally insulated case, the temperature rise is calculated if all this energy was absorbed in heating the polymer in the flight clearance.

Since the energy is developed within a layer of polymer a fraction of a millimetre thick, it is reasonable to neglect any thermal resistance for transfer of heat to the barrel. Within the barrel wall heat will be transiently conducted in three dimensions from the source. In the axial direction, at any instant each side of the flight abuts a portion of the screw channel where only a low intensity of shearing occurs, and thus heat will be conducted either side from the flight. However, at a slightly earlier or later instant, the situation will be reversed, when the flight at that axial position is slightly rotated and an axially adjacent portion of the flight occupies the corresponding circumferential position, so that the axial heat flow is equal and opposite. Thus the net axial heat flow is zero, but the instantaneous value follows an oscillating wave pattern. Similarly, there will be an instantaneous circumferential heat flow, foward and backward relative to the rotation, from the position of the flight, and at a fixed position on the barrel this also will be reversed as the flight moves to and from this circumferential position. These oscillating heat flows will attenuate slightly the temperature fluctuations calculated for one-dimensional radial heat flow. The latter therefore represents a 'worst case' (Fig. F1). Standard curves for one-dimensional transient heat conduction (e.g. Fig. 3.6 of Ref. 19 for negligible surface resistance) give the distribution of temperature as a function of time and position—Fig. F2. Equation (3.4) of Ref. 19 gives the heat absorbed by the barrel as a function of time; it also shows that at a given time it is directly proportional to the excess of surface temperature above the original (uniform) temperature of the barrel. This is calculated in Appendix F for the transit time of the flight and equated to the shear heat generated in the polymer to give the surface temperature T_a of the barrel as the trailing edge of the flight passes. The transient conduction calculation assumes that the surface temperature rises instantaneously from T_0 to T_a at time t_0 and then remains constant at T_a until $t_0 + dt$; in fact this temperature is indeterminate since it affects viscosity and hence shear heat generated and also the rate of heat diffusion through the barrel. However, the surface temperature is largely determined by the rate of heat diffusion in the barrel and for small temperature changes the final surface temperature and the total energy transferred will not depend critically on the time-variation of surface temperature. As shown by Table F1 the shear heat generated also varies slowly with temperature, so the assumption of a constant surface temperature equal to the final value T_a will lead to a small underestimate of energy and T_a. This energy of shear heating represents the mechanical power consumed in the flight clearance and by comparison with the isothermal value given by eqn (6.26) indicates how much mechanical power is reduced by finite heat diffusion into the barrel wall.

The remaining part of the problem concerns the rate of diffusion of heat in the barrel once the flight has passed. The very low thermal conductivity of the polymer gives a resistance ratio (McAdams m) for heat flow from barrel to polymer in excess of 100, so that, as shown by Figs 3.2 and 3.4 of Ref. 19, this surface may now be treated as insulated, and the heat stored in the barrel as diffusing further into the metal. As an alternative to a lengthy graphical construction, Appendix F approximates the temperature distribution at $t_0 + dt$ shown in Fig. F2 to a uniform temperature over a thickness estimated to give approximately the same energy content. For a shearing time of 0·1 s, this is approximately 2·6 mm at $(T_0 + T_a)/2$. This is then treated as diffusing heat into the remainder of the barrel initially at T_0, with a resistance ratio $m = k_{steel}/k_{steel} = 1$ for the remainder of the screw revolution (0·9 s at 1 rps). These results indicate that for a speed of 1 rps at $T_0 = 150°C$, the barrel surface temperature rises to approximately 151·5°C during passage of the flight and drops to 150·23°C before the flight again passes. In these conditions, the effect on power over the flight (second term of eqn (6.27)) is small, but the temperature does not return to T_0 before the next revolution, so a cumulative build up of temperature must be expected. This increased temperature must also affect viscosity near the barrel wall in the screw channel and will thus also reduce the channel power represented by the first term of eqn (6.27).

From Table F1 it is seen that the above conditions correspond to a melt viscosity in the flight clearance of the order of 100 Ns/m² and evidently the effects would be similar for a Newtonian fluid of this viscosity, which approximates to typical values at processing temperatures for nylon 66 and PET. However, a Newtonian fluid of similar viscosity to polyethylene at low shear rate—say 2000 Ns/m² would lead to a temperature rise of approximately 20°C and a reduction of some 20% in flight power; the reduction in channel power must also be significant. Calculations in Appendix F on the same assumptions but higher speeds of revolution show the temperature rise increasing to 2·6 and 4·6°C at 2 and 4 rps respectively, while the residual increase before the next revolution is 0·4 and 0·7°C, respectively, i.e. the temperature rises are almost proportional to speed. These rises will certainly give an important reduction of total power, and the proportion contributed by the flight clearance, at high rotational speeds and high melt viscosities. So far, the flight clearance has been considered in isolation, but must now be related to other heat flows in the melt pumping section.

In the example studied, the drag flow output is $1·283 \times 10^{-4}$ m³/s at 1 rps and the channel power at a viscosity of 2200 Ns/m² is 783 W/turn. The equivalent longitudinal temperature rise is:

$$\frac{783}{1·283 \times 10^{-4} \times 760 \times 2300} = 3·49°C/turn$$

due to shear heating. However, the distribution of shear heating in the channel is not uniform. In Section 2.2 (Fig. 2.9) this was considered for simple shear and capillary flow; now it will be derived for an isothermal Newtonian fluid for the screw channel where both drag and pressure flows co-exist, together with a transverse circulation. The transverse velocity at a position y from the screw root is given by:

$$u_y = \frac{Uy}{h}\left(\frac{3y}{h} - 2\right) \quad (5.25)$$

The tangential component of this velocity is:

$$u_{\text{tang}} = \frac{Uy}{h} \sin \phi \left(\frac{3y}{h} - 2\right) \quad (6.36)$$

The longitudinal velocity at y from the screw root is:

$$w_y = \frac{Wy}{h} + \frac{y^2 - yh}{2\eta} \cdot \frac{dP}{dz} \quad (A8)$$

and substituting for pressure gradient from eqn (5.62):

$$w_y = \frac{Wy}{h} + \frac{y^2 - yh}{2\eta} \cdot \frac{12\eta W}{h^2}\left(\frac{1}{2} - \frac{Q}{Wbh}\right)$$

$$= \frac{Wy}{h} + \frac{6Wy}{h}\left(\frac{y-h}{h}\right)\left(\frac{1}{2} - \frac{Q}{Wbh}\right) \quad (6.37)$$

and the tangential component is:

$$w_{\text{tang}} = \frac{Wy \cos \phi}{h}\left[1 + 6\left(\frac{y-h}{h}\right)\left(\frac{1}{2} - \frac{Q}{Wbh}\right)\right] \quad (6.38)$$

For constant Q/Wbh, combining eqns (6.36) and (6.38) and substituting $U = W \tan \phi$ from eqn (6.18), gives the total tangential velocity at y from the screw root:

$$\frac{Wy \cos \phi}{h}\left[\left(\frac{3y}{h} - 2\right)\tan^2 \phi + 1 + 6\left(\frac{y-h}{h}\right)\left(\frac{1}{2} - \frac{Q}{Wbh}\right)\right] \quad (6.39)$$

For brevity, putting

$$\tan \phi = T \quad (6.40)$$

and differentiating with respect to y to give the tangential shear rate:

$$\dot{\gamma}_{\text{tang}} = \frac{W \cos \phi}{h}\left[\frac{6yT^2}{h} - 2T^2 + 1 + \frac{6}{h}(2y - h)\left(\frac{1}{2} - \frac{Q}{Wbh}\right)\right] \quad (6.41)$$

The shear heating per unit volume is given by:

$$\tau\dot{\gamma} = \eta(\dot{\gamma})^2 \quad (2.43)$$

Substituting from eqn (6.41):

Shear heating per unit volume W/m³

$$= \frac{\eta W^2 \cos^2 \phi}{h^2} \begin{bmatrix} 1 - 4T^2 + 4T^4 - 12\left(\frac{1}{2} - \frac{Q}{Wbh}\right) + 24T^2\left(\frac{1}{2} - \frac{Q}{Wbh}\right) \\ + 36\left(\frac{1}{2} - \frac{Q}{Wbh}\right)^2 \\ + y\left\{\frac{12T^2}{h} - \frac{24T^4}{h} + \frac{24}{h}\left(\frac{1}{2} - \frac{Q}{Wbh}\right) \\ - \frac{120T^2}{h}\left(\frac{1}{2} - \frac{Q}{Wbh}\right) - \frac{144}{h}\left(\frac{1}{2} - \frac{Q}{Wbh}\right)^2 \right\} \\ + y^2\left\{\frac{36T^4}{h^2} + \frac{144T^2}{h^2}\left(\frac{1}{2} - \frac{Q}{Wbh}\right) + \frac{144}{h^2}\left(\frac{1}{2} - \frac{Q}{Wbh}\right)^2\right\} \end{bmatrix}$$

(6.42)

The shear heating per unit surface area (channel only) from 0 to y is obtained by integration; since shear heating is zero at y = 0, the constant of integration is zero.

$$\eta \int_0^y (\dot{\gamma})^2 \, dy = \frac{\eta W^2 \cos^2 \phi}{h} \begin{bmatrix} \frac{y}{h}\left\{1 - 4T^2 + 4T^4 - 12\left(\frac{1}{2} - \frac{Q}{Wbh}\right) \\ + 24T^2\left(\frac{1}{2} - \frac{Q}{Wbh}\right) \\ + 36\left(\frac{1}{2} - \frac{Q}{Wbh}\right)^2 \right\} \\ + \left(\frac{y}{h}\right)^2\left\{6T^2 - 12T^4 + 12\left(\frac{1}{2} - \frac{Q}{Wbh}\right) \\ - 60T^2\left(\frac{1}{2} - \frac{Q}{Wbh}\right) \\ - 72\left(\frac{1}{2} - \frac{Q}{Wbh}\right)^2\right\} \\ + \left(\frac{y}{h}\right)^3\left\{12T^4 + 48T^2\left(\frac{1}{2} - \frac{Q}{Wbh}\right) \\ + 48\left(\frac{1}{2} - \frac{Q}{Wbh}\right)^2\right\} \end{bmatrix}$$

(6.43)

Principles of Energy Balance

For computation this may conveniently be rearranged as:

Shear heating per unit surface W/m² from 0 to y

$$= \frac{\eta W^2 \cos^2 \phi}{h}$$

$$\times \begin{bmatrix} \frac{y}{h}(1-4T^2+4T^4)+\left(\frac{y}{h}\right)^2(6T^2-12T^4)+\left(\frac{y}{h}\right)^3(12T^4) \\ +\left(\frac{1}{2}-\frac{Q}{Wbh}\right)\left\{\frac{y}{h}(-12+24T^2)+\left(\frac{y}{h}\right)^2(12-60T^2)+\left(\frac{y}{h}\right)^3(48T^2)\right\} \\ +\left(\frac{1}{2}-\frac{Q}{Wbh}\right)^2\left\{\frac{y}{h}(36)+\left(\frac{y}{h}\right)^2(-72)+\left(\frac{y}{h}\right)^3(48)\right\} \end{bmatrix}$$

(6.44)

For drag flow, when $Q/Wbh = 0.5$, this reduces to:

$$\frac{\eta W^2 \cos^2 \phi}{h}\left[\frac{y}{h}(1-4T^2+4T^4)+\left(\frac{y}{h}\right)^2(6T^2-12T^4)+\left(\frac{y}{h}\right)^3(12T^4)\right]$$

(6.45)

The total shear heating in the channel $(y = h)$ is:

$$\frac{\eta W^2 \cos^2 \phi}{h}\left[1+2T^2+4T^4+12T^2\left(\frac{1}{2}-\frac{Q}{Wbh}\right)+12\left(\frac{1}{2}-\frac{Q}{Wbh}\right)^2\right]$$ (6.46)

Table 6.1 gives the *cumulative* shear energy from 0 to y in W/m² of the barrel inside surface area for values of $y/h = 0.1 \times 0.1$–1.0 and $Q/Wbh = 0 \times 0.1$–0.6. Table 6.1 also gives the percentage of the total from 0 to h, i.e. a uniform distribution would show 10, 20, 30%, etc., for $y/h = 0.1$, 0.2, 0.3, etc. The absolute values are plotted in Fig. 6.4, which shows how the total shear energy increases as Q/Wbh decreases and back pressure increases. As shown by eqn (6.54) and Fig. 6.13, the corresponding reduction in output means that the associated temperature rise will increase even more rapidly. The near-vertical portion of Fig. 6.4 for $Q/Wbh < 0.3$ indicates that very little shear energy is dissipated in the region $0.2 < y/h < 0.4$, reflecting the low shear rates in this range in Figs 5.7a and b. The steep slopes near the barrel wall for $Q/Wbh < 0.3$ show that most of the shear heating (approximately 85%) takes place in the region $0.6 < y/h < 1.0$, and even at $Q/Wbh = 0.4$, 25% occurs in 10% of the channel depth close to the barrel surface. In drag flow $(Q/Wbh = 0.5)$ the increase is non-linear due to

TABLE 6.1
Cumulative Shear Heating in Screw Channel with a Newtonian Fluid

y/h		\multicolumn{7}{c}{Q/Wbh}						
		0	0·1	0·2	0·3	0·4	0·5	0·6
0·1	W/m²	0·35417	0·17993	0·06424	0·00712	0·00855	0·06855	0·18711
	%	7·30	4·93	2·39	0·36	0·58	5·51	15·06
0·2	W/m²	0·50457	0·24500	0·07951	0·00811	0·03078	0·14753	0·35836
	%	10·40	6·71	2·96	0·41	2·07	11·86	28·85
0·3	W/m²	0·53853	0·25372	0·08124	0·02107	0·07322	0·23769	0·51448
	%	11·10	6·95	3·02	1·07	4·93	19·11	41·42
0·4	W/m²	0·54338	0·26457	0·10481	0·06408	0·14240	0·33975	0·65614
	%	11·20	7·25	3·90	3·26	9·59	27·32	52·82
0·5	W/m²	0·60645	0·33605	0·18566	0·15526	0·24487	0·45447	0·78407
	%	12·50	9·21	6·91	7·89	16·49	36·54	63·12
0·6	W/m²	0·81507	0·52665	0·35919	0·31270	0·38716	0·58258	0·90256
	%	16·80	14·43	13·36	15·90	26·07	46·84	72·66
0·7	W/m²	1·25657	0·89486	0·66083	0·55448	0·57581	0·72482	1·00151
	%	25·90	24·52	24·58	28·19	38·77	58·28	80·63
0·8	W/m²	2·01828	1·49917	1·12598	0·89871	0·81736	0·88193	1·09242
	%	41·60	41·07	41·88	45·69	55·03	70·91	87·95
0·9	W/m²	3·18751	2·39806	1·79005	1·36347	1·11833	1·05464	1·17239
	%	65·70	65·70	66·58	69·32	75·29	84·80	94·39
1·0	W/m²	4·85162	3·65004	2·68845	1·96687	1·48528	1·2437	1·24212
	%	100	100	100	100	100	100	100

the contribution of the transverse circulation. The latter will mix the various layers, tending to reduce differences in shear heating and consequent temperature rise, but Fig. 6.4 illustrates the non-uniformities due to shear heating, especially at high pressure gradients. It may be argued that if the average melt temperature is to be raised by shear heating (mechanical working), then this will lead to more uniform temperature at the die if the final screw section is working at an output $0·5 < Q/Wbh < 0·6$, i.e. a slightly falling pressure.

Pseudoplastic behaviour will tend to exaggerate the non-uniformity shown in Fig. 6.4 for a Newtonian fluid, whereas radial temperature gradients due to external heating (next paragraph) or the shear heating itself will tend to moderate the radial variation of shear energy.

A typical rating for resistance heaters is 3 W/cm^2 ($= 3 \times 10^4 \text{ W/m}^2$) of total outside barrel surface, which may be doubled at the inner surface of a small barrel say 50 mm inside diameter and 25 mm thick. Taking the thermal conductivity of steel as 50 W/m K (Table 3.1), the radial temperature gradient in the barrel $dT/dx = 3 \times 10^4/50 = 600°C/m$ or $0·6°C/mm$. If this external heating is considered as applied to the

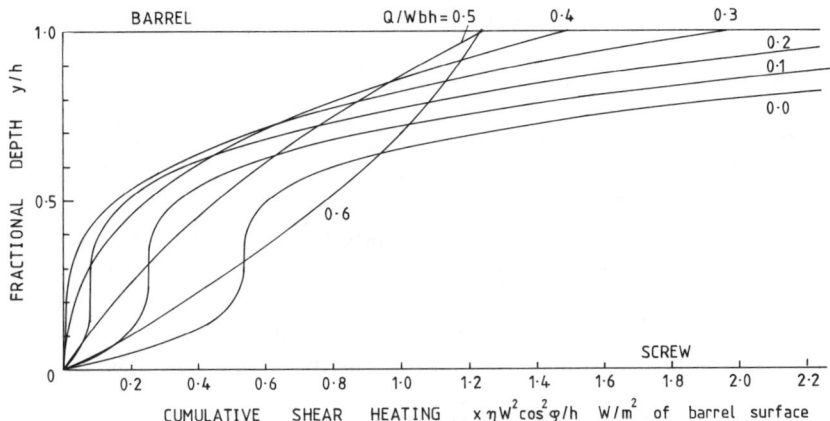

FIG. 6.4. Radial distribution of shear heat in combined drag and pressure flow.

channel flow, power input per turn $= 3 \times 10^4 \times \pi \times 0 \cdot 1 \times 0 \cdot 1 =$ 940 W/turn. The corresponding average temperature rise of the polymer will be:

$$\frac{940}{1 \cdot 283 \times 10^{-4} \times 760 \times 2300} = 4 \cdot 19°\text{C/turn}$$

due to external heating. Regarding the polymer in the channel as a semi-infinite slab at an initial temperature of $T_0 = 150°\text{C}$, heated on the outer surface only by the barrel at T_a with a resistance ratio $m = 0 \cdot 50/50 = 0 \cdot 01 \approx 0$, then from eqn (3.4) of Ref. 19, $X = \alpha\theta/r_m^2 = 1 \cdot 5 \times 10^{-7} \times 1/(0 \cdot 01)^2 = 1 \cdot 5 \times 10^{-3}$ (similar to calculations in Appendix F) and

$$Q = 940$$
$$= 0 \cdot 01 \times \pi \times 0 \cdot 1 \times 0 \cdot 1 \times 760 \times 2300 (T_a - T_0)$$
$$\times \left[1 - \frac{8}{\pi^2} (e^{-\pi^2 \times 1 \cdot 5 \times 10^{-3}/4} \cdots)\right]$$

and $T_a - T_0 = 27 \cdot 62°\text{C}$. Figure 3.6 of Ref. 19 shows that at $2 \cdot 0 = Z = x/2\sqrt{(\alpha\theta)} = x/2\sqrt{(1 \cdot 5 \times 10^{-7} \times 1)}$, i.e. $x = 1 \cdot 55 \times 10^{-3}$ m, the temperature will have changed only $0 \cdot 5\%$, so that the polymer more than $1 \cdot 5$ mm from the barrel is virtually unchanged in temperature. Thus, but for the transverse circulation (Fig. 5.8), this rate of energy input from the heaters would lead to highly non-uniform polymer temperatures in the channel.

Returning to the flight problem, superimposing on Fig. F2 the radial temperature gradient due to external heating shows that this is similar in magnitude, but opposite in direction, to that near the barrel surface due to shear heating in the flight. Thus this degree of external heating would severely reduce the diffusion of heat from the flight, increase the local temperature rise and reduce the power consumption. In contrast, severe barrel cooling will produce a temperature gradient in the same direction as diffusion from the flight, leading to a decrease in local temperature and increase in flight power consumption.

This preliminary study shows how sensitive flight power will be to barrel heating or cooling; indeed the observed effects on total power may well be primarily due to this (second term of eqn (6.27)), with important consequences for stable operation and melt temperature uniformity. Especially with viscous polymers, low melt temperatures, and high rotational speeds, substantial effects on performance may be expected, and the subject deserves further study.

6.6. EFFECT OF VARIABLES ON ENERGY BALANCE

The direct effects of dimensions and operating variables on power consumption with a Newtonian fluid ($n = 1$) have been discussed on pp. 161–163, and the modifications for a pseudoplastic fluid ($n < 1$) on p. 163. In Chapter 5 (eqn (5.43)) the drag flow was expressed in terms of dimensions D, h and N, and for a constant value of dimensionless output Q/Wbh, the total flow followed the same relation to diameter D, channel depth h and speed N:

$$Q \propto D^2 Nh \qquad (6.47)$$

For some purposes it is convenient to express energy per unit volumetric output E/Q, then combining eqns (6.29) and (6.30) with eqn (6.47):

$$\frac{E_{\text{channel}}}{Q} \propto \frac{\eta DNL}{h^2} \qquad (6.48)$$

and:

$$\frac{E_{\text{flight}}}{Q} \propto \frac{\eta DNL}{h\delta} \qquad (6.49)$$

Principles of Energy Balance

For some purposes it is useful to group independent variables such that the effect of varying shear rate is revealed. From eqn (5.98) wall shear rate:

$$\dot{\gamma}_h = \frac{W}{h}\left(4 - \frac{6Q}{Wbh}\right) \quad (5.98)$$

and for constant Q/Wbh and substituting eqn (5.4):

$$\dot{\gamma}_h \propto \frac{W}{h} \propto \frac{DN}{h} \quad (6.50)$$

The results are summarised in Table 6.2.

The important deduction at this stage from Table 6.2 is that every change of dimension or operating conditions causes different changes in power input E from those in output Q. Consequently, the specific power input E/Q changes, resulting in changes in energy balance. The implications for scale-up (change of D), for example from laboratory semi-technical machine to large-scale production will be studied in

TABLE 6.2
Output and Power Input as Functions of Dimensions (Constant Q/Wbh). From eqns (6.29), (6.30) and (6.47)–(6.50)

	Q	E		E/Q	
D proportional to:	D^2	D^3		D	
L	Constant	L		L	
		channel	flight	channel	flight
h	h	$\dfrac{1}{h}$	$\dfrac{1}{\delta}$	$\dfrac{1}{h^2}$	$\dfrac{1}{h\delta}$
η (e.g. changing temperature T)	Constant	η		η	
N	N	N^2	N^2	N	N
ηN Newtonian ($n=1$)	N	ηN^2		ηN	
ηN Pseudoplastic ($n<1$)	N	N^{n+1}		N^n	
Q/Wbh (e.g. changing pressure)	$\dfrac{Q}{Wbh}$	$4(1+\tan^2\phi) - \left(\dfrac{6Q}{Wbh}\right)$	Constant	$\dfrac{4(1+\tan^2\phi)}{Q/Wbh} - 6$	$\dfrac{1}{Q/Wbh}$

Section 8.5. Increase of melt length L produces no direct change in output Q but a proportional increase in both total power input E and specific power E/Q. Thus, other things being equal, a long machine will tend to run hotter than a short one, which may be an advantage for processes like paper-coating which require high melt temperatures but a disadvantage for blow moulding or for high viscosity melts such as high molecular weight UPVC and HD polyethylene and certain rubbers. If melt temperature (and I) is controlled to be constant, Fig. 6.5, lengthening the melt section will decrease the heater input to the polymer, although losses S will increase. In most cases the increased surface area will permit a decrease in heater rating (W/cm^2). If power input is high or melt temperature low such that barrel cooling is required, then as shown in Fig. 6.5, increasing length will increase the cooling required. As mentioned on p. 121, increasing screw speed tends to decrease melt length, and this interaction modifies, as above, the effects of screw speed as such.

Increasing the channel depth h increases the drag flow Q_D in proportion, and also the total output Q if Q/Wbh is held constant. Note that the latter condition implies a pressure gradient dP/dz varying inversely as the square of channel depth h (eqn (5.45)). However, because of the decrease of shear rate with increasing channel depth h

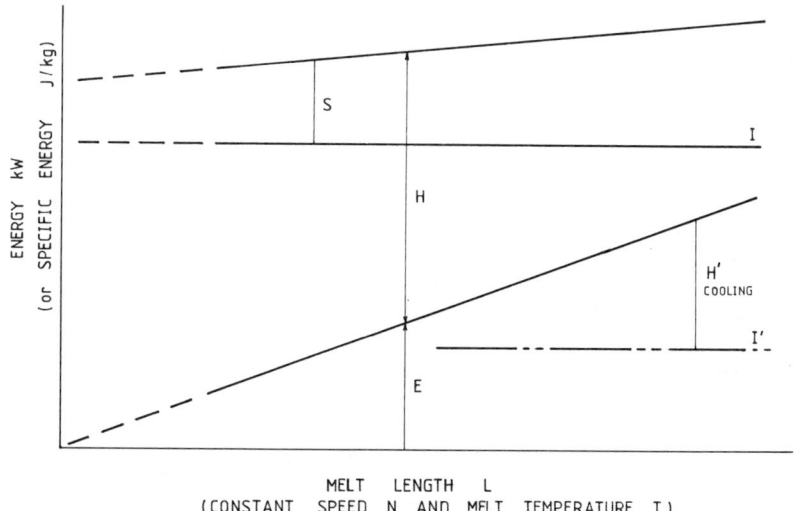

Fig. 6.5. Effect of melt length on heat requirements.

or flight clearance δ, the power input in channel and flight clearance is inversely proportional to h and δ, respectively. Thus with increase of channel depth h, the specific power input E/Q in channel and flight clearance decreases (inversely) as $1/h^2$ and $1/h\delta$, respectively. If for mechanical reasons, flight clearance δ is constant, the corresponding specific power E_{flight}/Q varies as $1/h$ and the total specific power inversely by a factor between $1/h^2$ and $1/h$. Equation (6.4) shows that at constant temperature the enthalpy increase I is proportional to output Q and therefore:

$$\frac{I}{Q} = \rho \int_{T_1}^{T_2} C_p \, dT - \text{a constant for given } dT \qquad (6.51)$$

In Fig. 6.6, output Q and enthalpy I are represented by straight lines, proportional to channel depth h. However, $E_{\text{channel}} \propto 1/h$. At constant temperature, the losses S will be constant, so as channel depth is increased, barrel heating H must be rapidly increased (or cooling decreased) to maintain temperature. Figure 6.7 represents the same situation in terms of energy per unit output and demonstrates not only that the total heater power increases with channel depth, but also the proportion of energy from the heaters increases and from the mechanical drive decreases. This means that low melt temperatures will be

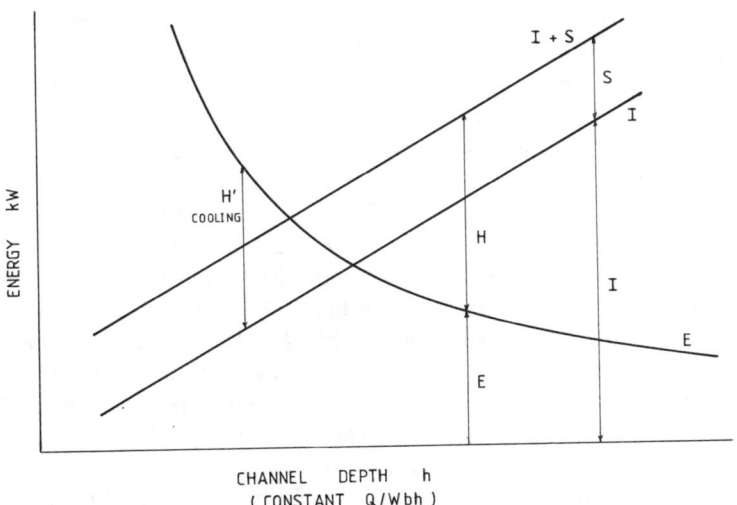

FIG. 6.6. Effect of channel depth on heat requirements.

178 Extruder Principles and Operation

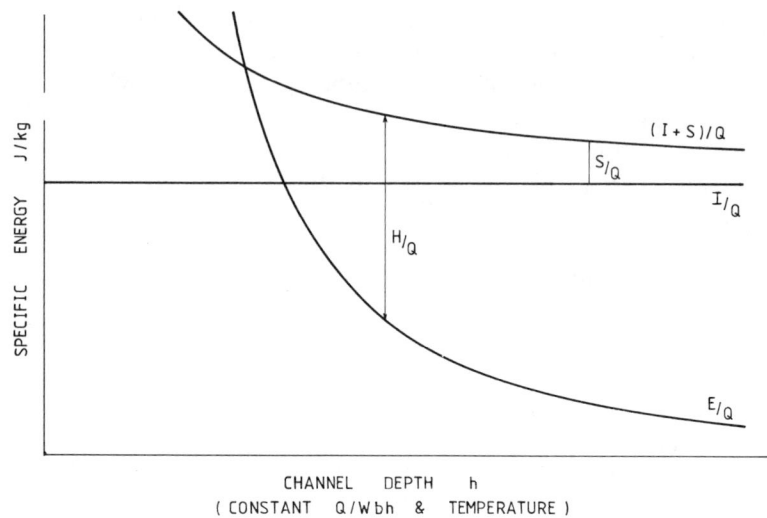

FIG. 6.7. Channel depth vs. specific energy.

easier and high melt temperatures may be more difficult to achieve, that steady conditions will be more dependent on the performance of the temperature controllers and less on a consistent viscosity (molecular weight, composition, etc.) of feed material, but that greater radial temperature differences may occur in the channel due to greater heat flow across a deeper channel. An increase of polymer molecular weight or decrease of melt temperature imply an increase of viscosity; at constant Q/Wbh, this has no effect on output Q, though eqn (5.45) shows it implies an increased pressure gradient dP/dz. However, both terms of the power equation (6.27) are increased in direct proportion, hence also is the specific power E/Q. Thus in Fig. 6.8, the power E decreases with increase of melt temperature T. However, increased temperature involves increased enthalpy I in the polymer, and also increased heat losses S from the machine, so heater power H increases rapidly with temperature, owing both to the increased heat requirements I and S and to the reduction in mechanical power input E. Since output is constant, it is only necessary to change the vertical scale for Fig. 6.8 to represent the specific power input E/Q and the balancing factors I/Q (eqn (6.51)), losses S/Q and heater input H/Q in $W/m^3/s$ or J/m^3.

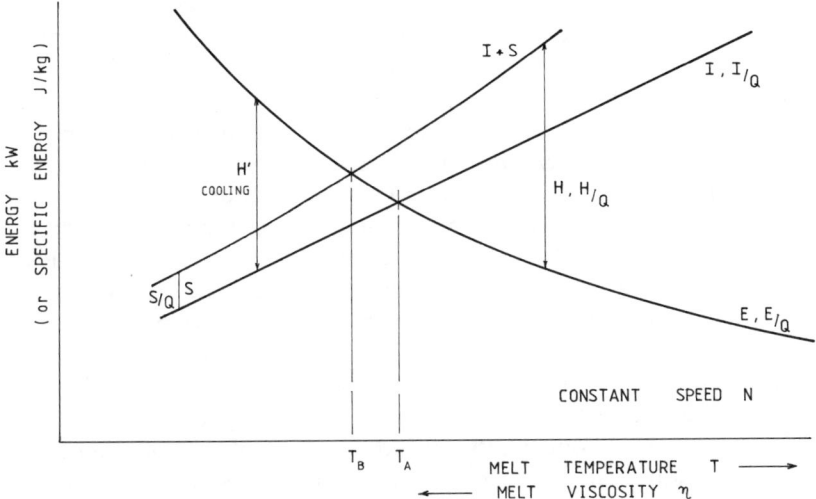

FIG. 6.8. Change of energy balance with melt temperature.

This shows that, as with increasing channel depth h, heater input not only increases but also becomes a greater proportion of the total energy; also it is difficult to operate at high temperatures, especially with large machines where the heater surface generally increases more slowly than output (see Section 8.5). It will be noted that at T_A, the drive motor supplies all the required increase in polymer enthalpy, leaving the heaters to supply only the heat losses; this is sometimes referred to as the 'adiabatic' temperature, since the polymer is neither giving heat energy nor receiving it from the surroundings. The actual value of T_A in a given case will depend on the output and the power input, which itself depends on screw dimensions, pressure (Q/Wbh), speed, and polymer type and molecular weight. At a somewhat lower temperature T_B, the drive motor also supplies the losses, so that the heaters may be switched off. This temperature T_B is also sometimes referred to as the adiabatic temperature, but the author believes is more correctly called 'autothermal' or 'autogeneous', since mechanical work is converted into heat in the polymer, some of which is then transferred to the barrel, etc., to supply heat losses. It has been held that this temperature is highly desirable for stable operation; it is true that temperature gradients due to external heat flows will be minimised, but other factors, e.g. shear rates, may not be optimised and, as discussed in Section 8.4, control problems may result. If a melt temperature below T_B is required, then barrel cooling must be applied, and Fig. 6.8 shows that this must increase rapidly as temperature is further reduced. However, barrel cooling necessarily affects most the polymer nearest the barrel surface, where, as shown in Section 6.5, shear rate and shear heating are generally a maximum, and these will be disproportionately increased by radial temperature gradients due to external cooling. Similarly, barrel cooling has a profound effect on the energy dissipated in the flight clearance. Practical experience shows that these effects cause a disproportionate increase in mechanical power and a corresponding increase in cooling requirement (Fig. 6.9) which may be beyond the possibilities of heat transfer to the barrel, especially with large machines. The author has experience of 200 and 250 mm diameter extruders in which barrel cooling caused overload of the drive motor without significant reduction in polymer melt temperature; the additional power only increased the losses to the cooling water, giving very inefficient operation. In such cases, lower melt temperature may only be possible by reducing speed and output, a deeper screw, and/or a larger machine. Screw cooling generally has a quite different effect and is discussed in Section 7.1. It will be noted that in Figs 6.8 and 6.9, the arrows relating to H (cooling) extend to the enthalpy line I rather than that for $I+S$, as in heating. This is

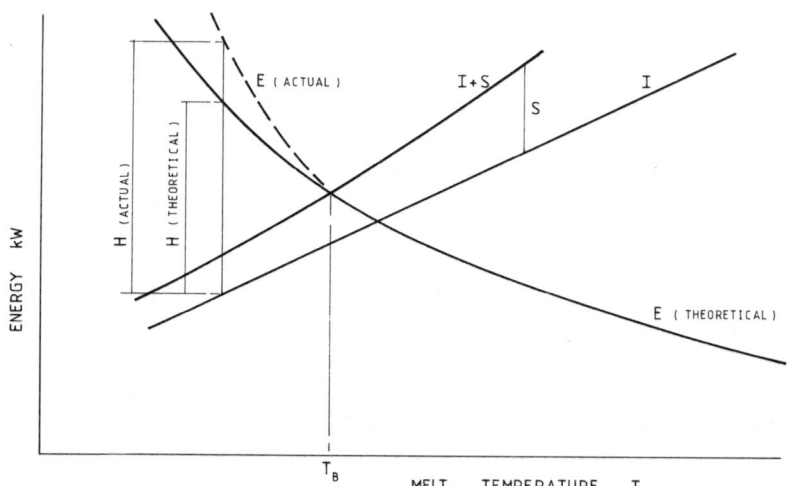

FIG. 6.9. Interaction between barrel cooling and mechanical power.

approximately correct because when the heaters are shut off and cooling applied, the external surface temperature of the barrel (but not the die) drops near to the coolant temperature and most of the natural cooling due to heat loss from the heaters disappears. This effect is minimised with induction heaters where most of the heat is generated by eddy currents within the barrel itself or with externally heat-insulated heaters where the losses are in any case much reduced.

At constant Q/Wbh, the effect of screw speed is simply that output Q is increased in direct proportion (eqn (6.47)), whereas both terms of the power equation (6.27) increase as the square of speed (eqns (6.30) and (6.31)). Thus the specific power input E/Q (per unit output) increases in proportion to speed. At constant melt temperature, the total enthalpy increase I is proportional to output Q, while the losses S are constant. Thus Fig. 6.10 represents the total energy balance and Fig. 6.11 the energy per unit output. Figure 6.10 shows that at standstill the heaters supply only the losses, but at low speeds must also supply an increasing deficiency from the drive motor. Due to the upward curve of the latter's contribution, a speed is reached at which the heater power is again reduced, until at N_A the motor E supplies all the enthalpy increase I in the polymer and the heaters again supply only the losses. As speed is further increased to N_B, the autogeneous speed, the heaters are shut off and the drive motor supplies energy to

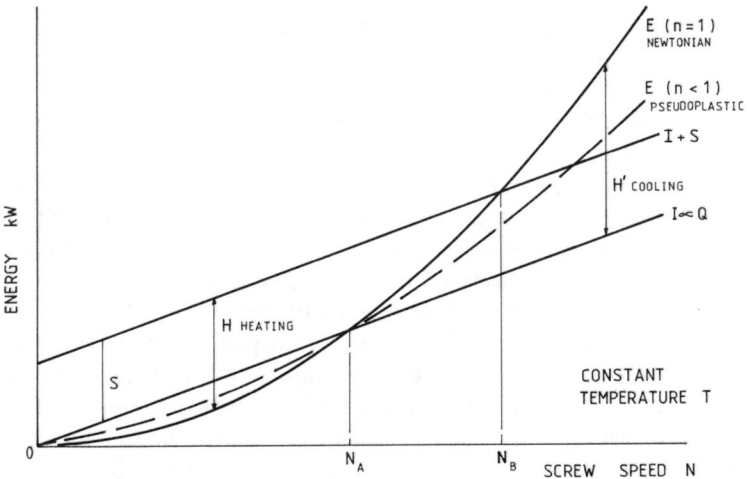

FIG. 6.10. Change of energy balance with screw speed.

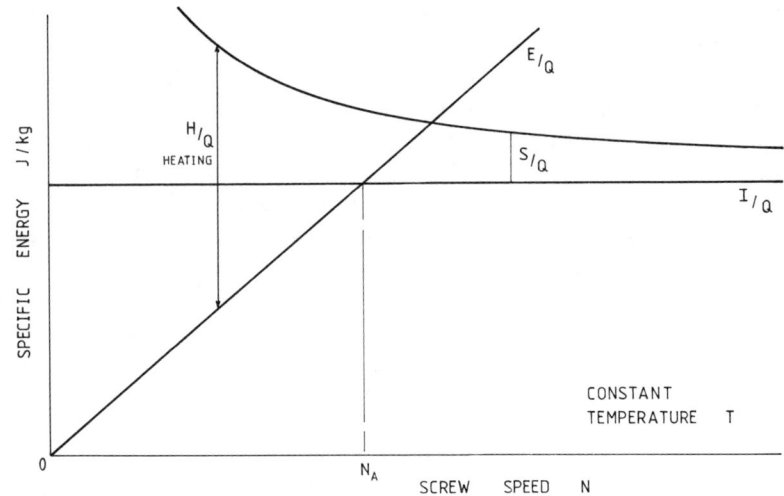

FIG. 6.11. Specific energy vs. screw speed.

meet both enthalpy increase in the polymer I and losses S. This point corresponds to the autogeneous temperature T_B, where N and T_B in Fig. 6.8 are, respectively, equal to N_B and T in Fig. 6.10. As speed is further increased at melt temperature T, the drive motor power increases more rapidly than the polymer enthalpy and cooling is required to maintain temperature T. As in Fig. 6.8, barrel cooling reduces the natural cooling represented by S and the heat to be removed by intentional cooling approximates to $E-I$ rather than $E-(I+S)$. As given by eqns (6.30) and (6.31), in the Newtonian case $E \propto N^2$; the numerical value of E at any speed, and hence the position of the intersections N_A and N_B, will depend on the dimensions of the screw and viscosity of the polymer, including in the latter the effect of temperature. For a given screw and polymer, Figs 6.8 and 6.10 may be regarded as mutual sections of three-dimensional surfaces of E, I, S and H representing together the energy balance.

In Fig. 6.11 the energy values are plotted per unit flow rate, when I/Q is constant at constant melt temperature and S/Q decreases with speed due to the increase of Q. Then E/Q, as given by eqns (6.48) and (6.49) is proportional to speed N. This again shows heater input H/Q decreasing with increase of speed, both in absolute value and as a proportion of total energy. This again shows that severe cooling, with its attendant problems, is necessary to avoid melt temperature rising with speed above N_B, just as Fig. 6.8 shows it necessary to reduce melt temperature below T_B at fixed speed N. The author recalls operating a 50 mm (2 in) diameter laboratory extruder with a 2:1 compression ratio screw on a K65 unplasticised PVC at a melt temperature of approximately 200°C; the autogeneous speed was reached at 40 rpm. With the same screw at similar temperatures, LD polyethylene of MFR 2·0 still required heating at 200 rpm. Since the enthalpies of the two polymers are not grossly different at this temperature, the difference in autogeneous speed N_B was a reflection of the very different melt viscosities and hence drive power required.

In the Newtonian case, viscosity is a function of temperature and not of speed, so that the effects of temperature and speed can be separated. As shown by eqn (6.52), in the pseudoplastic case, the combined effect of speed N on shear rate and viscosity gives:

$$E \propto N^{(n+1)} \qquad (6.52)$$

where the index $(n+1)$ is between unity and two. The dashed line in Fig. 6.10 shows this function compared with the full line for a Newtonian fluid ($n = 1$). It will be noted that as a result the heater power H required changes more slowly with speed and less cooling will be required at speeds above the autogeneous speed N_B. Note also that for purposes of comparison, this curve is shown for a pseudoplastic material having the same viscosity as the Newtonian material at temperature T and shear rate corresponding to screw speed N_A, so that at this speed the drive power is the same for both materials. Table 6.2 shows that the increase in enthalpy (or temperature) equivalent to the drive motor power increases as the n power of speed ($n < 1$) for a pseudoplastic material compared with direct proportionality ($n = 1$) for the Newtonian. Hence the achievable melt temperature will be less dependent on speed in the former case than in the latter. Although the effect of shear rate will be discussed further in Section 8.5 Scale-up, the effect of maintaining constant shear rate on heat balance (e.g. when varying more than one dimension) is mentioned briefly here. Dividing eqn (6.47) by eqn (6.50) gives output Q varying with Dh^2 at constant shear rate. Then for a Newtonian polymer, drive power in the screw channel varies as $\eta D^2 NL$, and over the flight the same with the addition of the ratio h/δ. The specific drive power E/Q, representing the temperature increase due to shearing increases as $\eta DNL/h^2$ and $\eta DNL/h\delta$, respectively. For a pseudoplastic polymer, the relations are similar, with ηN replaced by

N^n. The implications of this are that (i) an increase of viscosity, speed or dimensions will tend to increase the autogeneous temperature, i.e. the temperature below which cooling is required, (ii) a decrease of channel depth will cause a greater increase in specific power (proportional to $1/h^2$), and (iii) pseudoplastic rather than Newtonian behaviour of the polymer only reduces somewhat the effect of screw speed. These serve to emphasise that altering dimensions or operating conditions almost inevitably leads to changes in energy balance and the 'window' of satisfactory operation, which must be considered in adapting processing techniques and experience from one polymer, machine, output or scale to another. This will be a major aspect of operation included in Chapters 7 and 8.

So far comparisons from Table 6.2 have been made at constant Q/Wbh, i.e. with longitudinal pressure gradient as given by eqn (5.45). If this dimensionless output is now considered as a variable, then for given dimensions and speed, output Q is proportional to Q/Wbh. Drive power in the channel is proportional to $\left[4(1+\tan^2\phi)-6\dfrac{Q}{Wbh}\right]$, while that in the flight clearance is unchanged. Dividing out:

$$\frac{E_{\text{channel}}}{Q} \propto \frac{4(1+\tan^2\phi)}{\dfrac{Q}{Wbh}} - 6 \qquad (6.53)$$

TABLE 6.3
Variation in Mechanical Power with Back Pressure (kW/turn)

		\multicolumn{7}{c}{Q/Wbh}					
		0	0·1	0·2	0·3	0·4	0·5	0·6
Channel								
$n=1$	E	2·233	1·928	1·624	1·320	1·016	0·7122	0·4081
	E/Q_{rel}	∞	13·54	5·701	3·089	1·783	1·000	0·477
$n=0·5$	E	1·116	1·045	0·9712	0·8897	0·8037	0·7122	0·6452
	E/Q_{rel}	∞	7·336	3·409	2·082	1·411	1·000	0·755
$n=0·3$	E	0·846	0·819	0·7893	0·7603	0·7315	0·7122	0·7750
	E/Q_{rel}	∞	5·750	2·771	1·779	1·284	1·000	0·907
Flight								
$n=1$	E	←————————— 3·100 —————————→						
	E/Q_{rel}	∞	15·5	7·75	5·17	3·875	3·100	2·583
$n=0·5$	E	←————————— 0·428 —————————→						
	E/Q_{rel}	∞	2·14	1·07	0·713	0·535	0·428	0·357
$n=0.3$	E	←————————— 0·194 —————————→						
	E/Q_{rel}	∞	0·97	0·485	0·323	0·242	0·194	0·162

and:
$$\frac{E_{\text{flight}}}{Q} \propto \frac{1}{\dfrac{Q}{Wbh}} \tag{6.54}$$

Thus, as pressure gradient is increased, Q and Q/Wbh decrease, but channel power increases. The result, mainly due to decreasing output, is that specific power E/Q increases, tending to higher melt temperatures (compare Fig. 6.8).

The following example illustrates the variations of total and specific power (representing temperature rise due to shear heating) as functions of back pressure.

Diameter	D	100 mm
Channel depth	h	5 mm
Pitch	p	100 mm
Flight width	t	10 mm
Speed	N	1 rps
Viscosity (channel)		1000 Ns/m² at shear rate 59·87 s^{-1}
Pseudoplasticity index	n	1, 0·5, 0·3
Flight clearance		0·1 mm

Then
Channel width	$b = (p - t)\cos\phi = 0{\cdot}08576$ m
Helix angle	$\phi = 17{\cdot}567°$
Helical length	$dz = 0{\cdot}1\pi/\cos\phi = 0{\cdot}3297$ m (one turn)
Peripheral speed	$\pi DN = \pi 0{\cdot}1 \times 1 = 0{\cdot}3142$ m/s
Longitudinal velocity	$W = 0{\cdot}2994$ m/s
Transverse velocity	$U = 0{\cdot}0953$ m/s
Drag flow	$Q_D = 0{\cdot}2994 \times 0{\cdot}08576 \times 0{\cdot}005/2$
	$= 0{\cdot}642 \times 10^{-4}$ m³/s
Mass drag flow	$\rho \times Q_D = 0{\cdot}049$ kg/s if density is 760 kg/m³
Drag flow shear rate	$\dot{\gamma}_D = W/h$
	$= 0{\cdot}2994/0{\cdot}005 = 59{\cdot}87$ s^{-1}
factor $4(1+\tan^2\phi)$	$= 4{\cdot}4053$
Channel power	$dE = \dfrac{1000 \times (0{\cdot}2994)^2 \times 0{\cdot}08576 \times 0{\cdot}3297}{0{\cdot}005}$

$$\times \left(4{\cdot}4 - \frac{6Q}{Wbh}\right)$$

$$= 506{\cdot}8\left(4{\cdot}4 - \frac{6Q}{Wbh}\right) \text{ W/turn}$$

Flight power $= \dfrac{\eta \times (0\cdot 2994)^2 \times 0\cdot 01 \times 0\cdot 3297}{0\cdot 0001 \times \cos \phi}$

$= 3\cdot 10 \eta$ W/turn

The results for several values of Q/Wbh are given in Table 6.3 and plotted in Figs 6.12 and 6.13. These show that in the Newtonian case the channel power increases rapidly with increase of back pressure, but that the constant flight power is also large. With more highly pseudoplastic melts, the variation of channel power with back pressure becomes much smaller, as does the proportion of total power contri-

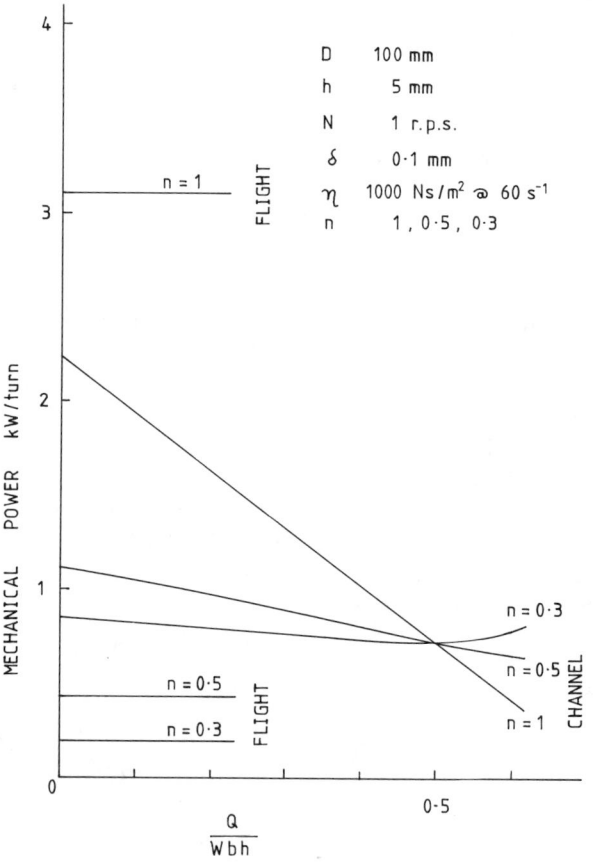

FIG. 6.12. Variation of mechanical power vs. pressure.

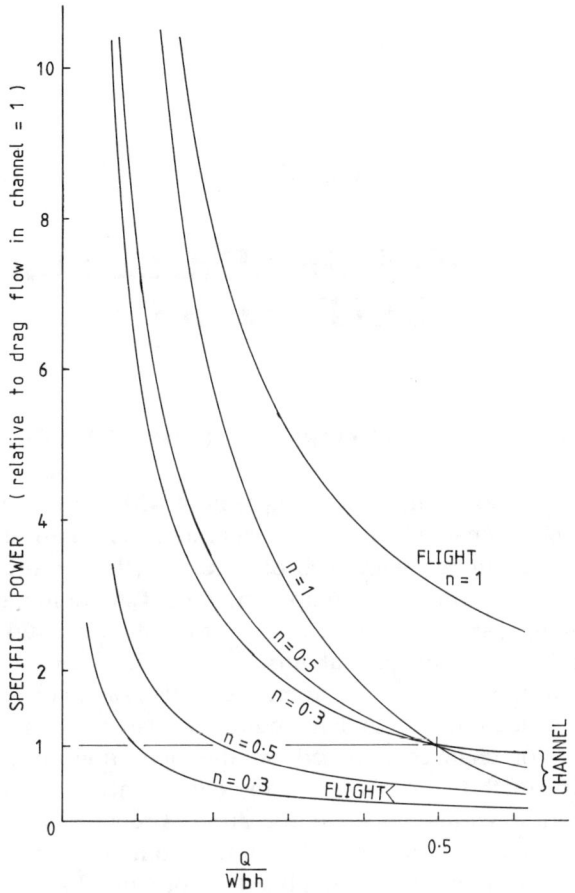

FIG. 6.13. Specific power vs. back pressure.

buted from the flight clearance. Due to the decrease of output Q with increase of back pressure, the specific power E/Q (plotted relative to a value of unity at $Q/Wbh = 1/2$ for the channel) increases even more rapidly, indicating severe problems of temperature rise at high back pressures and low outputs.

Chapter 7

OPERATION OF SINGLE-SCREW EXTRUDERS

7.1. OVERALL PERFORMANCE OF THE SCREW

The preceding chapters have dealt separately with mass and heat flow within the screw channel and flight clearance. Even in steady-state operation, these interact with each other and with pressure and temperature gradients to produce a complex pattern of simultaneous velocity, shear and temperature variations. Any attempt at a total representation of these in mathematical terms would lead to either a very generalised formulation or an extremely complex one, neither of which would readily lead to an understanding of the mechanisms or the consequences of changes imposed on the operating conditions. An understanding of the latter, which is the chief concern of this book, is essential for rational and effective operation. The alternative approach taken here is to use a qualitative description to modify and extend the approximate solutions of the simplified theory in Chapters 5 and 6. Practical experience suggests that these solutions are useful guides to operating strategies and the consequences of *changes* of conditions, as distinct from design for a specified performance. The following description refers to steady-state operation of a solid-fed single-screw extruder with fixed die, but without optional attachments or functions.

Material Flow
Steady-state requires that the net mass flow of material must be constant with time and the same at all points within the extruder and die, despite differing densities (due to temperature and change of state) and mechanisms of flow.
  Feeding of particulate solids into the screw and conveying along it

have been discussed in Section 5.10. The former, including flow in the hopper, is primarily by gravity with a possibility of entrainment by the moving screw surfaces. Bridging may restrict flow, at least intermittently, and segregation due to differing densities or particle shapes may affect uniform composition (Fig. 7.1). Experiments suggest that the loosely packed particles quickly compact, displacing entrapped air, to form an approach to a solid plug in which relative motion between particles is virtually inhibited. The increase in density implies a corresponding reduction in forward velocity and increased area in contact with the barrel for transfer of conducted heat. Theoretical studies suggest conveying is sensitive to relative coefficients of friction of polymer on screw and barrel/feed pocket and hence to relative surface temperatures; in practice cooling of the feed pocket is almost universal while cooling the feed end of the screw is often beneficial.

During melting, complex flows occur (Fig. 7.2), the main ones being 'collection' of a molten film near the barrel by the advancing screw flight, rapid forward flow of the resulting 'melt pool' and much slower flow of the remaining solid polymer, giving a wide range of residence times in this section. Other mass and heat flows are indicated in Fig. 7.2 which affect the overall process of melting. As the width of the melt pool progressively increases along the screw, the drag flow shape factor also increases;[31] both factors increase flow rate of melt so that by continuity the velocity of the remaining solid should appreciably reduce.

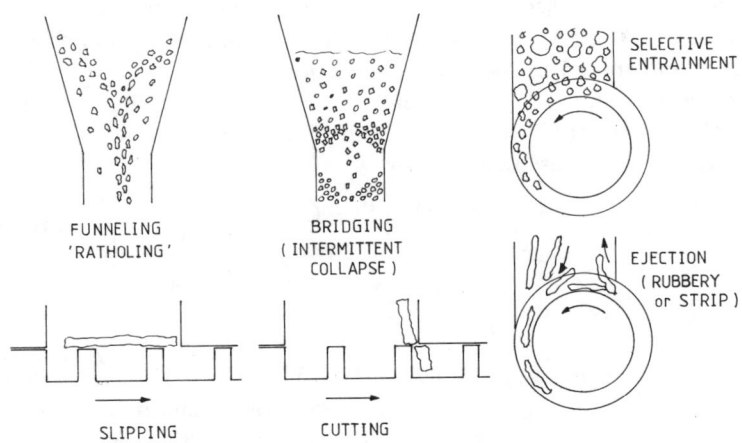

FIG. 7.1. Causes of feed segregation.

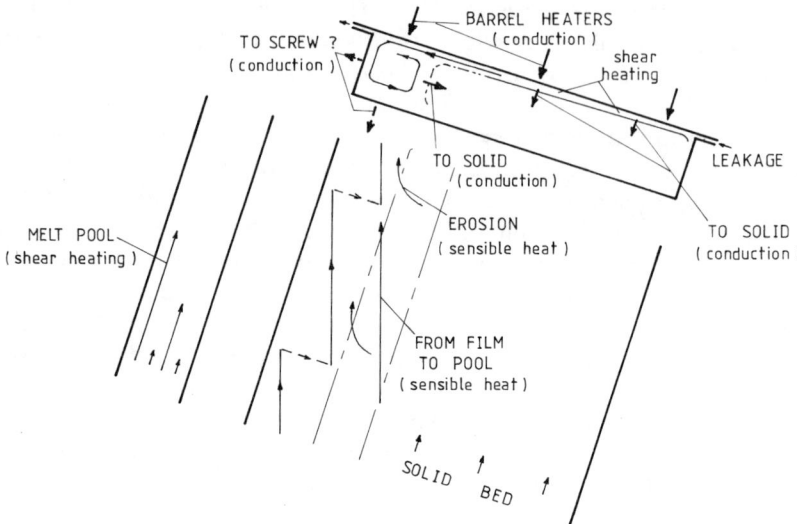

Fig. 7.2. Heat and mass flows in melting section (see also Fig. 5.36).

At some point, the remaining partly softened solid material will break up and the mechanisms of melting will change, as discussed in Section 5.9; the polymer will then behave as a suspension and pressure will commence to rise. The location of this change can be determined experimentally by pressure measurements, but the mechanisms controlling it are obscure; the general objective is to delay the bed break-up to minimise the size and extent of remaining particles to be melted and mixed in the pumping section.

Isothermal flow in the channel of the melt pumping zone is adequately described by eqns (5.96) and (5.25) for the down-channel and transverse velocities, respectively:

$$w = \frac{Wy}{h} + \frac{6W}{h^2}\left(\frac{1}{2} - \frac{Q}{Wbh}\right)(y^2 - yh) \qquad (5.96)\,(7.1)$$

$$u = \frac{Uy}{h}\left(\frac{3y}{h} - 2\right) \qquad (5.25)\,(7.2)$$

The former gives a net flow rate as in eqn (5.17) and the latter (neglecting leakage) gives zero flow rate represented by eqn (5.24). The resulting flow pattern is discussed in Section 5.2 (Fig. 5.9), producing both end-to-end and transverse mixing, the former (only) being influenced by longitudinal pressure gradient. The effect on pressure gradient of a screw of varying depth is discussed in Section 5.4 (Fig. 5.13), where it is seen that Q being constant by continuity, decreasing h implies an increase in the local value of Q/Wbh. If the position in the channel y is represented by extracting the fractional depth y/h from eqn (7.1), it is seen that the drag flow term changes only by the vertical (radial) scale of h, whereas the pressure flow term also decreases in magnitude as $Q/Wbh \to 1/2$. Thus in a tapered screw, the down-channel velocity distribution changes from Fig. 5.7b to c and eventually to Fig. 5.7d or e, though with a changing scale of h (Fig. 7.3). Equation (7.2) shows the transverse velocity distribution changes only with the scale of h. Thus the flow pattern along the length of a varying depth screw is continually changing even on an isothermal basis. The qualitative effects of non-Newtonian and non-isothermal flow have been discussed in Sections 5.7 and 5.8, respectively. Leakage, although small (Section 5.3), reduces net flow as given by eqn (5.17), re-entraining into the flow in the previous turn of the screw close to the barrel surface, where it is rapidly dispersed. Power consumption in the flight clearance is appreciable (Section 6.3) and consequent temperature rises will affect conditions in the adjacent channels (Section 6.5). The increase in specific volume due to longitudinal temperature gradients (as set indirectly on the barrel heaters) will increase mean velocity towards the die, aside from that due to any change in depth.

Varying and non-axial velocities from the screw may affect flow in a die close to the end of the screw, but are quickly rearranged in the

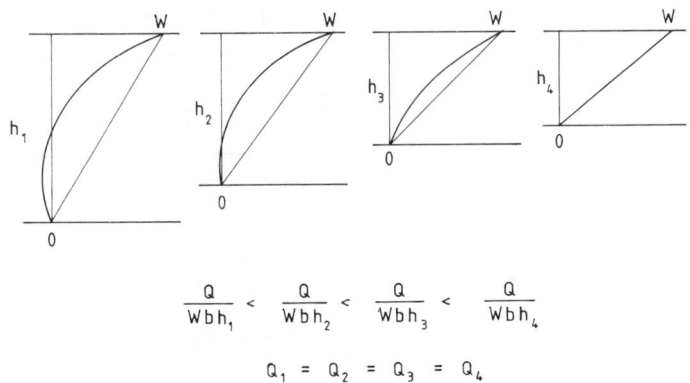

FIG. 7.3. Changing velocity profiles in a tapered screw.

usual die adaptor. The entrance to the die usually involves considerable increase in mean velocity; if this is too sudden dead spaces may be formed leading to decomposition and/or high shear stresses may arise leading to the possibility of melt fracture. In any case, elastic strain energy will be imparted to the polymer giving rise to die swell when the constraint is removed at die exit; in wide sheet or very large tube and film dies a diverging flow may lead to a retraction at the die exit in one dimension, e.g. width and diameter, respectively. In asymmetric dies, e.g. a thin rib on the side of a solid section, the increased L/D ratio of the narrow section will cause a lower mean velocity than in the wide section; attempts to correct this by reducing the land length in the narrow section lead (Fig. 7.4) to different longitudinal pressure gradients, which in turn cause a transverse component of flow due to lateral pressure differences. In die design, this can be resolved by a finite-difference network, but there are complex effects on elastic strain and swelling, and the transverse component must be suppressed by long die lands to avoid further flow distortion at the outlet. Since viscosity, elastic strain and recovery, etc., are non-linear functions, any solution, whether theoretical or empirical, is likely to be correct for

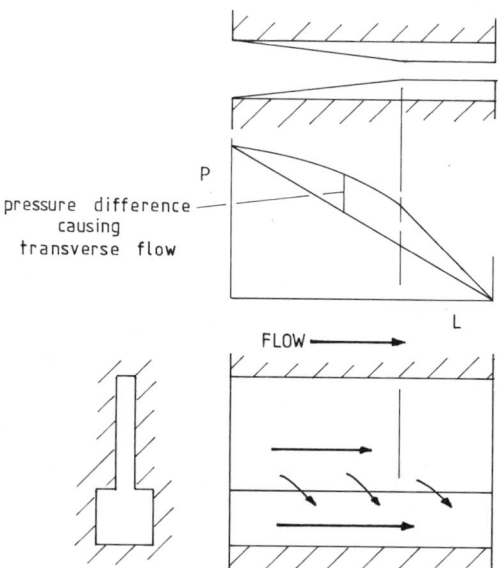

Fig. 7.4. Pressure gradients in dies.

only one combination of polymer, flow rate, temperature etc. Differences in viscosity and elastic properties due to temperature differences in polymer or die will also distort flow and therefore elimination of these temperature differences is especially important with wide sheet and tube dies (Section 4.1 p. 41 and Table 4.2).

Pressure

The effects of pressure are inseparable from those of flow. In the absence of force feeding, etc., pressure is negligible in the feed section, but the author's experiments[55] have shown that pressure generated by solids conveying leads to compaction of the solid particles and appears to be the principal cause of forward movement of the solid bed through the melting zone. Kosel[61] has experimented with feed sections designed to give substantial pressure in the melting zone and thus absorb considerable energy. It is claimed that this not only increases output like force feeding in the hopper (Section 5.10), but also dramatically increases the rate of melting, leading to more uniform melt temperature at the end of the screw. In this case there may be a positive pressure at the start of the melt pumping section, though with conventional feed and melting zones, experiments[55] have indicated the pressure to be zero at this point. The longitudinal pressure gradients in the melt pumping section, including the effects of back pressure and channel depth, have been dealt with in Section 5.4. These, combined with the transverse pressure gradient (rising from trailing to leading flight edge, right to left in Fig. 5.8), give rise to a periodically fluctuating pressure at a fixed point on the barrel, as indicated in practice by a pressure gauge. This fluctuation is normally damped out a short distance beyond the end of the screw. Pressure falls progressively through the adaptor and die, with the major drops through die lands and any restrictor bar/ring, which thereby control the flow. The possibility of lateral pressure difference is indicated in Fig. 7.4, while time-fluctuations will be discussed in Section 8.2.

Temperature

Solid feed to the extruder is usually at room temperature;† water cooling of the feed pocket is commonly used to maintain the feed in a

† The author has used air at up to 105°C in a partly fluidised bed to pre-heat LD polyethylene granules, which fed satisfactorily by gravity. Drive motor power was considerably reduced, but the effect on uniformity of melting could not be verified.

free-flowing state until it is beyond the feed opening—whether this also contributes to conveying along the screw is uncertain. Heat conducted from the first barrel heater may cause sticking in the feed, which sets an upper limit to the first zone temperature. The feed section of the screw also receives heat by longitudinal conduction, but its temperature is usually unknown and uncontrolled—an exception is referred to on p. 53. Within the barrel the polymer is heated by conduction from the barrel and possibly from the screw, leading to temperature gradients within the compacted plug; shear heating will also occur, mainly at the metal surfaces. Since the solid bed retains its shape, these temperature differences within it will tend to persist until it breaks up at the end of the melting section. As soon as an axial position is reached at which the barrel surface temperature is above the softening/melting point of the polymer, a film of molten polymer will form due to a combination of conducted and local shear heating. Initially this film will improve contact between the barrel and solid polymer particles and thus improve conducted heating, but possibly reduce frictional heating. As soon as the film thickness exceeds the flight clearance, it will tend to be collected by the flight, forming a 'melt pool' on the leading flight face. These mechanisms (Fig. 7.2) continue while the solid bed exists, though at rates dependent on the velocity of the solid bed, the thickness of the molten film and the local temperature of the barrel. The temperature of the melt pool will be governed by the balance between energy received by conduction from the barrel, sensible heat of the polymer collected from the film and internal viscous heating, and energy lost by conduction to the adjacent screw flight and interface with the solid bed. Heat conduction at the liquid/solid interface and their relative velocities will lead to melting and erosion (as observed experimentally[55]) in addition to that on the outer face of the solid bed close to the barrel. Temperature differences between liquid and solid are illustrated by the experimental data of Barnett et al.,[48] who showed that they tended to increase towards the melt pumping section. These workers also showed that temperature fluctuations increase with screw speed, and the position of maximum difference moves towards the die.

A relatively low barrel temperature in the melting section may be expected to suppress formation of the molten film near the barrel surface and enforce shearing within the solid bed, increasing mechanical work input and reducing heater input. The latter is confirmed by experience, which also shows output is not necessarily affected and final melt temperature may increase. In some cases it appears to improve temperature uniformity of the final melt and increase the limiting speed at which unmolten particles appear at the die. This emphasises the importance of correct setting and accurate control of barrel temperatures in the melting section.

The effect of screw temperature in the melting section has not been determined, but would be expected to affect the forward velocity of the solid bed and possibly the position at which the bed breaks up; especially with low barrel temperatures, screw cooling in this section only (to avoid disturbing the melt pumping section) might be expected to improve melting by delaying both solid bed movement and break-up.

It is a matter of common experience that uniform melting of powder in a single-screw extruder is often more difficult than with the same polymer in granule form; the author has found at high screw speed a stream of unmolten powder particles sheathed with a coating of molten polypropylene in conditions in which the same polymer fed in granule form gave a satisfactorily melted output. The reasons for this difference have not been established; it may be the smaller area of contact with the barrel of irregular powder particles, or that molten polymer at the barrel surface is more easily dispersed within the interstices of the powder instead of forming a coherent film, or that greater cohesion between powder particles reduces internal shearing. The melting section of powder-fed extruders is frequently extended by four to eight diameters to allow for slower melting. When the solid bed breaks up, softened particles become suspended in the melt, which follows velocity patterns similar to Figs 5.7 and 5.8 and begins to raise pressure. The mechanisms giving rather slow elimination of the solid particles have been outlined in Section 5.9 and in addition the velocity patterns will tend to reduce temperature differences arising in the melting section proper. The experiments of Barnett et al.[48] show these temperature differences decreasing from a maximum towards the die, but also that at high screw speeds significant temperature variations may persist at the die—these will be further discussed in Section 8.1—and it has been suggested that the elimination of these is a main function of the 'metering' section of the screw. Concurrently, the melt will be heated by internal shearing and frequently by heat conducted from the barrel; in Sections 3.3 and 3.2 it is shown that both will tend to produce non-uniform temperatures. It may be argued that these effects will be minimised by small pressure gradients (Fig. 5.7c or d) and small temperature changes, respectively, near the end of the melt pumping section; the former implies a tapered screw at modest die pressure (Fig. 5.13) and the latter a stepped set temperature profile (3 in Fig. 7.8), especially for high melt temperatures. Temperature variations from the screw will affect downstream operations such as drawing and cooling as well as distribution and flow in the die. Any attempt to correct the variations by heating in the die is undesirable since, in the absence of mixing, conducted heat will produce further variations. The purpose of die heating is primarily to reduce surface cooling and maintain the melt temperature; small upward adjustments affect the surface layers only and are used commercially to improve surface finish of tube, cast film, etc. A lowering of metal temperature tends to produce surface mattness or roughness, and this is often the first indication of the failure of a die heater.

Energy Transfer

Generation and transfer of energy are closely related to temperature changes; unbalance in the former appears as a change in sensible heat (internal energy) and temperature, while the latter affects viscosity and energy generated by shearing.

The energy required to cause flow in the hopper and into the screw comes from the potential energy represented by the static head. Little energy is required to convey solid particles along the screw (as seen in feeding experiments),[59] but as soon as resistance is encountered, causing consolidation and pressure against the barrel wall, work is done on the polymer and heating commences. As mentioned in Section 5.10, this may become excessive where conveying is very positive (e.g. a long cooled section and strip feeding), leading to rapid heating but high mechanical power input. Since it appears that little internal shearing occurs in the compacted bed, the energy dissipation and temperature rise are concentrated near the barrel surface, depending critically on the temperature of the metal surface, the remainder being heated mainly by conduction. As it develops, energy dissipation will also occur in the melt pool by viscous heating (Fig. 7.2). The forward flow of this melt pool assists in maintaining greater axial temperature gradients than when the screw is stationary (Section 6.2 Energy losses). Longitudinal conduction in the polymer is negligible, but in the metal barrel it occurs from (heater) zone to zone† and to the feed pocket.

† Note in Appendix C the changes in respective energy contributions from Zone 1 and 2 heaters as set temperature gradients are altered.

Due to the effect on viscous heating, the temperatures set on the controllers may be regarded more as an adjustment of energy input to the polymer, since the former only bears an indirect relation to the polymer temperature.

As melting continues, energy transfer to the solid bed is mainly by conduction through the molten film; if energy input from the barrel is increased, some will be absorbed in increasing the thickness and temperature of the molten film, so that heating of the solid bed is not necessarily increased in proportion. As speed and output are increased, residence time in a given length of screw is decreased and thus either melting becomes less complete or occupies a greater length. With semi-crystalline polymers, additional energy is required at the melting point to supply the energy released on crystallisation, though no melting 'plateau' of temperature is observed since melting is progressive with time and position. When the solid bed breaks up, the particles are melted mainly by conduction from the melt. In the melt pumping section, as mentioned under 'Temperature', conduction from the barrel and internal shearing increase the energy content of the melt (non-uniformly) while circulation tends to reduce temperature differences due to these and to the mechanisms of melting. The gross mechanical power input will be influenced by temperature and hence by barrel heating, while increased back pressure (eqn (6.22)) increases the specific power input, mainly by decreasing output, and hence forms a second means of control of final melt temperature and balance between mechanical and heater energy inputs.

Barrel cooling is considered separately from heating since its effects are not simply negative; it produces a reversed temperature gradient through the channel depth, but the lowest temperature and highest viscosity now occur in the region of highest shear rate, tending to modify the latter. In pressure (capillary) flow, boundary cooling decreases wall shear rate (Fig. 3.1c), and steepens the velocity profile, which here occurs asymmetrically (Fig. 7.5a), regarding the screw as thermally neutral. The drag flow will also be distorted by cooling (Fig. 7.5b), so although superposition of flows is no longer strictly valid, the resultant profile (Fig. 7.5c) will be distorted from the isothermal condition. Since velocity at the barrel remains constant and viscosity increases, the power dissipated in the channel (eqn (6.22)) must increase, as commonly observed. The effect of barrel cooling on power absorbed in the flight clearance (eqn (6.26)) will be much greater, so that both total power and the proportion from the flight clearance will

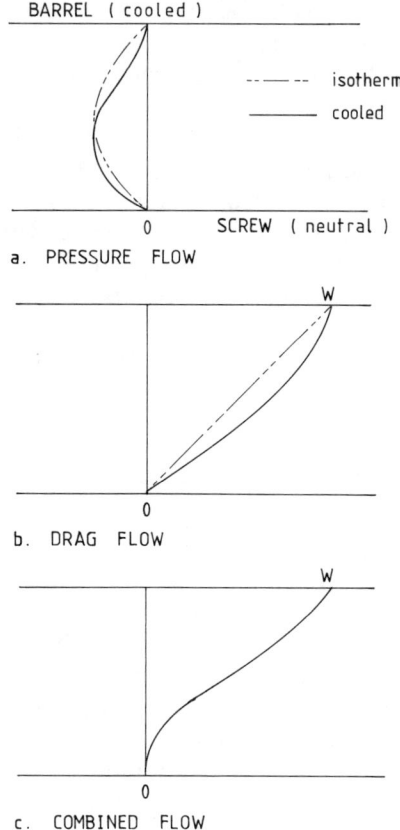

Fig. 7.5. Velocity profile with cooled barrel.

increase, tending to increase the melt temperature. In the limit, barrel cooling will freeze a layer of polymer and stall the motor, but if heat removal is inadequate, cooling will increase mechanical power but also increase bulk mean temperature and probably increase spatial variation of temperature. In large machines, the greater channel depth permits greater temperature differences and also the surface area for cooling is limited in comparison to the heat and mass flows. The author recalls a large hot melt-fed extruder in which barrel cooling dramatically increased the mechanical power input, but the final melt temperature tended to rise rather than fall.

Screw cooling depends on quite different factors from barrel cooling.

Its possible use in feeding and melting sections has been mentioned on pp. 189 and 195. Due to low thermal conductivity of the melt and low relative velocities near the screw root (Figs 5.7 and 5.8), screw cooling in the melt pumping section has little direct effect on conditions near the barrel wall. Considering the screw as an entity, it is virtually surrounded by polymer, so in the absence of cooling it can otherwise only lose heat through its rear end to drive shaft and bearings. With conventional screw proportions this is limited by the length and cross-sectional area of the screw root, as is internal conduction from (say) pumping section to melting section, which would tend to promote melting. In the absence of intentional cooling, it is reasonable to treat the screw as thermally neutral (insulated), though Gale's[54] results suggest this may not always be correct in the melting section. A small degree of screw cooling, preferably directed to the extreme tip of the screw, is frequently used with heat-sensitive polymers such as UPVC to reduce the temperature of the screw tip and hence build up of degraded polymer at this point.† Larger flows of air or water or lower inlet temperatures produce significant reductions in output with frequently some increase in mechanical power input and melt temperature—similar effects to those of reducing channel depth at constant screw speed (Fig. 6.6). The effect in the melt pumping section of screw cooling indeed appears to be mainly that of reducing channel depth and is conventionally used for this purpose when the available screw is too deep. However, this is not always immediately reversible; the author made a momentary application of water cooling to the screw of a 50 mm extruder running on UPVC but 8 h after removing cooling, output and power had not returned to the values before applying cooling. For this reason, and because of the sensitivity of operation to changes in channel depth (pp. 90, 177, and Fig. 6.6), screw cooling must be gradually applied and accurately controlled for flow rate and temperature. In the die, apart from the conversion of pressure energy to frictional heat (see Section 6.1) energy transfer should be minimal (p. 196).

Mixing

Mixing is associated with the mechanisms of flow and energy transfer and has been included under these headings. Here it is only necessary

† The author found in extended running a tendency to form a discóloured 'pipe' down the centre of the extrusion; in addition to cooling, a pointed screw nose (Fig. 7.6) and radiusing spanner flats minimised the fault.

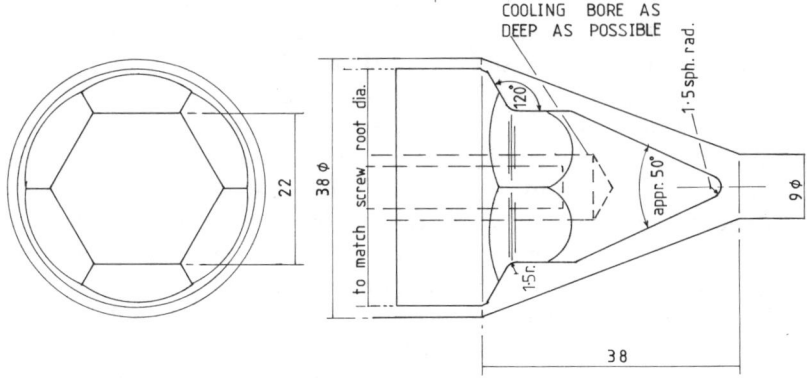

Fig. 7.6. Experimental screw nose.

to summarise the main features. Little mixing occurs while the polymer is solid; indeed avoidance of segregation is probably the chief concern. The mechanism of melting leads to a degree of end-to-end mixing as well as transverse mixing in the molten portion. The unmolten material is only mixed as it is incorporated in the melt, and thus the last portion, which eventually breaks up into semi-molten particles, undergoes least mixing in this section and has least time for subsequent mixing. Until these semi-molten particles melt, they cannot mix into the bulk and remain a source of thermal inhomogeneity; however, despite mixing in the melt due to both transverse circulation and longitudinal back mixing due to pressure, non-uniform temperature is frequently encountered at the end of the screw. At least four sources of inhomogeneity can be listed, though their relative importance is difficult to quantify and will change with operating conditions. These include:

(i) lack of radial mixing in the channel; the transverse circulation in Fig. 5.8 tends to be laminar, with an almost static 'core', although the latter has a longitudinal velocity (except at $Q = 0$);

(ii) non-uniform shear heating in the channel, as discussed under 'Energy transfer' and in Section 6.5, leading to non-uniform temperature increases;

(iii) barrel heating or cooling, producing temperature gradients between barrel and screw root;

(iv) incomplete mixing in the channel of polymer undergoing drastic shearing and heating in the flight clearance.

Mixing in the melting section will assist uniformity of composition (additives, etc.), and distributive mixing continues in the screw channel in the melt pumping section, though with the limitation in (i) above. Shear stresses high enough for significant dispersive mixing occur only in the flight clearance, they apply to only a small proportion of the total flow, and non-uniformly within that proportion. Back pressure improves distributive mixing both by changing the longitudinal velocity distribution (Fig. 5.7) and by increasing mean residence time through reduced output.

7.2. EFFECTS OF CONTROLLED VARIABLES

The foregoing description of the simultaneous mechanisms in the extruder gives the qualitative background against which operating strategies and decisions must be made. The extruder operator is usually faced with the problem of operating an existing set-up, including extruder, screw, drive motor and gearing, heating/cooling, control systems and die to make a saleable product or intermediate of sufficient quality and uniformity at maximum output and efficiency. Only when this is impossible, or only possible at uneconomic output, can changes in the machine be considered, though since the latter are expensive in terms of cash and loss of production, it is as well to know what changes are required and what they can achieve. Frequently, the set-up, or some part of it, was designed for a different polymer, product or output rate and compromises must be made, at least until new equipment can be obtained or justified by the market.

The machine variables under the operator's control are usually screw speed, back pressure, heating and temperature profile. The general effects of these will first be summarised, together with those of melt temperature, polymer properties and screw profile, which are important though not usually under direct control. In making any such summary, it is important to define the conditions, and in this case it will be that generally other independent variables are held constant. Here the die geometry will be regarded as fixed, so that back pressure becomes a variable dependent on output and melt temperature, and with a constant controlled (heater) temperature profile, energy balance and melt temperature become dependent variables.

Screw Speed

It is shown in Section 5.5 that for a fixed die, the output of an isothermal Newtonian fluid is directly proportional to screw speed (eqn (5.87)) and that for a pseudoplastic polymer it is approximately proportional (p. 103). Thus at any point in the screw the dimensionless flow rate Q/Wbh is approximately constant. Equation (6.27) shows that for a Newtonian fluid at constant Q/Wbh, mechanical power in both channel and flight clearance increases in proportion to the square of speed (eqn (6.33)), whereas for a pseudoplastic it increases in proportion to an index of speed between unity and two ($n+1$ for a power law fluid where $0 < n < 1$ eqn (6.34)). Consequently, the mechanical power input per unit volume output, which equals the shear heating per unit volume, is proportional to speed N for Newtonian, and N^n for pseudoplastic fluids (Table 6.2). At constant temperature, Fig. 6.10 shows that as a consequence the energy required from the heaters first increases and then decreases with increasing speed, before ultimately going negative (cooling). If at any point heating or cooling is inadequate to maintain the set temperature, then final melt temperature will fall or rise accordingly. As screw speed increases, the melting section will occupy a greater length of the screw so that the effective melt point and start of pressure rise will move forward. This reduces melt length Z so that B in eqn (5.80) must increase, Q_{Tot} in eqn (5.86) will decrease and P must decrease (eqn (5.81)) compared with these factors for the original melt length, i.e. the previous statement (eqn (5.87)) must be modified so that output and pressure increase, but somewhat less than the increase in speed. In practice, this effect is usually small and depends on the screw profile and where the melt point occurs—Fig. 5.13 shows that in a tapered screw, the first portion of the melt length usually contributes a comparatively small proportion of the total pressure. A more important effect is that the distance is reduced for melting the particles remaining after the solid bed breaks up. This, together with a shorter residence time in the melt pumping section for mixing (for composition and temperature) results in greater temperature variations and poorer mixing at the die. The more common limits on speed increase are:

(i) restriction of solid feed—in hopper or conveying section of screw;
(ii) unmolten polymer at the die;
(iii) excessive temperature variation at the die;

(iv) inadequate mixing for composition at the die;
(v) inadequate cooling to maintain melt temperature *or* excessive rise of melt temperature;
(vi) inadequate motor power.

Back Pressure

This may be increased by increasing the land length of the die lips, partially closing a restrictor bar or ring before the die lips or inserting additional screens or breaker plate. Here the conditions are assumed to be constant speed and set temperatures. Equations (5.17), (5.22), (5.51), (5.59) and (5.60) show that for all common types of screw, increasing pressure will decrease output. This is also shown from A to C in Fig. 5.20, but note that the pressure drop through the die decreases from A to A', so that the pressure restriction must raise the pressure on the screw from A' to C. The mechanical power absorbed in the channel increases with increase of pressure (decrease in Q/Wbh) according to eqn (6.22), though the effect seems small in practice, but the decrease in output means that the specific power E/Q increases significantly and temperatures tend to rise (eqn (6.53)). The power consumed in the flight clearance is nominally unchanged (eqn (6.26)). The wall shear rate increases (eqn (5.98)) and longitudinal mixing is increased (Fig. 5.7b and c), but transverse flow is nominally unchanged (eqn (5.24)). There is more time for melting because of the reduced output, and experience shows melting to be more complete, although temperature variations at the die may increase, possibly due to increasing non-uniformity in the radial distribution of shear heating (Section 6.5). Incidentally, increase of back pressure increases the load on the thrust bearing ($\pi D^2 P/4$).

Heating

Raising the set temperatures on the barrel heaters will of course tend to increase the heater energy to the polymer and also the heat losses. By reducing the viscosity near the barrel, it will also decrease the mechanical power input in the channel, and to a greater extent in the flight clearance. This leads to a change in the energy balance in the melt pumping section (to the right in Fig. 6.8) and a higher final melt temperature. The higher temperatures in screw and die will give lower pressures, but assuming the temperature coefficients of viscosity are similar at the different shear rates in screw and die, the output will be little changed since, if suffices 1 and 2 refer to screw and die: for the die

$$Q_{\text{Tot}} = \frac{KP}{\eta_2} \qquad (5.81)$$

and for pressure flow

$$Q_P = \frac{BP}{\eta_1} \qquad (5.78)$$

and

$$B = \frac{bh^3}{12Z} \qquad (5.80)$$

At higher temperature (Q', P', η', etc.)

$$\eta_2' = C\eta_2 \qquad (7.3)$$

and

$$\eta_1' = C\eta_1 \qquad (7.4)$$

Then

$$Q'_{\text{Tot}} = \frac{KP'}{\eta_2'} = \frac{KP'}{C\eta_2} \qquad (7.5)$$

and

$$Q'_P = \frac{BP'}{\eta_1'} = \frac{BP'}{C\eta_1} \qquad (7.6)$$

By rearrangement of eqn (5.17):

$$Q_D = Q_{\text{Tot}} + Q_P \qquad (7.7)$$
$$Q_D' = Q'_{\text{Tot}} + Q_P' \qquad (7.8)$$

Substituting eqns (5.81) and (5.78) in eqn (7.7) and eqns (7.5) and (7.6) in eqn (7.8):

$$Q_D = \frac{KP}{\eta_2} + \frac{BP}{\eta_1} \qquad (7.9)$$

$$Q_D' = \frac{1}{C}\left(\frac{KP'}{\eta_2} + \frac{BP'}{\eta_1}\right) \qquad (7.10)$$

But from eqn (5.10), and independent of temperature:

$$Q_D' = \frac{Wbh}{2} = Q_D \qquad (7.11)$$

therefore

$$\frac{P'}{C}\left(\frac{K}{\eta_2} + \frac{B}{\eta_1}\right) = P\left(\frac{K}{\eta_2} + \frac{B}{\eta_1}\right)$$

or

$$\frac{P'}{C} = P \tag{7.12}$$

Then from eqns (7.5) and (7.6):

$$Q'_{Tot} = \frac{KP}{\eta_2} = Q_{Tot} \tag{7.13}$$

$$Q'_P = \frac{BP}{\eta_1} = Q_P \tag{7.14}$$

This is represented graphically in Fig. 7.7. From eqn (7.13):

$$\frac{Q'_{Tot}}{Wbh} = \frac{Q_{Tot}}{Wbh} \tag{7.15}$$

and since Q/Wbh is constant, the bracketed term in eqn (6.22) is constant, and the channel power is also decreased only by the change in viscosity, i.e. by the factor C in eqn (7.4). If viscosity in the flight clearance changes by the same factor C, the total power given by eqn (6.27) also changes by factor C. Constant Q/Wbh (eqn (7.15)) also implies that the longitudinal velocity profile (Fig. 5.7) is constant, giving similar mixing, but the change in the balance of thermal and

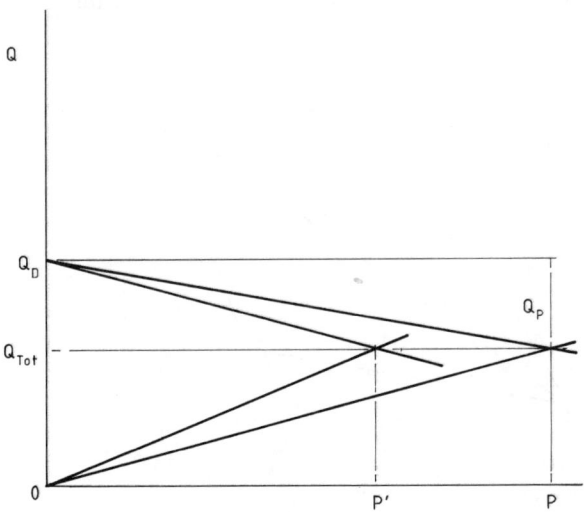

FIG. 7.7. Output and pressure vs. temperature.

mechanical energy input is likely to create additional temperature gradients, leading to greater variations in final melt temperature.

Additional heating in the melting section may promote faster melting, but may also lead to earlier break-up of the solid bed and hence poorer temperature uniformity of the melt, so that, as discussed on p. 195, temperature control in this section is of great importance. Barrel and screw cooling have been discussed previously on pp. 197–9.

Temperature Profile

This refers to the intentional variation of set temperatures along the barrel, for which it is difficult to give general rules since it depends on screw profile and polymer, on the rate of melting relative to the length of screw and especially to the level and uniformity required of final melt temperature. If the first heating zone temperature is too high, it may cause slippage in the solids feeding section; if it is too low it will delay melting and may cause problems due to entrained air in the feed: however, the latter is sometimes used to reduce the effective length of the screw and limit power input. As remarked earlier, this set temperature is more closely related to the heater power input than to the polymer temperature in this zone. One policy (Profile 1 in Fig. 7.8) is to set the temperature as high as possible without affecting feeding; this will tend to promote rapid melting and heating of the polymer

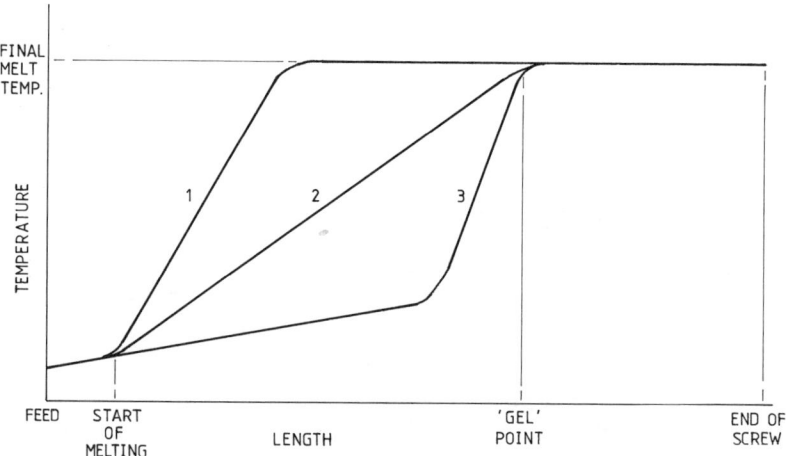

FIG. 7.8. Idealised barrel set temperature profiles.

before the melt pumping section and will minimise the mechanical power input for a given screw design and speed. However, this leads to conductive rather than internal shear heating and may tend to give early break-up of the solid bed and large temperature fluctuations at that point. This profile is thus possibly worthwhile with high melting point polymers such as nylon 66 and PET which are difficult to shear in solid form and which usually use screws with a long parallel or slightly tapered section, said to allow for adequate heating. It is also a possible strategy where motor power is limited or maximum melting rate is required, but melt temperature variation is not critical, or where screw length is barely sufficient for the required output.

A common strategy is to use a uniform gradient of *set* temperatures in the melting section (Profile 2); this is likely to lead to a slower initial melting rate but raise the temperature of the polymer at the end near to that required in the melt pumping section. However, it may also lead to early break-up of the solid bed. A third policy (Profile 3) is to maintain as low a temperature as motor power, screw strength, etc., allow, but sufficient to allow melting, which may now require an undesirably large part of the screw length, but delay solid bed break-up and give more uniform heating of the polymer in this section, so that when break-up occurs the temperature is fairly low but uniform, and less time is required for melting the remaining solid particles. This may be the ideal for low and uniform final melt temperature, but impractical for higher temperatures in view of limited length of screw.

It is generally agreed that in the melt pumping section the melt temperature should be raised as quickly as possible to the final value required and then be held constant to enable variations in temperature to be minimised or eliminated, so a constant set temperature profile is usually employed in this section. This may seem to favour the 'ramp' Profile 2 in the melting section since a step change in barrel temperature is impractical; a compromise between Profiles 2 and 3 is probably preferable, bearing in mind that the position of the melt point is usually unknown, and changes with screw speed.

Melt Temperature
Although this is strictly a dependent variable, largely determined physically by heating and temperature profile, its value must be correct for the polymer and downstream processes of shaping, drawing, quenching, etc., so it is often a matter of adjusting temperature profile to give the desired melt temperature. The effects of an increase on

other factors are similar, e.g. lower pressure in screw and die with similar output, lower mechanical power input but greater heater input (Fig. 6.8), and possibly less mixing and poorer temperature uniformity; this assumes a constant screw design and speed.

7.3. POLYMER PROPERTIES

The most important property in this context is the melt viscosity, which depends on the type of polymer and molecular weight, as well as temperature and shear rate. Operating temperatures are usually adjusted to give melt viscosities in the range 100–1000 Ns/m^2 at low shear rates; however, limits for thermal degradation give viscosities near the upper value for UPVC, while extrusion grades of nylon 66 have viscosities nearer the lower value. For a given polymer and temperature, viscosity is approximately inversely proportional to melt flow rate (BS 2782:720A) since the latter represents mass flow rate at a given pressure drop (compare eqn (5.81)), and increases with molecular weight. The variation of viscosity with shear rate, the pseudoplasticity index n (eqn (2.7)) is important both in giving the relation between viscosity in the screw channel (low shear rate) and die (high shear rate) and that in the flight clearance. The former relation affects the necessary depth and length of melt pumping section and the latter influences the proportion of total power absorbed over the flight clearance. In Section 6.4 the variation of mechanical power with screw speed is shown to depend on n (eqn (6.34)). Polymers including the nylons and PET have a value of $0.8 < n < 0.9$, i.e. they are nearly Newtonian, whereas polyolefines and polystyrene have values of n in the range 0·2–0·5 and for polyacetal and unplasticised PVC, $0.4 < n < 0.5$. However, non-Newtonian behaviour is also a function of molecular weight distribution, a broader distribution generally giving a lower value of n, but a smaller dependence on temperature.[12]

The softening or crystalline melting temperature is mainly dependent on the polymer type, differing little with molecular weight in the commercial range. A high melting point tends to require a long length of screw for the melting section and a high heater power, but in the cases of nylon 66 and PET melt viscosities are fairly low at temperatures little above the melting point, so substantial heating after melting is rarely required. Another factor is the high rigidity of these two polymers in their solid form, which means that they are less easily heated by internal shearing before melting; this is given as a reason for the common use of stepped screws with long first sections of constant or near-constant depth rather than because of their high melting points.

Other polymer factors such as thermal expansion, melt specific heat, thermal conductivity and temperature coefficient of viscosity tend to vary little within a given polymer type. There seems little point in analysing their effects to explain differences in the behaviour of different polymer types; the latter are well known commercially and probably result from a combination of a number of factors. Swell ratio (recoverable strain) varies considerably with polymer type, molecular weight distribution, temperature, shear rate, fillers and, at least with LD polyethylene, very considerably on immediate shear history. This is a matter of polymer rheology which hardly concerns performance of the extruder, but is of course of vital importance in die design and dimensional control of the product.

7.4. SCREW DESIGN

Other factors being constant, residence time in the screw is proportional to its effective length and inversely proportional to speed (output). Thus factors dependent on residence time such as melting, heating between melting and final temperatures, distributive mixing and reduction of solid particles and temperature variations in the melt pumping section will all be similarly influenced, and a longer screw will tend to offset limitations occurring with increased speed. A longer melt pumping section permits higher pressures at the same output or a deeper screw (and higher output) with similar response to pressure and pressure changes. A short screw will show limitations in situations requiring high output, high temperature, high pressure, but small change in output, etc., with pressure, also for uniform mixing or uniform melt temperature; operating changes tending to improve one factor will probably cause deterioration in at least one other (see Table 7.2).

The author would regard a total length of 15 diameters as a minimum for thermoplastics, although $12D$ is used for cold-fed rubbers, and $20D$ is normal. If high speeds, high pressures, or high melt temperatures or very stable output and uniform temperature are required, then $24D$ or even $28D$ would be recommended. If venting is required, an additional $6D$ or preferably $8D$ should be provided; a vented screw of less than $28D$ total length would almost certainly be limited in one of the foregoing aspects, especially if the first section is working against zero back pressure. These lengths are only rule-of-thumb regarding the limits within which it is suggested the operator should work with an existing machine; for design it would be preferable to determine the total length by summing the lengths required for feeding, melting and pumping/mixing. It must be remembered that a longer screw will generally absorb more power, so a limit is placed by torsional strength at the feed end. Where low melt temperatures are required, these also set limits to practicable length, which are likely to be different for various polymer types. For example, unplasticised PVC at low outputs may require a screw of $15D$ or even less, whereas nylon 66 at high output may require a minimum of 24 diameters screw length.

Compression ratio is a term commonly used, but for this discussion channel depth in the melt pumping section is a better basis of comparison. As shown by eqn (5.10), a shallow screw will have a low drag flow rate, but eqn (5.15) shows the pressure flow is also small, being proportional to the cube of h and, as shown in Fig. 5.18, at high die pressures may have a greater total output than a deeper screw at the same speed. Figure 5.18 also shows that the output changes much less with a change in pressure than in a deeper screw, and in this sense the shallow screw will tend to be more stable. Appendix B compares pressure gradients in the two sections of a stepped screw, i.e. where speed and output are equal. On this basis (point A in Fig. 5.18) it is shown that except when Q/Wbh_2 is close to 0·5, i.e. pressures are low, the pressure gradient in a shallow screw is always greater than in a deep screw, so that a given die pressure may be reached in a shorter melt pumping section. However, at a given output, when h is reduced, Q/Wbh is increased so that the velocity distribution resembles Fig. 5.7c rather than Fig. 5.7b, indicating a reduction in longitudinal mixing; on the other hand, the smaller depth will give smaller temperature differences due to a given heat input and might be expected to give reduced variations due to different rates of internal shear heating. The wall shear rate and mechanical power absorbed in the channel will increase slightly due to a reduction in depth (Table 2 of Ref. 55) so that somewhat higher temperatures will be achieved with a given heater input, or at a fixed temperature the heater input will become smaller. As mentioned on p. 199, similar effects may often be obtained by screw cooling, which appears to reduce the effective depth of the melt pumping section. In the melting section, a reduction in depth will reduce the residence time and may well lead to a deficiency in melting rate and of course a reduction in feed section depth may lead to reduction in feed rate (Fig. 5.11). However, reduction in output rate from either of these causes is of less importance than the unstable operation which is likely because both the mechanism of solid feeding and of melting are more subject to random disturbance than melt pumping. So far the comparison has been made on the basis of identical output and pressure from shallow and deep screws. More frequently a shallow screw gives a lower output at the same screw speed (e.g. A and B in Fig. 5.20), or a higher speed is used to give the same output. In the case of a lower output, mechanical power input will not be greatly different but will be absorbed in a smaller flow of polymer so that specific power input and temperature rise will be greater, less dependent on external heating, and mixing will be improved. The lower output will also tend to result in a more uniform final melt temperature. If output is restored by increasing screw speed, then the effects will combine to give a higher temperature or less heat input (i.e. approach the autogeneous point in Figs 6.8 and 6.10). However, increase of speed will tend to increase temperature inhomogeneity and melting limitations may occur; the product uniformity may be either better or worse than with the deeper screw at lower speed.

The screw profile, i.e. the variation in depth along the length, is the subject of strongly held views, especially in respect of certain polymers and product types. It is often stated for instance that certain nylons will hardly extrude at all through taper or taper-parallel screws, while these are almost universally used for unplasticised PVC. It seems logical that polymers of low melt viscosity such as nylons and PET are usually extruded with screws having shallow and comparatively long final sections for melt pumping, in order to achieve useful pressure gradients (eqn (5.17)) and stability against pressure fluctuations. Applications for these polymers, e.g. film and fibres, require high pressures and stable outputs, again leading to long, shallow 'metering' sections. Polymers of high melt viscosity such as natural rubber and UPVC tend to be extruded with deeper and shorter melt pumping sections, which achieve sufficient pressure without excessive power input (proportional to dz in eqn (6.27)) and temperature rise; the latter being important since these polymers are usually processed at temperatures

TABLE 7.1
Examples of Screw Types (90 mm diameter, single start, pitch = diameter (helix angle 17·6°))

Purpose	Polymer	Depth (mm)			Length (diameters)				Compression ratio[a]
		Feed	Transition	Meter	Feed	Trans	Meter	Total	
High output compounding	LDPE	15	15–6·5 T[b]	6·5	6	12	6	24	2·1:1
General purpose	LDPE / HDPE / PS	15	15–5·5 T	5·5	4	10	6	20	2·4:1
Film	LDPE	15	15–3·0 S	3·0	10	1	9	20	4·3:1
High output compounding	PP powder	—	15–7·9 T	7·9	—	18	2	20	1·8:1
General purpose	PP	—	15–6·0 T	6·0	—	16	—	16	2·2:1
General purpose	PP	—	12·7–5·6 T	5·6	—	10·9	9·1	20	2·1:1
General purpose	POM	15	15–5·0 T	5·0	4	12	4	20	2·7:1
General purpose	PMMA	15	15–4·0 S	4·0	14	1	5	20	3·3:1
Compounding	Nylon 66	—	11·1–9·6 T/S	2·8	—	$12+\frac{1}{2}$	7·5	20	3·6:1
Film	PET	15	15–3·0 S	3·0	14	1	9	24	4·3:1
Film	UPVC	15	15–7·0 T	7·0	4	12	4	20	1·9:1
Low output compounding	UPVC	15	15–9·0 T	9·0	4	7	4	15	1·5:1

[a] Cross-section basis
[b] T = tapered; S = stepped

Note: The screw lengths recommended on p. 209 are generally greater than listed here, due to development in design and requirements since these screws were commissioned.

near to the limit of thermal degradation in order to reduce their viscosities. Other polymers of limited thermal stability such as polypropylene and polyacetal (POM) also tend to be processed with deeper screws, but often of considerable 'metering' length. Evidently a deep feed section is desirable to avoid limitations from solid feeding, and, except with very large machines, feed depth is limited by torsional strength of the screw. To assist in feeding, many screws have a parallel section of 4 to 6 diameters (including the feed opening). The controversy therefore centres around a uniform taper or a parallel followed by a step, in the melting section. As stated earlier, the argument for the latter seems to be that as rigid solids the polymers used with stepped screws are not readily sheared to assist melting and require a long time to reach their high melting points by conducted heat, leading to long parallel or near-parallel sections which are usually continuations of the feed section. It is claimed that the rapid decrease of channel depth ('step') in one turn or less acts as a barrier to forward flow of unmolten polymer and thereby 'fixes' the gel point. As shown in Section 5.4, most of the pressure will be raised in the shallow section and considerable movement of the gel point (point of initial pressure rise) rearward will give little change in pressure (or output). The case for a uniform taper is less clearly stated though it arguably maintains the down-channel velocity of the unmolten portion of the melting bed and reduces stagnation which might form a hold-up point for degradation. It also means that matching of the screw profile to the actual point of pressure rise is less critical, so might be expected to perform more flexibly, both in respect of an initial mismatch and the changes which take place in operation, especially with increasing screw speed.

It is impossible to generalise further on screw dimensions, leaving aside special types for venting, compounding, etc. It may be useful to give some examples (Table 7.1) which have operated successfully; they are given in relation to a 90 mm diameter screw with a 15 mm deep feed section, though as discussed in Section 8.5 channel depths usually vary somewhat less than in direct proportion to screw diameter. The terms used refer to the geometric portions of the screws, not the functional sections for feed, melting and melt pumping used in the remainder of this book to describe regions of particular mechanisms.

7.5. OPERATIONAL STRATEGIES

The preceding sections have described: (7.1) the overall performance of the extruder in respect of flow, pressure, temperature, energy transfer and mixing; (7.2) the effects of changes in controlled variables of screw speed, back pressure, heating, temperature profile and melt temperature. Sections 7.3 and 7.4 have considered the effects of polymer properties and screw design, respectively. This section takes the opposite view, in showing what operational changes may be made to achieve a desired objective, and what other consequences may follow, so that the operator may choose which alternative to apply, depending on other constraints of the individual situation. For clarity this is given in tabular form (Table 7.2). To avoid unnecessary reference back, the basic equations for flow and energy are reproduced here; Figs 5.20 and 6.8 are also relevant.

Flow in a channel of constant depth:

$$Q_{\text{Tot}} = \frac{Wbh}{2} - \frac{bh^3}{12\eta} \cdot \frac{P}{Z} \qquad (5.22)$$

Mechanical power in channel only:

$$E_{\text{channel}} = \frac{\eta W^2 b \, dz}{h} \left[4(1 + \tan^2 \phi) - \frac{6Q}{Wbh} \right] \qquad (6.22)$$

Mechanical power in channel and flight clearance:

$$E_{\text{dz}} = \frac{\eta W^2 b \, dz}{h} \left[4(1 + \tan^2 \phi) - \frac{6Q}{Wbh} \right] + \frac{\eta W^2 t \, dz}{\delta \cos \phi} \qquad (6.27)$$

In the foregoing table, some of the statements, especially on the effects of reduction in screw channel depth, may appear contradictory. At constant output and appreciable back pressure, the velocity distribution in the channel will change from Fig. 5.7b with a deep screw to Fig. 5.7c with a shallow screw, but with the vertical scale reduced so that wall shear rate and stress are increased (unless the situation of Fig. 5.7e obtains); however, the reverse-flow component giving longitudinal mixing is reduced or eliminated. Reference to Fig. 5.18 shows that for two screws of different depths, running at the same speed to give the same output with the same die, the operating point must be at A, and the die line of Fig. 5.19 must also pass through this point; this is the basis of the comparisons made at constant output. Any other depth of screw will require some adjustment of speed to give the same operating point. In other cases the comparison is made on a basis of a fixed die characteristic (Q_{Die} in Fig. 5.20) and constant speed, where decreasing the screw channel depth decreases output and pressure,

TABLE 7.2
Operational Strategies

Required	Action	Response	Reference	Limitations
Increase output	Increase speed (fixed die)	Output Q and pressure P increase approx. proportional to speed N	eqn (5.87)	Feed restriction Maximum haul-off and cooling speeds
		Mechanical power increases approx. proportional to $N^{1.5}/N^2$	Fig. 6.10	Maximum motor power Maximum heater/cooling power
		Heater power increases/decreases	Fig. 6.10	Thermal degradation
		Melt temperature T_m rises (heaters constant)	p. 183	Unsatisfactory flow from die (e.g. drawing, blowing, quenching)
		Gel point moves towards die	pp. 132, 202	Incomplete melting and surging
		Temperature variations increase	p. 202	
		Residence time (mixing) decreases	p. 202	Unstable flow and poor distribution in die Non-uniform composition and product properties
	Increase 'metering' depth (fixed speed and die)	Q increases at low P only	Fig. 5.18	Feed restriction
		Greater fluctuation in Q if P changes	Fig. 5.30	Product breaks in haul-off, especially with drawdown
		Mechanical power E decreases slightly	eqn (6.29)	
		Melt temperature T_m may fall	Fig. 6.6	
		Gel point moves towards die	p. 134	Incomplete melting and surging

TABLE 7.2—contd

Required	Action	Response	Reference	Limitations
Increase pressure	Add screens to breaker plate (fixed die)	Temperature variations *may* increase	p. 210	Unstable flow in die
		Residence time (mixing) increases but mixing uniformity *may* decrease	p. 211	Non-uniform product properties. Non-uniform composition and product properties.
		At constant speed, output Q from screw decreases, requiring higher speed N (as above) to restore Q and P in die.	Fig. 5.17	Output cannot be restored by increasing speed or screw depth
		Mechanical power E increases	Fig. 6.12 p. 164	Maximum motor power Maximum torsional strength of screw
		Heater power H decreases	Analogous to Fig. 6.8 eqn (6.53) p. 203	Heaters off (no cooling) Maximum cooling (if cooled)
		Melt temperature T_m rises (heaters constant)	p. 203	Thermal degradation Unsatisfactory flow from die
		More complete melting	p. 203	—
		Temperature variations *may* increase	p. 203	—
		Residence time (mixing) increases Mixing uniformity increases		—
Increase melt temperature	Increase barrel set temperatures	Output Q approx. constant Pressure reduced (fixed speed and die)	eqn (7.13) Fig. 7.7	— —

Operation of Single-Screw Extruders

Action	Effect	Reference	Comment
Decrease set temperature in melting section *only*	Mechanical power E decreased considerably	eqn (6.27)	—
	Heater power H increased	Fig. 6.8	Maximum heater power
	Melt temperature T_m rises if E was previously small, e.g. deep, short screw	Fig. 6.8	As required—degradation, etc., will be anticipated
	Melt temperature *falls* if E was previously large, e.g. shallow screw, high pressure	p. 195	*Contrary to requirement*
	Melting *may* be more complete, if output Q and mechanical power E low	p. 256	
	Melting *may* be less complete, if output Q and mechanical power E are high	p. 195	—
	Temperature variations may increase/decrease, depending on melting	p. 206	Incomplete melting
	Mixing decreases, especially dispersive, due to lower shear stresses and power	p. 206	*If* 'increase', unstable flow in die
	Output Q and pressure P approx. constant (fixed die)	p. 240	Non-uniform product properties
	Mechanical power E increased considerably	p. 280	Non-uniform composition and product properties
	Heater power H decreased	p. 239	—
	Melt temperature T_m rises if mechanical power E was previously high	p. 205	Maximum motor power
		p. 194	Maximum torsional strength
		p. 207	—
		p. 206	As required—may give shear degradations, e.g. dulling of pigments
		p. 195	

TABLE 7.2—contd

Required	Action	Response	Reference	Limitations
		Melt temperature T_m falls if mechanical power E was previously low or if considerable heating took place after melting		Contrary to requirement
		More complete melting—gel point may move towards feed	p. 135	
		Break-up of solid bed may be delayed	p. 206	—
		Temperature variations may decrease due to higher temperature at start of melt pumping section, giving improved flow in die	p. 195 p. 208	—
		Mixing may increase due to longer time after melting, giving improved uniformity in product	p. 200	
	Commence or increase screw cooling	Effective channel depth in pumping section is reduced	p. 199 p. 210	
		If pressure is low, output Q and pressure P will be decreased	A to B in Fig. 5.20	Uneconomic output
		If pressure is high, output Q and pressure P will increase slightly	C to D in Fig. 5.20	
		Mechanical power E increases slightly	p. 199	—
		Melt temperature T_m increases due to smaller Q	Fig. 6.6	As required

Action	Effect	Reference	Note
Increase screw length	Greater power input *may* give more complete melting	p. 195	—
	Temperature variations *may* increase due to greater difference in screw and barrel temperatures and shorter residence time	p. 211	Unstable flow in die Non-uniform product properties
	Mixing *may* decrease due to longitudinal shear gradient, though radial mixing distance reduced	p. 200 p. 211	Possible non-uniform composition
	Pressure gradient decreased, so output Q and pressure P increase (fixed die)	Fig. 5.20 p. 209	—
	Mechanical power E in pumping section increased	eqn (6.27)	Maximum motor power
	Heater power H (and losses S) increased	Fig. 6.5	Maximum heater power
	Melt temperature T_m increased, especially if previously close to melting point	Fig. 6.5	As required—degradation, etc., will be anticipated
	Melting more complete if reduced pressure gradient allows length for melting to increase, despite greater output Q	p. 209	—
	Melt temperature variations decrease due to more complete melting and greater pumping length for mixing	p. 209 p. 239	—
	Mixing increases due to greater pumping length	p. 209	—

TABLE 7.2—contd

Required	Action	Response	Reference	Limitations
Reduce melt temperature	Reduce barrel set temperatures	Output Q approx. constant		
		Pressure P increased (fixed speed and die)	eqn (7.13)	Safe working pressure of barrel and die
		Mechanical power E increased considerably	Fig. 7.7	Maximum motor power
		Heater power H reduced	eqn (6.27)	—
		Melt temperature T_m falls if E was previously small, e.g. deep, short screw	Fig. 6.8	As required
		Melt temperature T_m rises if E was previously large, e.g. shallow screw, high pressure	Fig. 6.8	
			p. 195	*Contrary to requirement*
			p. 256	
		Melting *may* be less complete if output Q and power E previously low (mainly conducted heating)	p. 195	Incomplete melting
			p. 206	
		Melting *may* be more complete if output Q and power E previously high (mainly shear heating)		—
		Temperature variations *may* increase or decrease, depending on melting	p. 206	*If* increase, unstable flow in die. Non-uniform product properties
			p. 240	
		Mixing increases, especially dispersive, due to higher shear stresses and power	p. 208	—
			p. 239	
	Increase set temperatures in	Output Q and pressure P approx. constant (fixed die)	p. 194	
			p. 205	

Condition	Effect	Reference	Result
melting section *only*	Mechanical power E decreased considerably	p. 194	—
	Heater power H increased	p. 207	Maximum heater power
	Melt temperature T_m falls if mechanical power E was previously high	p. 206	As required
	Melt temperature T_m rises if mechanical power E was previously low	p. 195	*Contrary to requirement*
	Less complete melting—gel point may move towards die	p. 135	Incomplete melting
	Break-up of solid bed *may* be accelerated	p. 206	
	Temperature variations *may* increase due to lower temperature at start of melt pumping section	p. 195 p. 208	Unstable or non-uniform flow in die Non-uniform product properties
	Mixing *may* decrease due to shorter time after melting	p. 200	Non-uniform composition and product properties
Decrease screw length (melt pumping section)	Pressure gradient increased so output Q and pressure P decrease (fixed die)	Fig. 5.20	Uneconomic output
	Mechanical power E (in pumping) decreased	eqn (6.27)	—
	Heater power H (and losses S) decreased	Fig. 6.5	—
	Melt temperature T_m decreased, especially if considerably above melting point	Fig. 6.5	As required Minimum processing temperature will be anticipated

TABLE 7.2—contd

Required	Action	Response	Reference	Limitations
		Melting less complete if higher pressure gradient tends to reduce length for melting, despite lower output Q	p. 209	Incomplete melting
		Melt temperature variations increase due to poorer melting and shorter pumping length	p. 209	Unstable or non-uniform flow in die
			p. 239	Non-uniform product properties
		Mixing decreases due to shorter pumping length	p. 209	Non-uniform composition and product properties
	Increase channel depth in melt pumping section	At low pressures, output Q and pressure P will increase	B to A in Fig. 5.20	—
		Mechanical power E in channel will decrease	eqn (6.22)	
		Constant flight power becomes greater proportion of total power E	eqn (6.27)	
		Melt temperature T_m decreases due to lower specific power E/Q	Fig. 6.6	As required
				Minimum processing temperature will be anticipated
		Melting *may* be less complete due to deeper channel and higher output	p. 134	Incomplete melting
		Temperature variations *may* increase due to greater output Q and greater channel depth h	p. 210	Unstable flow in die
		Mixing *may* increase or decrease depending on melting and effect of higher output	p. 211	Non-uniform composition

Operation of Single-Screw Extruders

Action	Sub-action	Effect	Reference	Consequence/Limit
Reduce mechanical power input	Reduce screw speed	Output Q and pressure P decrease approx. in proportion to speed N	eqn (5.87)	Uneconomic output. Minimum haul-off and draw-down speeds
		Mechanical power E decreases *more* than in proportion to speed, so also reduces torque, where this is limited by the drive	Fig. 6.10	As required
		Almost instantaneous effect		
		Melt temperature T_m decreases	p. 183	—
		Melting more complete—longer residence time	p. 202	—
		Mixing more complete—longer residence time	p. 202	
	Increase set temperatures in melting section only	Output Q and pressure P approx. constant (fixed die)	p. 194	
		Mechanical power E decreased considerably	p. 205	As required
		Heater power H increased	p. 194	Maximum heater power
		Melt temperature falls if mechanical power E was previously high	p. 207	Softening/melting point reached
		Melt temperature rises if mechanical power E was previously low	p. 206	Thermal degradation
		Less complete melting—gel point moves towards die	p. 195	
		Break-up of solid bed *may* be accelerated	p. 135	Incomplete melting
			p. 206	

TABLE 7.2—contd

Required	Action	Response	Reference	Limitations
	Increase *all* set temperatures	Temperature variations *may* increase due to lower temperature at start of melt pumping section	p. 195	Unstable or non-uniform flow in die
		Mixing *may* decrease due to shorter time after melting	p. 208	Non-uniform product properties
		As above, but further decrease in power E and increase in melt temperature T_m due to melt pumping section	p. 200	Non-uniform composition and product properties
		Further heating in melt pumping section *may* lead to greater temperature variations	eqn (6.27) Fig. 6.8	Thermal degradation
	Decrease screw length (melt pumping section)		p. 195 p. 240	Non-uniform flow and product properties
		Pressure gradient increased so output Q and pressure P decreased (fixed die)	Fig. 5.20 p. 209	Uneconomic output
		Mechanical power E (in pumping section) decreased	eqn (6.27)	As required
		Heater power H (and losses S) decreased	Fig. 6.5	—
		Melt temperature T_m decreased, especially if considerably above melting point	Fig. 6.5	Softening/melting point of polymer reached
		Melting less complete if higher pressure gradient tends to reduce length for melting, despite lower output Q	p. 209	Incomplete melting

Operation of Single-Screw Extruders

Action	Effect	Reference	Notes
	Melt temperature variations increase due to poorer melting and shorter pumping length	p. 209	Unstable or non-uniform flow in die
	Mixing decreases due to shorter pumping length	p. 239	Non-uniform product properties
Increase channel depth in melt pumping section	At low pressures, output Q and pressure P will increase	p. 209	Non-uniform composition and product properties
	Mechanical power E in channel will decrease	B to A in Fig. 5.20 eqn (6.22)	—
	Constant flight power becomes greater proportion of total E	eqn (6.27)	As required
	Heater power H increases	Fig. 6.6	Maximum heater power
	Melt temperature T_m decreases due to lower specific power E/Q	Fig. 6.6	Softening/melting point reached
	Melting *may* be less complete due to deeper channel and higher output Q	p. 134	Incomplete melting
	Temperature variations *may* increase due to greater output Q and greater channel depth h	p. 210	Unstable flow in die
	Mixing *may* increase or decrease depending on melting and effect of higher output	p. 211	*If* decrease, non-uniform composition
Reduce die restriction or remove screens	At constant speed, output Q increases with decrease in (total) back pressure	C to A in Fig. 5.20 & p. 203	Maximum haul-off and cooling speeds
	Mechanical power in channel E decreases with decrease of pressure (increase of Q/Wbh)	eqn (6.22) Fig. 6.12	As required. NB Small effect in practice, perhaps due to reduction in T_m

TABLE 7.2—contd.

Required	Action	Response	Reference	Limitations
		Heater power H increases	Fig. 6.8	Maximum heater power
		Melt temperature T_m reduced due to lower specific power E/Q	eqn (6.53) p. 203	Softening/melting point reached Viscosity rises, partially offsetting power reduction
		Melting *may* be less complete due to higher output Q	p. 134 p. 203	Incomplete melting
		Temperature variations *may* increase due to greater output Q	p. 203	Unstable flow in die
		Mixing *may* increase or decrease depending on melting and effect of higher output	p. 203	Non-uniform composition
Reduce melt temperature variations	Reduce screw speed N (fixed die)	Output Q and pressure P reduced approx. in proportion to speed N	eqn (5.87)	Uneconomic output
		Mechanical power E reduces approx. proportional to $N^{1.5}/N^2$	Fig. 6.10	
		Heater power H increases/decreases depending on high/low speed	Fig. 6.10	Maximum heater power Maximum cooling or heaters off
		Melt temperature T_m falls (heaters constant)	p. 183	Softening/melting point reached
		Gel point moves towards feed	p. 132	—
		Melting more complete—longer time	p. 202	—

Operation of Single-Screw Extruders

Reduce barrel set temps	Temperature variations decrease	p. 202	As required
	Residence time for mixing increases	p. 202	—
	Output Q approx. constant	eqn (7.13)	—
	Pressure P increased (fixed speed and die)	Fig. 7.7	Safe working pressure of barrel and die
	Mechanical power E increased considerably	eqn (6.27)	Maximum motor power
	Heater power H reduced	Fig. 6.8	—
	Melt temperature T_m falls if E was previously small	Fig. 6.8	Softening/melting point reached
	Melt temperature T_m rises if E was previously large	p. 195	Effect on temperature variations—see below
	Melting *may* be less complete if output Q and power E previously low (mainly conducted heating)	p. 256	Incomplete melting
	Melting *may* be more complete if output Q and power E previously high (mainly shear heating)	p. 195, p. 206	—
	Temperature variations *may* decrease if melting improves, and longer time in pumping section if temperature rise in latter is small, i.e. T_m is low	p. 206	As required
	Temperature variations *may* increase if melting deteriorates, and/or temperature rise in pumping section is large, i.e. T_m is high	p. 240	*Contrary to requirement*

TABLE 7.2—contd.

Required	Action	Response	Reference	Limitations
	Increase back pressure by die restriction or adding screens (constant speed)	Mixing *may* improve if T_m decreases due to longer residence time and higher shear stress in pumping section	p. 208 p. 239	—
		Output Q decreases and total pressure P increases	A to C in Fig. 5.20 eqn (6.22)	Uneconomic output Safe working pressure of barrel
		Mechanical power in channel E increases		Maximum motor power
		Heater power H decreases	Analogous to Fig. 6.8	—
		Melt temperature T_m rises due to increased specific power E/Q (heaters constant)	p. 203	Thermal degradation Unsatisfactory flow from die Effect on temperature variations—see below
		More complete melting due to longer residence time and increased specific power	p. 134	—
		Temperature variations *may* decrease due to improved melting and longer time in pumping section, if temperature rise in pumping section is small, i.e. T_m is low	p. 203	As required
		Temperature variations *may* increase if temperature rise in pumping section is large, i.e. T_m is high/increases		*Contrary to requirement.*
		Mixing improves due to longer residence time in melt pumping section	Fig. 5.7 p. 203	—

Operation of Single-Screw Extruders

Action	Effect	Reference	Notes
Increase screw length (melt pumping section)	Pressure gradient decreased so output Q and pressure P increase (fixed die)	Fig. 5.20, p. 209	—
	Mechanical power E (in pumping) increased	eqn (6.27)	Maximum motor power
	Heater power H (and losses S) increased	Fig. 6.5	Maximum heater power
	Melt temperature T_m increased especially if previously close to melting point	Fig. 6.5	Thermal degradation increased. Effect on temperature variations—see below
	Melting more complete if reduced pressure gradient allows length for melting to increase, despite greater output Q	p. 209	—
	Melt temperature variations decrease due to more complete melting and greater pumping length for mixing, despite greater output Q	p. 209, p. 239	As required
	Mixing increases due to greater pumping length	p. 209	—
Change depth of melt pumping section	At constant screw speed, output will usually increase with screw depth	Fig. 5.18	
	Theoretically mechanical power E will decrease with increase of screw depth, but in practice changes appear small	eqn (6.22), eqn (6.29)	

TABLE 7.2—contd

Required	Action	Response	Reference	Limitations
		Melt temperature will decrease with increasing screw depth due to smaller specific power E/Q	Fig. 6.6	Softening/melting point reached
		Melting *may* be less complete with deeper screw and higher output Q	p. 210	Incomplete melting
		Temperature variations—experiment suggests medium depth screws give more uniform temperature than very shallow or very deep—in the former case this would be exaggerated by the additional speed to restore output. However, at high output melting tends to be incomplete in deeper screws; mixing distances in the pumping section are greater and shear stresses less, whereas longitudinal mixing tends to be greater	p. 244	This is complex and depends on previous values of speed, pressure gradient (Q/Wbh), power input, melting length, screw channel depth, etc. Also depends on whether comparison is made at constant speed or constant output
Improve melting	Reduce screw speed N	Output Q and pressure P reduce approx. in proportion to speed N	eqn (5.87)	Uneconomic output
		Mechanical power E reduces approx. in proportion to $N^{1.5}/N^2$	Fig. 6.10	—

Action	Effect	Reference	Limiting factor
	Melt temperature T_m falls due to reduced specific power (heaters constant)	p. 183	Softening/melting point reached
	Gel point moves towards feed	p. 202	—
	Melting more complete—longer residence time for conducted heat	p. 132	As required
	Temperature variations decrease due to more complete melting and longer time/greater length in pumping section	p. 202	—
	Residence time for mixing increases	p. 202	—
Increase back pressure (constant speed)	Output Q decreases and total pressure P increases	A to C in Fig. 5.20 eqn (6.22)	Uneconomic output
	Mechanical power E increases	Analogous to—Fig. 6.8	Maximum motor power
	Heater power H decreases		—
	Melt temperature T_m rises due to greater specific power E/Q (heaters constant)	eqn (6.53)	Thermal degradation
	More complete melting		—
	Temperature variations *may* increase or decrease	p. 203	
	Mixing improves due to longer residence time and greater longitudinal mixing	Fig. 5.7 p. 203	
Reduce barrel set temperatures in melting section only	Output Q and pressure P approx. constant (fixed die)	p. 194	—
	Mechanical power E increased considerably	p. 205 p. 207	Maximum motor power Maximum torsional strength
	Heater power H decreased	p. 206	—

TABLE 7.2—contd

Required	Action	Response	Reference	Limitations
		Melt temperature T_m rises if mechanical power E was previously high	p. 195 p. 206	Thermal degradation
		More complete melting due to greater shear stresses, smaller temperature differences between solid and molten portions, possible delay of solid bed break-up	p. 135 p. 195 p. 206	As required
		Temperature variation (at die) *may* increase due to lower average temperature at start of melt pumping section	p. 195 p. 208	Unstable flow in die Non-uniform product properties
		Mixing *may* increase due to longer time after melting, and greater shear stresses at lower temperatures	p. 200	—
	Increase screw length	As previously, lower pressure gradient tends to increase length for melting	p. 209	
		Greater length also allows more gradual temperature gradients in melting section		
		Greater length also allows greater time in melting section for elimination of solid particles and temperature variations	p. 209 p. 239	

Action	Effect	Reference	
Improve mixing	—	—	Uneconomic output
Increase back pressure (constant speed)	Output Q decreases and total pressure P increases	A to C in Fig. 5.20 eqn (6.22)	Uneconomic output
	Mechanical power E increases	Analogous to—Fig. 6.8	
	Heater power H decreases	eqn (6.53)	
	Melt temperature T_m rises due to greater specific power E/Q (heaters constant)	p. 203	
	More complete melting		
	Temperature variations *may* increase or decrease		
	Mixing improves due to longer residence time and greater longitudinal mixing	p. 203 Fig. 5.7	
Reduce screw speed (fixed die)	Output Q and pressure P reduced approx. in proportion to speed N	eqn (5.87)	Uneconomic output
	Mechanical power E reduces approx. proportional to $N^{1.5}/N^2$	Fig. 6.10	—
	Melt temperature T_m falls (heaters constant)	p. 183	—
	More complete melting	p. 132	—
	Temperature variations decrease	p. 202	—
	Mixing improves due to longer residence time, lower temperatures and more complete melting	p. 202	—

TABLE 7.2—contd

Required	Action	Response	Reference	Limitations
	Reduce barrel set temperatures	Output Q approx. constant	eqn (7.13)	—
		Pressure P increased (fixed speed and die)	Fig. 7.7	—
		Mechanical power E increased considerably	eqn (6.27)	Maximum motor power
		Melt temperature T_m falls if E was previously small	Fig. 6.8	Softening/melting point reached
		Melt temperature rises if E was previously large	p. 195	Thermal degradation
			p. 256	Effect on temperature variations
		Melting may be more/less complete if mechanical power was previously high/low	p. 195	
		Break-up of solid bed *may* be delayed	p. 206	

	Temperature variations *may* increase/decrease depending on changes in melting, in melt temperature T_m, and whether latter is much above melting point	p. 195 p. 206 p. 240
	Mixing will probably improve due to more complete and uniform melting, and higher shear stresses in pumping section, but may be offset if melt temperature is raised appreciably in pumping section	p. 208 p. 239
Increase screw length	As previously, lower pressure gradient tends to increase length for melting	p. 209
	Greater length in pumping section allows more time for mixing	p. 209 p. 239
Surging	See Section 8.2	p. 245

though increasing the value of Q/Wbh (from A to B in Fig. 5.20). Then if the operational requirement is for a constant output (as at A in Fig. 5.20), the speed of the shallower screw must be raised until the operating point B coincides with A. As shown in the first section of Table 7.2, the consequences of increasing speed will frequently be contrary to those of other changes, especially reducing channel depth, and the total effect may be improvement or deterioration depending on the relative magnitudes of the changes and their effects in a particular situation.

Chapter 8

EXTRUDER OPERATION AS PART OF A TOTAL PROCESS

In Chapter 4, the principles of several commercial extrusion processes have been described together with broad indications of the performance each requires from the extruder. In Chapter 7 the performance of a single-screw extruder has been discussed together with the changes expected from intentional changes in the operating conditions. Since some form of die is always involved, this has been assumed to be fixed in dimensions by the remainder of the process. Note that if comparisons between machines or conditions are made on the basis of, say, constant pressure, very different conclusions will result. In Chapter 7, the discussion centres on steady-state conditions and parameters such as output, pressure, energy, temperatures and mixing; and small changes in these which may be brought about by altering the running conditions. In this chapter, an attempt will be made to see how these may be applied to meet the conditions and requirements of complete commercial processes, and this introduces additional factors such as stability and fluctuations, transients when conditions are altered, and methods of control. Scale-up is also considered in some detail, being an important step in establishing a new process.

8.1. QUALITY

Up to this point attention has been concentrated on physical quantities which can more or less readily be measured. For commercial products, however, it is not sufficient to obtain a given mass flow rate at given conditions of temperature, pressure or energy input; the concept of 'quality' is rather intangible but very necessary in determining satisfactory operation. It cannot be overemphasised that one or other aspect of quality in the product is almost inevitably the limiting factor on output, rather than physical limits as such. Many theoretical studies neglect quality and so are of limited value. Concepts of efficiency and optima which ignore the limits imposed by quality are also largely irrelevant. The theory presented in Chapters 5 and 6 is deliberately simplified to reveal changes in the overall performance until more exact mathematical models are interpreted in an equally comprehensive representation of the real situation.

So what is quality? In a finished product it involves uniform composition of polymer and additives on both microscopic and macroscopic scales; the required macromolecular structure, including crystalline form and extent, orientation in direction(s), distribution and extent, and residual strain; correct and uniform colour, appearance, surface texture and optical properties; required combination of mechanical, electrical and chemical properties and their distribution; absence of visual and mechanical defects such as die lines, ripples, sinks, voids, cracks, crazing, warping, residual stress concentrations, dirt, gels, 'fisheyes', and pigment agglomerates. Even though not all of these may apply in a particular product, satisfactory quality represents a difficult target which may inadvertently be lost through process changes. Certain properties will depend largely on correct and consistent feed material, e.g. mean molecular weight, mean pigment and additive levels. Others, given correct feed material, depend heavily on post-extrusion operations such as drawing and cooling; these include crystallinity, orientation and surface texture. It is therefore desirable to define quality at the die, where it is largely a function of the extrusion process. These properties may be modified subsequently, but correct and uniform post-extrusion processes will depend on properties, e.g. uniform flow velocity and temperature, at the die. The following definition of quality of extrusion is proposed:

Constant flow rate Q and pressure P
Uniform and constant melt temperature especially the absence of unmolten material
Uniform and complete mixing of polymer constituents and additives
Correct shear history,
   e.g. maximum for film grade LDPE
      medium for degradation of PP
      minimum for compounding of PVC
      minimum for reducing elastic defects.

Mechanisms of mixing in the screw have been discussed previously (Sections 4.2 and 7.1). In summary, the scale of mixing is limited by the rather small volume of the screw channel, and the scale will further decrease with the residence time as output is increased. A fair degree of distributive mixing, e.g. of masterbatch and base polymer having similar density and viscosity, is achieved in the screw channel, though not uniformly. Increased back pressure increases relative mixing velocities while reduced temperature increases the resultant shear stresses. Increasing speed reduces the time for mixing. High shear stresses necessary for dispersion of pigment agglomerates or mechanical chain scission only occur over the flight clearance, involving randomly a small proportion of the flow and thus likely to be incomplete and energy-inefficient.

Temperature Variations

Temperature variations have been included in Table 7.2, based largely on a qualitative understanding of changes in the melting process and the mixing effect in the melt pumping section. It is a matter of experience that as screw speed is increased, a point is reached beyond which the flow in the die becomes increasingly unstable, also that this is less likely to occur, or only at higher speed, with longer or shallower screws, other set conditions being equal. With increasing length the effect is likely to be small while with a shallower screw it is almost certainly at a lower output per revolution. The instability may be due to melt fracture or surface defects arising in the die or due to the onset of surging in the screw (see Section 8.2). In the limit, the outline of granules may be observed, or, with powder feed a stream of opaque unmelted or partly melted powder may be encapsulated in a sheath of translucent melt. Such conditions will obviously give unsatisfactory drawdown and mechanical weakness, especially brittleness, in the product, but the unmolten material may exist along with molten polymer whose average temperature is well above the softening or crystalline melting point. As discussed in Sections 5.9 and 7.1, the solid material usually results from a gross break-up of the solid bed in the melting section and the subsequent mechanism of melting in the pumping section (which is almost pure conduction) is slow. However, the molten portion may simultaneously be heated both by conduction and internal shearing, giving considerable spatial differences in temperature of the polymer emerging from the die. Even at lower speeds or higher back pressures where unmolten material does not appear at the

240 Extruder Principles and Operation

die, temperature variations may be sufficient to distort flow in the die and give unstable conditions in post-extrusion operations. Possible origins of these temperature variations include:

(i) Non-uniform temperature at end of melting section (break-up of solid bed) within melt and semi-molten solid
(ii) Conductive heating from/to the barrel wall
(iii) Internal shear heating within melt pumping section
(iv) Heat transfer to the screw
(v) Heat and mass transfer from polymer sheared in flight clearance.

The mechanisms tending to reduce temperature variations include:

(i) Transverse circulation in channel (Fig. 5.8)
(ii) Relative longitudinal velocities in channel (Fig. 5.7)
(iii) Thermal conduction within polymer in melt pumping section (including remaining solid particles)
(iv) Local multiaxial and overall longitudinal thermal diffusion in barrel and screw metal
(v) Preferential shear heating (as origins (iii) above) of more viscous element where shear *rate* is defined.

Considering the possible origins first:

(i) At the break-up of the solid bed this should occupy a relatively small proportion of the channel cross-section, leaving the melt in the remainder to follow flow patterns (Fig. 5.36) similar to Fig. 5.7d, since little pressure is generated, and Fig. 5.8, and undergoing considerable mixing. In the melting region, the screw temperature is unlikely to be above the melting point or the solid bed would become detached, nor is it likely to be much below the melting point or the initial film of melt would re-freeze when collected by the flight, and so fail to pump forward; hence little heat transfer is likely to the screw in the melting section. The barrel temperature is also unlikely to be below the melting point or a molten film would not form. Thus it would appear that any temperature differences within the molten portion will tend to be removed by mixing and that they will be minimised by keeping the barrel temperature in the melting section as low as possible without 'freezing' the polymer and causing excessive torque. In this case the mean temperature of the molten portion will also be as close as possible to that of the remaining solid portion. Low barrel temperatures will tend to minimise temperature differences due to conducted heat flow H_M, which will also be small, giving slow melting; much of the energy for melting will come from internal shearing E_M. The melting process is therefore likely to occupy a greater length of screw, giving a lower but more uniform temperature $\simeq T_M$ at the end of the section and probably delay break-up of the bed, leaving smaller solid particles nearer to their melting point T_M (see p. 195).

(ii) The total conductive heating H_P from the barrel (Fig. 7.2) will be a function of the increase of polymer temperature from melting $\simeq T_M$ to the end of the pumping section T_m and the radial temperature gradients in the polymer to achieve this will be inversely proportional to the length of the melt pumping section L, i.e. low final melt temperature and long pumping sections will favour temperature uniformity.

(iii) Internal shear heating and changes in its distribution with back pressure have been discussed in Section 6.5 (Fig. 6.4). It will be noted that the maximum shear heating occurs near the barrel, where the downchannel velocity w is greatest, giving minimum residence time, but the transverse velocity u, giving mixing, is also greatest. Severe shear heating, e.g. with high back pressure, will therefore tend to increase temperature variations, though this will be minimised by low viscosities implied by high final melt temperatures.

(iv) Heat transfer from the polymer to the screw will tend to lower the polymer temperature near the screw, adding to the variations caused by shear heating. Note that the polymer velocities along and across the channel are small close to the screw, so little mixing will occur. If forced cooling of the screw is used, e.g. by circulating water, it will cause local cooling of polymer which cannot be dissipated by mixing. Further, unlike against the barrel, freezing can occur without direct effect on screw torque, and the indirect effects of this on output, velocity distribution, shear rates and energy input will be much greater than the direct effect on temperature variations.

(v) As discussed in Section 6.5, considerable energy is dissipated in shearing a small volume of polymer in the flight clearance; this will be affected by screw speed and shear rate/viscosity characteristics of the polymer. However, it will also be dramatically affected by barrel temperature; if the latter is high, the shear energy will be low but so will local heat diffusion into the barrel ((iv) of reducing mechanisms). If barrel cooling is applied, local heating will diffuse within the barrel wall but also the local viscosity will be maintained which causes heat generation. Heat diffused within the barrel will raise the average wall temperature in the region and also be conducted longitudinally to cooler parts of the barrel (and possibly the die). Due to shearing in the flight clearance, this small proportion of the polymer will have a higher than average temperature which when mixed with the polymer in the channel will raise the latter's temperature, especially near the barrel where the leakage flow enters. As observed under (iii) this will tend to increase temperature variations before the die.

Of the mechanisms tending to reduce temperature variations:

(i) Transverse circulation (Fig. 5.8) will tend to reduce temperature variations by mass transfer, and since velocities are not uniform, by mixing, i.e. bringing closer elements which because of their initial position have different temperatures, and so promoting conduction between them. As mentioned earlier, note that there is a position in the channel depth where there is no lateral velocity, and Fig. 5.7b shows that at high back pressures this position has only a small down-channel velocity, and so will experience little mixing.

(ii) The relative longitudinal velocities, especially at high back pressure (Fig. 5.7), provide a degree of material mixing. In the absence of longitudinal temperature gradients this will not reduce lateral variations in temperature. In this respect it assists the transverse circulation in transmitting heat conducted from the barrel and generated by shear near the barrel throughout the channel cross-section, but as noted in Section 5.2, the resultant velocity is never rearward in respect of the screw axis, and therefore does not convey heat in this direction, as does longitudinal conduction within the barrel wall.

(iii) Thermal conduction within the polymer is the most obvious mechanism for reducing temperature variations, but depends on these variations for its existence, so is self-extinguishing. Evidently the effect will increase with length of melt pumping section, with reduction in channel depth and with increase in residence time (reduction in output).

(iv) As discussed, local diffusion in the barrel will diminish extreme local temperature variations in space and fluctuations in time, e.g. due to movement of the screw flight. Overall longitudinal conduction will tend to reduce axial temperature gradients in the barrel, and to some extent in the polymer, compared with the set heater temperatures, so in a heating situation the polymer will tend to be heated earlier, with smaller temperature gradients towards the die.

(v) In a system subject to constant shear stress, such as a suspension or a simple lateral temperature difference (eqn (2.43)) the greater shear rate and hence shear heating occurs at the position of lowest viscosity (e.g. highest temperature) tending to reinforce the temperature difference. However, where the shear *rate* is defined, as in the melting section (neglecting the molten film) where two streams of different viscosity are side by side, the greater shear stress in the stream of higher viscosity gives greater shear heating tending to reduce the temperature difference. This is unlikely to occur in the melt pumping section.

Thus it appears that conditions tending to minimise space-variations in final temperature are long screws and low outputs and to some extent shallow channels; on balance, small longitudinal temperature gradients and low melt temperatures are probably beneficial, whilst the lowest feasible barrel temperatures in the melting section will tend to minimise the input temperature variations. This suggests a set barrel temperature profile similar to 3 in Fig. 7.8, especially if final melt temperatures much above the softening/melting point are required. The step in temperature will be difficult to achieve, as would a form of construction including longitudinal insulation to assist the temperature step; this emphasises the importance of knowing the position at which the polymer becomes effectively a melt (by pressure measurement—p. 120) and recognising that this position changes with output, back pressure, etc.

Experiments carried out by the author to determine factors affecting temperature variations, admittedly not precisely controlled for operating conditions, used a 50 mm extruder with a 50 mm diameter barrel extension beyond the screw leading to an adjustable die. The first stage was to assess the reliability of temperature measurements; a 3 mm Pyrotenax[64] thermocouple was placed at the centre of the extension and a simple thermocouple was welded to the metal wall. The extruder was run in conditions expected to give considerable radial temperature variations and readings were taken with various levels of heating to the barrel extension. It was found that at 25 mm distance, the immersed thermocouple tended to follow indications of the wall thermocouple, rather than the melt. When the barrel extension was unheated but lagged externally, relying on heat conducted axially from adjacent sections, its temperature appeared to be unrelated to that of the immersed thermocouple, either higher or lower depending on extrusion conditions; while not a proof of accurate absolute indication of temperature, the differences caused by changes in conditions should be at least approximately correct. An array of thermocouples 50 mm

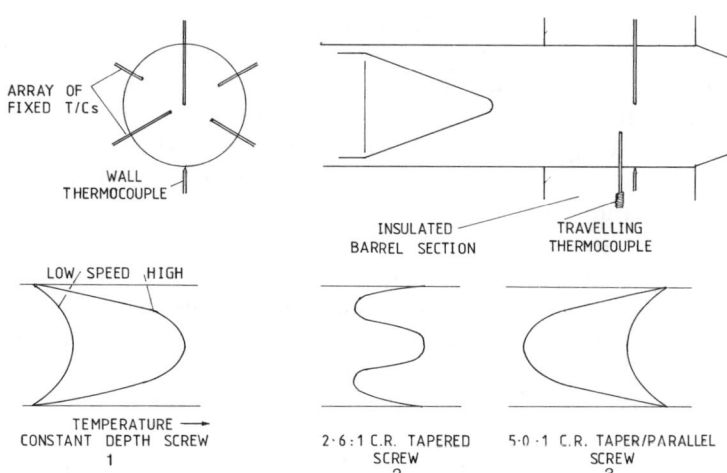

Fig. 8.1. Radial temperature profiles at end of screw.

downstream of the screw tip was then arranged at various radii, together with a single couple adjustable over the full range (Fig. 8.1). The extruder was operated on LD polyethylene at screw speeds from 20 to 100 rpm. Three screws of 16:1 L/D ratio were used: (1) constant depth 6·3 mm ($\frac{1}{4}$ in.); (2) uniform taper from 9·5 to 3·2 mm ($\frac{1}{8}$ in.)— 2·6:1 compression ratio; (3) 10 diameters from 9·5 to 1·6 mm ($\frac{1}{16}$ in.) plus 6 diameters at 1·6 mm ($\frac{1}{16}$ in.) deep—5·0:1 compression ratio. It was found that, when running, the adjustable thermocouple agreed at all times within 1°C with the fixed couple at the same radius, including that at the wall. Each screw was found to give a characteristic profile of temperature, that for the first screw being hotter in the centre with a gradual fall to the wall, and for screw 3 being hotter at the wall and falling to the centre. Screw 2 showed a 'moustache' profile with higher temperatures at both wall and centre and a lower temperature at mid-radius. At 20 rpm the variations with screws 1 and 3 were about 5°C and with screw 2 about 2°C. As speed was increased, all three screws retained their characteristic profiles, but increasing in magnitude until at 100 rpm screws 1 and 3 showed overall variations of as much as 50°C, with about 30°C for screw 2. Because of the mass and response time of the thermocouples, these readings represent time-averaged temperatures; a colleague obtained readings on a similar machine with fast response couples and recorder and showed rapid time-fluctuations of temperature of the same order of magnitude in apparently steady operation (see also Ref. 49 and p. 248). These results must be treated with caution; in some respects such as the effect of speed they agree with the foregoing theory but it is noticeable that variations with the shallowest screw (3) are greater than with screw 2, possibly due to excessive shear heating in the metering section.

Marshall et al.[47] have also detected time variations of temperature, but show clearly that variations arise in the melting section and tend to be reduced in the melt pumping section. They also show that these variations increase considerably with screw speed and the subsequent reduction takes a greater distance, resulting in an increased residual variation at the end of the screw. These results do not of course separate the effects of the various mechanisms discussed in this chapter.

8.2. STABILITY

In addition to the limits on useful output set by inadequate mixing and excessive temperature variations, a limit is frequently imposed by unsteady flow or 'surging' causing variations in product dimensions, drawdown, cooling and consequently in product properties. These usually occur, or only become significant, at high screw speeds and outputs. Slow variations may arise outside the machine itself due to bridging in the feed hopper, erratic filling of the screw or switching of heater controls; the last occurring especially with time-proportioning controllers, characteristically over a period of 10–20 min. Modern proportional controllers avoid this, but inspection of the 'fast cycling' control suitable for resistive heating, will often show a periodic variation in the power input over periods usually less than 1 min. This can be minimised by adjusting the 'power' limit until the load indicator light flashes at a constant frequency. Excessive deadband or overshoot in the control or heat capacity in the barrel can increase the temperature oscillations. Autogeneous operation might be expected to improve temperature stability, since heat flows through the barrel are at a minimum. However, the reversal of temperature gradients and the heat capacity of the barrel material, as well as the difficulty of controlling liquid cooling at low levels, must modify this. One approach is the energy-inefficient method of maintaining more or less constant air cooling and switching heaters to balance heat requirements. With correctly placed thermocouples and modern controllers a good compromise is to work slightly below the autogeneous speed or above the autogeneous temperature so that a small but continuous heat input is required.

It will be noted that the equations for flow and pressure (eqns (5.17), (5.81), (5.86)) are linear and thus their solution, represented by the operating point, e.g. A or B in Fig. 5.20, is single-valued. The equation (6.27) for mechanical power input E is linear in viscosity η (temperature) and output Q, and thus single-valued (Fig. 6.8). Equation (6.53) shows that as back pressure is increased, the specific power input E/Q is a linear (reciprocal) function of Q/Wbh and therefore single-valued. It might appear from eqn (6.27) that if Q is regarded as constant, two values of speed (proportional to W) will correspond to a given value of power E; however, Q and W are related (by the solution to eqns (5.17) and (5.81)) and a single value appears if W and Q/Wbh are treated as independent variables, the latter being determined by pressure gradient dP/dz, or in a specific case by the die constant K. A double value of speed N occurs at which the heater input H has a given value (Fig. 6.10); it would appear that if the heater input was predetermined, an instability in speed N (and with it output Q and pressure P) could occur. In practice, the power/speed characteristic of the drive motor would inhibit this and the author has not observed this type of oscillation in experimental machines with constant heater loading. In the usual case in which heater input is controlled to give the required barrel temperatures there is thus no mechanism in the melt pumping section to cause instability. Thus the origin of surging is usually attributed to the melting section. It has previously been observed that at given temperatures, the length required for melting increases with speed N (and output Q). A reasonable deduction from this is that, at a given speed and temperature, a shorter melting length will correspond to a lower rate of melting. For the melt pumping section, Fig. 8.2, which is derived from Fig. 5.20 and eqn (5.17), shows that as pumping length Z increases, so do output Q and pressure P. In stable conditions, melting length plus pumping length is constant and $Q_{\text{melting}} = Q_{\text{pumping}}$. However, supposing for a short time the pumping rate exceeds the melting rate; the difference can only be supplied by partially emptying the screw channel at the start of the pumping section (see Section 5.6). Where the channel is no longer full, no pressure can be raised, and the start of the effective pumping section moves forward towards the die. As it does so, the pumping rate (and

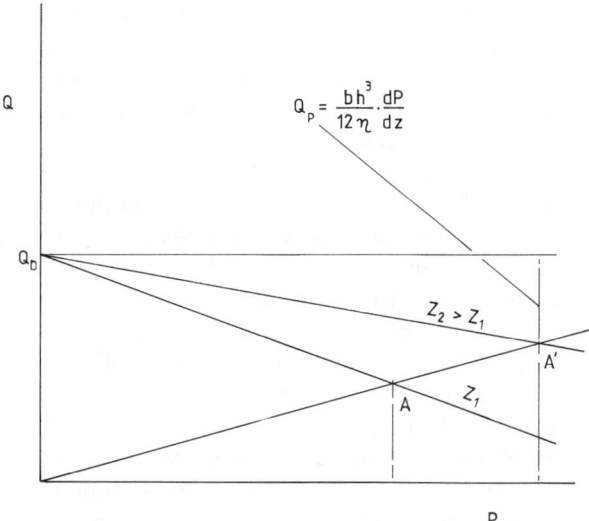

Fig. 8.2. Effect of melt length on output.

pressure) will fall, in some cases to zero, so that it is less than the melting rate. The excess melt now begins to refill the channel, and the greater length available increases the melting rate. As the channel is refilled with melt, the pumping length again increases, as does the output and pressure, so reducing the length for melting and hence the rate. When the pumping rate again exceeds the melting rate, emptying of the channel will recommence and the cycle be repeated indefinitely. Because of the melt capacity of the screw channel and the time required for additional melting to refill it, this cycling takes a finite time. It tends to increase in both frequency and severity with increasing screw speed and the author has found the period to be typically 5–15 revolutions of the screw, which corresponds roughly to the volume of the melt pumping section; in extreme cases the output has temporarily ceased completely, but recommenced with apparently completely melted polymer. It is claimed that with stepped screws, the gel point is virtually predetermined by the rapid change of channel depth; it is true that if the pumping length exceeds the 'metering' length, it will contribute little to the total pressure (Fig. 5.12 at low Q/Wbh) and therefore an approximate maximum output and pressure is fixed; however, this will reduce the length for the melting process and at the other extreme the effective length and output of the metering section can still be reduced by partial emptying. Surging can occur with stepped screws and this is colloquially explained as the metering section pumping itself empty and then waiting until the melting process (in this case said to be mainly by conduction) refills it so that pumping can recommence. Evidently in this case melting rate will be a positive function of the available length (barrel surface) and of the temperature difference between barrel and polymer, suggesting high temperatures in the melting section. Where melting is substantially by mechanical shearing, as is believed to be the case with the polymers for which tapered screws are suitable, low barrel temperatures and high mechanical power input should increase the rate of melting (see Table 7.2), delay break-up of the solid bed, and reduce the incidence of this type of surging.

Kirby[65] has deduced an equation for the stability of the melt pumping section to an external disturbance and Fisher[5] concludes that an increase in die pressure or increase in melt pumping length will improve stability. While agreeing with these conclusions, Appendix D suggests Kirby's equation also indicates that a decrease in channel depth will improve stability; these changes do not of course remove the cause of instability. As mentioned in Sections 5.5 and 5.6, the second portion of conventional vented screws is often subject to surging, due to the non-linear change in pumping rate with degree of filling and also possibly due to intermittent flow of polymer in the section under the vent, and here also a decrease of final depth of channel appears to reduce surging.

Tadmor and Klein[66] quote interesting experimental results on time-fluctuations of temperature and pressure and remark that 'a good correlation between pressure and temperature fluctuation implies that temperature fluctuation is the source of the problem and flow rate may be steady, while if a pressure fluctuation exists without a temperature fluctuation, it is almost certain that it will be accompanied by a flow rate fluctuation'. The suggestions under the heading 'Improve melting' in Table 7.2 may therefore help to minimise this source of surging, while the factors mentioned above should reduce the consequences; increasing length in particular will also help reduce temperature fluctuations and variations at the end of the melt pumping section (see pp. 209 and 243).

Although strictly outside the extruder, the stability of the drive system will also affect the extruder performance. Figure 6.10 and eqns (6.22) and (6.34) show that at constant temperature the power absorbed in the melt pumping section of the screw increases somewhat more than in proportion to its speed, depending on the shear rate/viscosity characteristics of the polymer. Thus torque on the screw from this source, proportional to E/N, will increase with speed, linearly for a Newtonian fluid. The total torque will therefore also increase with speed, since that due to feeding and melting will at least be constant (power proportional to speed). Except at low speeds (e.g. with AC commutator motor), the torque/speed characteristic of electric motors usually droops, while the torque of hydrostatic motors is essentially constant. Thus if speed falls below the set value (Fig. 8.3), an excess torque is available to accelerate, while a speed increase due to say a temporary load reduction leads to a torque deficiency tending to reduce speed, creating a stable system. With the usual high-speed motor driving through reduction gearing, there is appreciable rotational inertia tending to damp out such speed fluctuations. For applications requiring very stable speed, despite variations in extruder demand, thyristor controls for DC shunt motors with tachogenerator feed-back are available from fractional to several hundred kilowatt power and with increasing degrees of sophistication (and cost) for improved speed holding down to 0·25% from zero to full load.

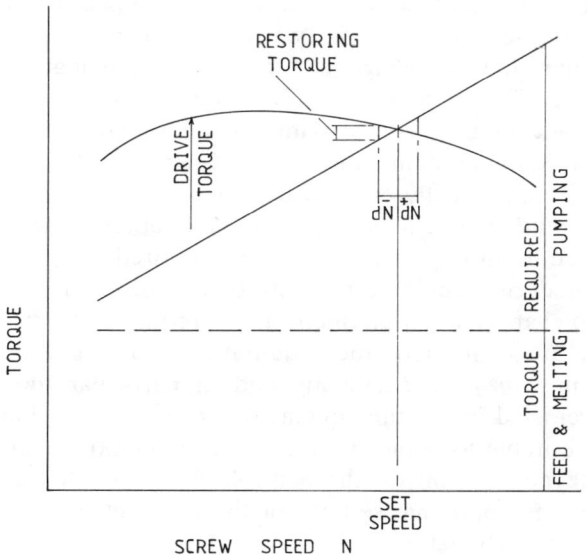

FIG. 8.3. Speed stability of drive.

Start-up and Condition Changes

So far stability has been considered in nominally steady conditions. The speed-holding mentioned above would not be achieved instantaneously for such a large change in load. Thus at start-up or when deliberately changing speed, short-term transients may cause overload; provided they are of short duration the motor will not be harmed and flexibility in couplings and belt drives will cushion the effects, but even momentary excess torque may damage the screw or drive shaft, for instance if solid material jams in the feed.

Gradual start-up and changing of conditions is discussed in Section 9.3; here it is only necessary to mention that changes in pressure influence output almost instantaneously and power a little slower. Temperature equilibrium may be reached in 10–20 min on a small machine, but takes an hour or more on a larger or slow-running machine. Response to speed changes is only slightly slower than to changes of pressure, but the major changes in energy balance shown in Fig. 6.10 mean that temperature-equilibrium may take longer than with change of pressure. Changes in set temperatures are usually the slowest to take effect and stabilise, partly because of thermal capacity of the extruder structure, partly due to changing ambient losses and partly due to changes in the mechanisms of melting. Limited heater power will slow temperature increases and temperature controllers must compromise between rapid warm-up and minimum overshoot. However, the slowest change is often in longitudinal temperature gradients, e.g. if the set point of one heating zone is changed in relation to neighbouring zones. With experience, such changes can be achieved quickly with minimum transients, but it must be realised that large changes will usually cause a temporary unbalance in the many mechanisms and interactions described in Chapter 7, and these will sometimes be in an opposite sense to that desired.

While modern temperature controllers respond quickly and smoothly to changes in requirement, they are limited by the system to which they are connected; the requirements for rapid start-up and stable running may be conflicting and in particular the maximum heater power used for heating up may be excessive for stable control— it may be desirable to reduce power when running (by manual adjustment of the power limit on the controller) or consult the controller manufacturer for optimum settings of the differential term (for overshoot) and proportional band (for power limitation).

8.3. SHEAR HISTORY

The remaining criterion of quality at the die (p. 238) is the correct shear history. This may broadly be defined in terms of the units of shear strain experienced by the polymer, although it has been remarked previously (pp. 51 and 239) for certain processes, such as dispersion of pigment agglomerates and polymer chain scission, that a threshold level of shear stress is required to produce any significant effect, regardless of the time of shearing. In such cases the definition might be modified to 'the units of shear strain at or above the threshold stress'. The shear strain is the integral of shear rate × time:

$$\text{shear strain} = \int_{t_1}^{t_2} \dot{\gamma} \, dt \qquad (8.1)$$

Ideally this will be the same for each element of volume, so that even though the shear rate and the time for which it is experienced may both vary with position in the screw, all elements are subjected to the same pattern. This is evidently not true in the single-screw extruder; consider the longitudinal velocity w, which in drag flow varies uniformly from zero at the screw to W at the barrel (Fig. 5.7) so that shear rate dw/dy is constant. However, neglecting movement in the y direction (radially), the residence time in a given melt length will be Z/w_y, varying from Z/W to infinity. Thus at y from the screw root:

$$\text{shear strain} = \dot{\gamma}_y \, dt = \frac{dw}{dy} \cdot \frac{Z}{w_y} \qquad (8.2)$$

and this has the hyperbolic distribution:

$$\frac{Z}{h} < \frac{WZ}{hw_y} < \infty \qquad (8.3)$$

Figure 5.8 shows that both transverse velocity and shear rate are non-linear; at the position of maximum (rightward) velocity, the shear rate is zero so that at this position no transverse shearing is experienced, though presumably the same volume element undergoes considerable shear when moving leftward near the barrel. Thus the polymer in the channel is subjected to a wide spectrum of shear strain; if shear degradation is intentional, then some average is relevant, assuming that the product is mixed to give acceptably homogeneous properties. Where degradation is to be minimised, then the maximum shear strain would be the correct criterion, but Figs 5.7 and 5.8 show that adjacent to the screw root the polymer has theoretically neither forward nor transverse velocity, indicating an infinite residence time; screw cooling will tend to increase the thickness of this stagnant layer. Traditionally, high back pressures are used to improve mixing, giving the maximum relative velocities in the longitudinal direction and greater residence time due to lower output; however, as pressure increases and $Q/Wbh \rightarrow 0$, the longitudinal velocity distribution of Fig. 5.7a is approached where a second point of zero velocity in both directions occurs. Thus at high back pressures, a second layer experiences considerable shear rate in both directions, but a very long residence time. Further scatter of shear history is due to the small proportion of the flow subjected to very high shear rates for a short time in the flight clearance. This emphasises that the single-screw extruder by itself is not very satisfactory for intentional shearing; the best conditions are evidently a long shallow metering section, low speed and fairly high back pressures giving low outputs. For heat- or shear-sensitive materials, such as rubbers and UPVC, short deep screws running at modest speed with low back pressures (which also minimise power input) are desirable and are in fact commonly used. The fact that the single-screw extruder performs reasonably well as a distributive mixer suggests that there is some mechanism of mixing (not just circulation as at the flight faces, Fig. 5.8) in the y (radial) direction not revealed by the simplified flow analysis. If this mechanism is weak or does not exist, as suggested by striations found in microtome sections of extrudate, there would be a case for radial mixing devices within the melt pumping section; otherwise it would be preferable to minimise shear strain in the screw and attach a mixing device, preferably giving uniform shear and residence time, and designed for the particular purpose and polymer in view.

Where severe shear is imposed by the extruder, this is dissipated in heat which increases the polymer temperature, reducing shear stresses and pressure gradients. If cooling is used to reduce temperature rises, this tends to increase power input and shear stresses and so be partly self-defeating. It also leads to distortion of temperature and velocity profiles (Section 3.3 and Fig. 7.5) which militate against both cooling and homogeneity. If shear heating can be accommodated by temperature rise in the polymer, the extruder is a simple means of achieving it, but cooling of highly viscous materials is in any case difficult to achieve uniformly and the single-screw extruder is no exception. Where only a modest degree of shear is required, it must be accepted that this will not be uniformly distributed through the polymer mass and an acceptable level can only be achieved by trial. Table 7.2 indicates how conditions favouring mixing and uniform temperatures can be obtained.

8.4. CONTROL

In the discussion on vented (two-stage) screws and combinations of screw extruder with gear pumps (Section 5.5) it was pointed out that since an increase of pressure difference or pressure gradient reduces the output of the melt pumping section, it will automatically adjust to accommodate small changes in output or pressure. Referring to Fig. 5.20, if a disturbance increases the output and pressure through the die from A to A'', the extruder output will decrease from A to A_{n}. Thus the supply of polymer to the die decreases and the die pressure tends to fall again. Similarly, if die output falls below A, the pressure will also fall, increasing the flow from the extruder, tending to restore output and pressure to point A. It will be noted that if the same screw was operating with a smaller die, as at C, the fluctuation in output corresponding to a given pressure variation would be larger than with a shallower screw. In this sense, the shallower screw can be said to be more stable, but for a given hold-up volume between screw and die, the greater output discrepancy between the die and the deeper screw could be expected to give a more rapid restoration of the original conditions. The single-screw extruder running at constant speed and feeding a fixed die is thus a self-regulating system as regards output and pressure. Figure 6.8 shows that, at constant speed, a change in final melt temperature involves a substantial change in mechanical power input and even greater (absolute) change in heat supplied. Thus a temporary reduction in temperature will lead to an increase in mechanical power and a reduction in losses, tending to raise the temperature to the original level. If the reduction in temperature is sensed by the heater controllers, they also will tend to restore temperature by additional heat input, but, due to the thermal capacity of the machine, this will occur over a much longer time scale. Thus the extruder tends to be self-regulating for temperature, irrespective of the heater controllers. It was shown in Fig. 8.3 that the usual 'regulation' curve of electric motors tends to make the drive self-stable in respect of speed. However, Fig. 6.10 shows that as speed changes, the difference between energy required by the system and that supplied by the mechanical drive changes relatively slowly and indeed at the point of maximum difference the lines are parallel; this difference represents the energy required from the heaters, which in short times is fixed by the heat capacity of the machine. Thus an instability in the extruder is possible, only restrained by the drive characteristics. Therefore in many respects the single-screw extruder is self-stable against short time-fluctuations in that these will be damped out more or less quickly, although transient effects cannot be avoided. This is provided that constant speed is assured. It is therefore unwise to impose any short-time external control on the extruder, apart from speed, which may react with the self-regulating system and even cause loss of control. For the same reason, any long-term adjustments should be made slowly and progressively.

The speed control provided by a thyristor drive is ideal in that the voltage output of the tachogenerator on the motor shaft responds virtually instantaneously and there is negligible lag in the thyristor control to provide the motor current necessary to maintain the speed. Provided that the speed regulation permitted by the control is small, the mechanical inertia of the motor and drive will also have negligible effect, giving a rapid and precise control of speed in response to load changes. However, when substantial changes in speed setting are imposed, mechanical inertia will cause a more delayed response and the possibility of slight overshoot. The speed regulation may be minimised by more sophisticated and expensive control circuits, which cannot always be readily adapted after initial installation; they are likely to be economically justified for products requiring minimum output fluctuations, such as film and fibre, and where drawdown in minimal, e.g. for large tube. It is recommended that to assist in re-setting conditions a digital setting should be provided for speed and for long-term stability this might operate through switching a stepped potentiometer with shunt-connected fine adjustment (Fig. 8.4). Of the older types of drive, the AC commutator motor with induction regulator is probably the most satisfactory, having small regulation up to about 20% overload and a fairly constant efficiency around 85% between 50 and 120% load. However, the efficiency drops considerably at speed settings substantially above or below the synchronous speed with large circulating currents in the secondary circuit. The movable-brush AC Schrage motor is capable of sustaining considerable overload torque for long

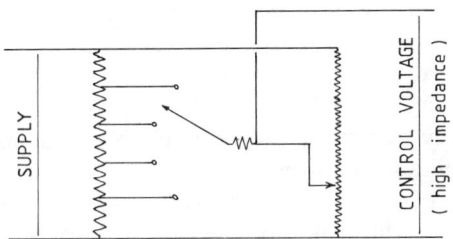

FIG. 8.4. Proposed digital speed control.

periods, but its speed-holding is poor, having characteristics similar to the series motor. The AC/DC Ward–Leonard drive is expensive and cumbersome, incorporating three rotating machines, which also limit the efficiency; the performance is fairly good. The AC induction motor with magnetic slip coupling is subject near full load to the fairly large speed regulation of the induction motor and at low speeds to low efficiency due to losses in the coupling.

In production, output cannot be measured directly and often only remote from the extruder, e.g. by continuous measurement of product thickness at constant haul-off speed in wire covering, tube and sheet extrusion. As a result, there is a time delay and the possibility of errors due to 'storage' capacity in the system between extruder and point of measurement, making output unsuitable as an indication for control. Pressure can be measured on-line and at the end of the screw, with fast response and good accuracy. Unfortunately, aside from possible temperature errors, these instruments seem liable to damage in a production environment; the force-balance design originated by ICI Petrochemicals and Plastics Division and marketed by Bowyer Engineering Ltd,[67,69] was intended to minimise this problem and gives a 20–100 kN/m^2 (3–15 psi) pneumatic signal suitable for control. Pressure is therefore the most rapid and suitable indication of changes in the process; however, direct control of pressure by adjustable valves or restrictors is slow and mechanically unreliable in the environment of hot, highly viscous polymers. In addition, it has been shown earlier to be highly dependent on temperature and effective melt pumping length, as well as output, so is not a good medium for control, though as mentioned on pp. 99 and 120 is an essential indication of conditions in the extruder.

Final melt temperature is both an important indication of operating conditions and a vital factor to be controlled for subsequent processing and product properties. It can be measured roughly by a portable thermocouple inserted manually in the stream of polymer emerging from the die; the thermocouple must be robust and therefore slow in response, so cooling the polymer which tends to stick to it. Repeated insertion is usually necessary to get a reliable (maximum) indication, and so doing inevitably disturbs downstream operations. More reliable indication will be obtained from a thermocouple permanently installed just before the die, which like the pressure transducer requires careful handling during cleaning, especially if it is made small to give rapid response. It will not generally respond as rapidly as pressure gauges, and in view of the space variations and time fluctuations mentioned earlier (see p. 244) will not show a true average value; to minimise the former it should always be located in the same position, and preferably in a fairly narrow flow channel, though this will accentuate errors due to heat from the walls. Even so, the indication of a melt thermocouple forms an important and useful signal for long-term control of barrel temperatures, especially with changes in speed, output, etc. Melt temperature cannot of course be controlled directly, but, with a given screw design and speed, indirectly through barrel temperatures, and thus relatively slowly. Note from Table 7.2 that increases in set temperature at different points along the screw length will have different, and even opposite effects on the final melt temperature. In this connection the ammeter on the motor drive or other torquemeter is a rapid indication, e.g. if the effect of an increase in a set temperature is a marked decrease in drive torque, the melt temperature is also likely to be increased. Thus the barrel temperature *profile* should be regarded as a control input distinct from the average or final set temperature.

Various temperature profiles were discussed in connection with Fig. 7.8 and in summary the author suggests the temperature of the first barrel zone determines the point at which melting commences; the set temperatures in the melting section determine the energy added by heat conduction, and by difference that from the mechanical drive; the set temperatures in the melt pumping section determine how far the final melt temperature is above the softening/melting point of the polymer.

Modern temperature controllers for plastics processing machines perform two main functions with a high degree of sophistication and reliability, (a) to give rapid start-up and change of conditions with minimum overshoot, (b) accurate maintenance of set temperatures

Extruder Operation as Part of a Total Process

despite wide changes in heat demand by the process. The details of operation are beyond the scope of this book, but the instrument manufacturers are knowledgeable of the requirements of the plastics industry and very helpful in establishing the instrument settings (time constants, integral and derivative terms, power level, proportional band) to obtain the best performance in a particular situation. For reasons of fast and unambiguous response, the control thermocouple should be close to the source of heating/cooling, and this is the temperature the controller attempts to maintain. As mentioned above, this is indirectly related to the desired polymer temperature, and subject to time delays, so the operator must establish the relation between set and melt temperatures and the appropriate strategy for control settings. In maintaining set temperatures, the controllers will automatically restore the energy balance (as represented by Figs 6.8 and 6.10) after a disturbance due to changes in polymer, speed, pressure or temperature, but due to interactions within the extruder this will take time, up to several hours, and this action must be recognised as accompanying any deliberate alteration in another set condition. During this stabilisation time there is the probability of off-standard product with the attendant material, energy and labour costs and loss of production capacity. In extreme cases it may be more economical to stop the downstream processes and either stop the screw or reduce output to a minimum until temperatures reach approximately the required values. When speed is restored, the consequences of this on energy balance must take place, but acceptable product is likely to be achieved quicker and with less waste of material, etc. With experience, it may be possible to reduce the time to reach equilibrium by judicious adjustment of speed, pressure or heater controls, but reaching correct temperatures is likely to remain the limiting factor on time and it must be remembered that any change of conditions will require time to stabilise, so inadvised adjustment is likely to extend the time to reach steady conditions.

Of the remaining parameters, a constant-torque control on the drive is undesirable, since it would lead to changes in speed and temperature, with consequent settling times; like pressure, motor amperes are useful for rapid indication but not control. Degree of mixing and melt temperature variations are difficult to measure on-line and are not directly controllable; the factors listed in Table 7.2 should be adjusted first, possibly with some sacrifice in output, and only then should changes in screw design be considered.

As discussed on p. 199, screw cooling requires very careful adjustment and reliable instantaneous indication of coolant flow rate and temperature; in practice it is usually under manual control as the operator's 'last resort'.

One possibility which the author has not seen discussed, but which should aid stable operation is measurement of the effective 'gel' point by extrapolation of two or more pressure readings and using this signal to control the melting process by heat input and screw cooling in this section *only*.

Each of the preceding control functions is capable of overriding or hierarchical control and integration with downstream functions such as cooling water temperature and haul-off speed to give computer control of the whole process. The subsidiary controls of motor speed and individual barrel temperatures would remain, with the set points being automatically adjusted in accordance with an overall strategy. This strategy, however, requires accurate modelling of the interactions, e.g. between adjacent heating zones and between the melting process and melt temperature variations, which must be determined experimentally for the individual polymer, machine and process. The model must also allow for the inherent time lags so that hasty control action is not defeated by the inherent characteristics of the process.

Control for Specific Processes

General conditions for control of the single-screw extrusion process have so far been discussed in this chapter. Because of the wide range of polymers, machines and products involved, this has necessarily been qualitative. In Chapter 4 the principal extrusion processes were outlined and their requirements in terms of process variables, e.g. output, pressure, melt temperature, mixing, etc., were summarised, both as to magnitude and accuracy, in Table 4.2. Details of this are open to criticism, especially as each process category covers a wide range of polymers, sizes, linear speeds and product specifications, e.g. in relation to homogeneity and dimensional tolerance. However, it distinguishes the critical process factors and illustrates how these differ between processes. In Table 8.1 the author puts forward suggestions as to how these requirements might be met in terms of screw type and operating conditions. The table also indicates the accuracy of control which is desirable to achieve acceptable product quality and uniformity, as far as this is determined by the extruder itself. Naturally, control of the haul-off speed, especially in relation to screw speed, is

TABLE 8.1
Suggested Conditions for Individual Processes

Process	Screw diameter (relative)	Screw length L/D ratio	Screw type	Metering depth	Screw speed (relative)	Melt temperature	Degree of Control			Comments
							Speed	Temperature	Mechanical[a] power	
Rod/section profile	High	20/24 PE, PP 15/20 PVC 20/24 Nylon	T T S	Shallow	Low	Medium	××	××	×××	Output relative to haul-off speed, for dimensions. Minimum elastic memory. Medium temperature for rigidity while cooling. Add screens to promote mixing.
Tube	High	20/24 PE, PP 15/20 PVC 20/24 Nylon	T T S	Medium	Medium	Medium except die	××	××	××	Minimum elastic memory. High die temperature for gloss. Medium temperature for drawdown of small tubes.
Wire covering/sheathing	Low	20/24 PE, PP 15/20 PVC 20/24 Nylon	T T S	Shallow	High	High	××	×	×	High temperature for good drawing and avoiding melt fracture. High pressure for thin insulation. Output relative to haul-off speed for thickness.
Sheet/laminating, (including continuous vacuum forming)	High	20/24 PE, PP 15/20 PVC 20/24 Nylon	T T S	Medium	Medium	Medium	××	××	××	Medium speed is compromise for high output and uniformity of Q & T. Steady output to avoid ripples. High output for thick sheet. Uniform temperature for die distribution. Medium temperature for drawdown.

TABLE 8.1—contd.

Process	Screw diameter (relative)	Screw length L/D ratio	Screw type	Metering depth	Screw speed (relative)	Melt temperature	Degree of Control			Comments
							Speed	Temperature	Mechanical power[a]	
Flat film	Medium	24/28 PE, PP 20 PVC	T	Shallow	Medium	High	×××	×××	×××	High temperature for drawdown. High pressure for die. Medium speed for uniform elastic memory. Uniform output for thickness.
Paper coating	Medium	24/28 PE, PP	T	Shallow	Medium	High	×××	×××	×	Output relative to drum/hauloff speed for thickness. Medium output for rapid cooling. Very high temperature for flow and surface finish.
Tapes	Medium	20/24 PE, PP	T	Medium	Medium	Medium	×××	×××	××	Medium temperature for crystallisation and drawdown. Uniform output for dimensions and crystallisation. Uniform temperature for dimensions.
Filament/fibres	Medium	20/24 PE, PP	T	Shallow	Medium	High	×××	×××	×	High temperature for drawing High pressure for die.
Tubular film	Medium	20/24 PE, PP	T	Shallow	Medium	Medium	×××	×××	×××	Medium temperature for drawing and blowing. Medium speed for uniform elastic memory.

Process	Output	L/D	Polymer	Screw type	Channel depth			speed	temperature	Notes
Foam	Medium	20/24		T	Medium	Low	Low	×	××	Temperature depends on blowing agent. Low shear heating for decomposition. Medium pressure and mixing for uniform distribution & pore size.
Compounding/blending	Medium/high	24/28 20 20/24	PE, PP PVC POM PMMA Nylon	T T T S S	Deep	Medium	Low	×	××	Deep screw at high pressure for mixing. Large diameter & medium speed for high output. Low temperature for mixing and heat history.
Degrading	Medium	24/28	PP	T	Shallow	High	High	×	××	
Blow moulding	High	20/24	PE, PP	T	Medium	Medium	Low	××	×××	No haul-off. Melt temperature and swell ratio e.g. controlled by compromise with output of speed and diameter.
Injection moulding	High	24/28 20 20/24	PE, PP PVC PS	T	Medium/ shallow	Medium	Medium/ high	×	××	No haul-off. Melt temperature control important. Back pressures low in practice.

Notes:
The screw type and dimensions are largely dependent on the polymer and temperatures.
Screw diameter and length are a compromise with capital cost.
Screw speed is a compromise between output and quality.
The length/diameter ratio and channel depth depend largely on the polymer used.
'Temperature' under 'Control' refers to set temperatures in the melt pumping section.
[a] Mechanical power input is controlled mainly by temperatures in the melting zone. Screw cooling is excluded since it is always very sensitive, requiring precise control.

Suggested *running* limits are:

speed	×	±1%	temperature	×	±2°C	mechanical power ×	±3%
	××	±0.3%		××	±0.5°C	××	±1%
	×××	±0.1%		×××	±0.1°C	×××	±0.4%

vital in determining lateral dimensions in processes such as profile, tube and wire covering where drawdown is minimal, and important in sheet, film and fibre extrusion where substantial drawdown is common. Control of coolant temperature and flow rate, including their relations to melt temperature and extruder output rate, is important in controlling distortion in shape and size, distribution and types of crystalline structure; these and similar control requirements are additional to those mentioned in Table 8.1.

One of the most common causes of fluctuations in extruder conditions and product properties arises from variations in properties of the polymer feed. This may arise from incorrect blending of constituents, especially re-fed scrap or deblending in storage or feeding; this is under the processor's control and must be monitored. Variations may arise from changes in size, shape or surface friction of solid feed particles; larger particles may be marginally poorer in packing and melting, but generally undue sensitivity to particle size, shape, etc., indicates poor design or conditions in the feed section so that potential solid feed rate is not sufficiently in excess of melt feed rate. A third source of fluctuations is due to changes in the melt viscosity and elasticity arising from changes in molecular weight, molecular weight distribution, chain branching, previous shear and heat history, etc. Polymer manufacturers are understandably reluctant to impose on themselves restrictions in terms of melt processing properties additional to those existing in respect of mechanical, electrical and environmental properties in the finished product. Such restrictions may impose undesirable economic limitations in respect of polymer intermediates, polymerisation processes and operating conditions. The processor must therefore determine for himself the melt properties and permissible variations necessary for his particular process and product. He may be able to agree with the polymer supplier a more detailed purchase specification, but in any event must have equipment to monitor melt viscosity, swell ratio, and their variations with shear rate and temperature. A recent paper on LLDPE by Teh et al.[68] describes equipment, methods and presentation of data, though the present author does not suggest that such comprehensive work is necessary for day-to-day control. However, a model for on-line computer control of the process (see p. 258) must contain an analogue of these flow properties as well as an input based on previously determined characteristics of the actual feed material. In the absence of the latter, the control system must have means of distinguishing changes due to polymer properties from those due to operation, e.g. set temperatures (see next paragraph).

It would be tedious to detail all the effects of changes in melt properties, but from eqn (5.81) it is seen that an increase in viscosity gives a proportional increase in pressure at the same output, while eqn (5.17) shows that such proportional changes in pressure and viscosity cause no change in extruder output, so output is constant while pressure rises—the same effect as of a reduction in final melt temperature. Equation (6.27) shows that at constant output Q, the mechanical power input increases in proportion to viscosity and, according to Fig. 6.8, either melt temperature will increase or heat input will decrease or require cooling. On operational grounds it may be argued that temperature should be allowed to increase until viscosity and pressure are restored to their original values, but this requires a pre-determined adjustment of heater set temperatures on the basis of either constant power or heater input; in any case the subsequent cooling conditions will be disturbed and thermal stability of the polymer may be affected.

A change in the non-Newtonian response to shear rate represented by the pseudoplasticity index n would alter the relation between viscosity in the die (high shear rate) and in the screw channel (low shear rate). For instance, a narrowing of the molecular weight distribution causing an approach to Newtonian behaviour ($n \to 1$) would give, for a fixed viscosity in the screw, an increase in the viscosity in the die. By eqn (5.81) this is equivalent to a decrease in the value of the die constant K, and Fig. 5.20 shows that the operating point will move, e.g. from A to C, or B to D, giving a lower output at higher pressure. This is also illustrated by eqn (5.86) where a decrease in K gives an increase in the factor $(1+B/K)$, or on p. 103 an increase in η_{Die} increases the factor $(1+B\eta_{\text{Die}}/K\eta_{\text{Screw}})$ and hence reduction in output at a given speed. The effect on mechanical power is shown by eqn (6.22), i.e. that a decrease in output causes an increase in power input, and since this is absorbed by a smaller flow of polymer, temperatures will tend to rise. This is analogous to the effects of increased back pressure discussed on p. 186. The increased viscosity at high shear rates will also increase the power absorbed in the flight clearance of the screw, and hence the total power and melt temperature. However, it also increases the *proportion* of total power contributed by the flight clearance and the resulting temperature effect on channel power (see Section 6.5). It is also likely to increase temperature variations at the die and consequently decrease product quality. Changes in melt elasticity will alter the swell ratio, modified by any accompanying changes in shear history in the extruder, leading to changes in the drawdown behaviour, or where drawdown is negligible, to changes in product dimensions. The foregoing illustrates the effects of changes in melt properties of the polymer, the necessity for prior knowledge of such changes by material testing, and the difficulties of incorporating the effects in the process algorithm necessary for on-line computer control. To add relevance to Table 8.1, suggested limits are included for the various degrees of control of speed, temperature and mechanical power input. It will be appreciated that these and other data given in Table 8.1 are in the nature of a general guide and will need to be modified in particular cases; the author trusts that it will not prove too far from processors' own experience and will be useful at least as a starting point.

8.5. SCALE-UP

Although scale-up borders on the province of the designer, it is included here because of operational factors which do not seem to have been fully considered in works on design. Also, these factors frequently become evident only when the equipment has been manufactured and the operator is then expected to obtain the best performance from this equipment. In particular, the author has on several occasions been faced with a demand for a changed screw design; this might take 6–12 months to manufacture during which production would be less than optimum, and then costly losses in production would be incurred while it is installed. In one instance, involving a number of very large identical machines, the author felt obliged to persuade management on theoretical grounds that such expense and delay was likely to leave the problem still unsolved; instead the time and money was spent on gearing of a different ratio so that at full output the maximum motor power was available.

The scale-up problem may occur in the progressive increase of production capacity, leading to larger machines; this may often be carried out successfully by empirical methods, and if the increments are modest any problems may be overcome by minor adjustments of machine dimensions or operating conditions. However, in the case of a new polymer, product or process, the initial development is necessarily done at small scale usually using extruders of 60 mm diameter or less. Economic production may require outputs more than ten times that of the laboratory scale so that the method of scale-up becomes important, not merely to achieve the output but to proportion correctly the drive motor, thrust bearing, heaters, screw, etc., and to obtain correct conditions of pressure, temperature, mixing, shear history, etc. A number of different strategies are possible, and several will be examined to show the likely results and where they might be employed.

The following analysis is based on the simplified Newtonian equations developed in Chapters 5 and 6 for output and mechanical power, modified where necessary for pseudoplastic flow. The deductions are thus not exact for design purposes, but enable the effects on a number of important operational parameters to be clearly seen, and the correct general method of scaling be selected.

In addition to output and mechanical power (giving the rating of the drive motor), the specific power input E/Q will be deduced, since this represents the adiabatic temperature, a key point on the energy balance. In order to operate at any other temperature, energy must be added by heaters or removed by cooling in proportion to the output, which if it increases as a greater power than D^2, requires a higher heating density (W/cm^2) since surface area increases as D^2 (for constant L/D ratio). If the adiabatic temperature represented by E/Q changes with scale-up, then, as shown by Fig. 6.8, the heat requirement per kilogram of output may change dramatically, even from heating to cooling or vice versa, and the total will be amplified by the increased output. However, of greater significance is that a change in the balance of energy input between mechanical E and heat H will alter the extruder performance in respect of melting, mixing, temperature variations at the die, stability, etc., so that the downstream processes may be affected and the quality of the product will almost certainly differ from that determined in the laboratory. The results are summarised in Table 8.2.

It has been seen earlier that the assumption of constant Q/Wbh simplifies comparison as well as ensuring similar velocity distributions in the screw channel. Equation (5.45) gives the implication for pressure gradient and this will be evaluated in each case. Mixing in the screw cannot be evaluated simply, but maximum shear rate at a fixed Q/Wbh is given by eqn (5.98) and is proportional to the drag flow shear rate W/h, proportional to DN/h. The mean residence time in the

TABLE 8.2
Scale-up of Single-Screw Extruder—Summary of Effects of Various Scaling Strategies with Q/Wbh as constant

	Volumetric output, Q	Mechanical power input, E ($n=1$)	Mechanical power input, E ($n<1$)	Specific mechanical power ($n=1$)	Adiabatic temperature, E/Q ($n<1$)	Pressure gradient, dP/dz ($n=1$)	Die pressure, P ($n=1$)	Wall shear rate, $\dot\gamma$ ($n=1$)	Mean residence time, t_{res}	Shear strain (mixing), $\dot\gamma t_{res}$	Specific heating surface, A/Q
Proportional to	D^2Nh	$\dfrac{D^3N^2L}{h}$	$\dfrac{D^{n+2}N^{n+1}L}{h^n}$	$\dfrac{DNL}{h^2}$	D^nN^nL	$\dfrac{DN}{h^2}$	$\dfrac{DNL}{h^2}$	$\dfrac{DN}{h}$	$\dfrac{L}{DN}$	$\dfrac{L}{h}$	$\dfrac{DL}{Q}$
Equation	(8.2)	(8.5)	(8.8)	(8.6)	(8.9)	(5.45)	(5.50)	(5.98)(5.4)	(8.1)	(5.98)(8.1)	—
Scale-up strategy											
Case 1 Constant depth h	D^2N	D^3N^2L	$D^{n+2}N^{n+1}L$	DNL	D^nN^nL	DN	DNL	DN	$\dfrac{L}{DN}$	L	$\dfrac{L}{DN}$
Case 2 Geometric similarity $D \propto L \propto h$	D^3N	D^3N^2	D^3N^{n+1}	N	N^n	$\dfrac{N}{D}$	N	N	$\dfrac{1}{N}$	Constant	$\dfrac{1}{DN}$
Case 2a Constant speed N	D^3	D^3	D^3	Constant	Constant	$\dfrac{1}{D}$	Constant	Constant	$\dfrac{1}{h}$	Constant	$\dfrac{1}{D}$
Case 3 Constant peripheral speed $D \propto \dfrac{1}{N}$	Dh	$\dfrac{DL}{h}$	$\dfrac{DL}{h^n}$	$\dfrac{L}{h^2}$	$\dfrac{L}{h^{n+1}}$	$\dfrac{1}{h^2}$	$\dfrac{L}{h^2}$	$\dfrac{1}{h}$	L	Constant	$\dfrac{L}{h}$
Case 4 Constant shear rate $N \propto \dfrac{h}{D}$	D^3N^2	D^2NL	D^2NL	$\dfrac{L}{DN}$	$\dfrac{L}{DN}$	$\dfrac{1}{h}$	$\dfrac{L}{h}$	Constant	$\dfrac{L}{DN}$	$\dfrac{L}{DN}$	$\dfrac{L}{D^2N^2}$
	$\dfrac{Dh^2}{N}$	$\dfrac{DhL}{N}$	$\dfrac{h^2L}{N}$	$\dfrac{L}{h}$		$\dfrac{1}{h}$	$\dfrac{L}{h}$		$\dfrac{L}{h}$	$\dfrac{L}{h}$	$\dfrac{L}{h^2}$

Case 4a)										
$h \propto D$	D^3	D^2L	D^2L	$\dfrac{L}{D}$, $\dfrac{L}{h}$	$\dfrac{1}{D}$, $\dfrac{1}{h}$	$\dfrac{L}{D}$, $\dfrac{L}{h}$	Constant	$\dfrac{L}{D}$, $\dfrac{L}{h}$	$\dfrac{L}{D}$, $\dfrac{L}{h}$	$\dfrac{L}{D^2}$, $\dfrac{L}{h^2}$
N constant										
Case 4b)										
$L \propto h$		Dh^2	Dh^2	Constant	$\dfrac{1}{h}$	Constant	Constant	Constant	Constant	$\dfrac{1}{h}$, $\dfrac{1}{L}$
$L/D \propto h/D \propto N$		DL^2	DL^2							
Case 4b(i)										
$L \propto h \propto \sqrt{D}$		D^2	D^2	Constant	$\dfrac{1}{\sqrt{D}}$	Constant	Constant	Constant	Constant	$\dfrac{1}{\sqrt{D}}$
$N \propto 1/\sqrt{D}$										
Case 4b(ii)										
$L \propto h \propto D^{1.5}$		D^4	D^4	Constant	$\dfrac{1}{D^{1.5}}$	Constant	Constant	Constant	Constant	$\dfrac{1}{D^{1.5}}$
$N \propto \sqrt{D}$										
Case 5)										
$L \propto D$		$D^{2.5}N$	$D^{3.5}N^2$	$D^{n/2+3}N^{n+1}$	DN	$D^{n/2+\tfrac{1}{2}}N^m$	N	DN	$N\sqrt{D}$	$\dfrac{1}{N}$	$\dfrac{1}{N\sqrt{D}}$
$h \propto \sqrt{D}$										
Case 5a)										
$N \propto 1/\sqrt{D}$		D^2	$D^{2.5}$	$D^{2.5}$	$\sqrt{D}$	$\sqrt{D}$	$\dfrac{1}{\sqrt{D}}$	$\sqrt{D}$	Constant	$\sqrt{D}$	$\sqrt{D}$	Constant

melt pumping section is given by volume/Q or:

$$\text{Mean residence time} = \frac{bhZ}{Q} = \frac{bhL}{\left(\dfrac{Q}{Wbh}\right)\pi DN \cos\phi\, bh \sin\phi}$$

$$\propto \frac{L}{DN} \tag{8.4}$$

for constant Q/Wbh and helix angle ϕ. Shear rate and residence time together give a semi-quantitative measure of total shear strain on which distributive mixing depends. Output is given by a modification of eqn (5.43) for a fixed value of Q/Wbh:

$$Q \propto D^2 Nh \tag{8.5}$$

Mechanical power in the channel is given by eqn (6.22) which can be represented by:

$$E \propto \frac{\eta W^2 bZ}{h} \tag{8.6}$$

for constant Q/Wbh and helix angle ϕ.

For mechanical reasons, the flight clearance δ is approximately proportional to diameter D. Noting that h is usually increased as D increases, though not necessarily in proportion, if we take:

$$\delta \propto h \tag{8.7}$$

and the ratio $\eta_{\text{flight}}/\eta_{\text{channel}}$ is taken as constant (since shear rates are proportional to W/δ and W/h respectively), then eqn (8.6) also represents the total power given by eqn (6.27). Substituting in eqn (8.6) for W (eqn (5.4)), b (eqns (5.2) and (5.3)) and Z (eqn (5.1)) we can write:

$$E \propto \frac{D^3 N^2 L}{h} \tag{8.8}$$

in the Newtonian case. Dividing eqn (8.8) by eqn (8.5) gives:

$$dT \propto \frac{E}{Q} \propto \frac{DNL}{h^2} \tag{8.9}$$

For a pseudoplastic fluid, from eqn (2.7):

$$\eta \propto (\dot{\gamma})^{n-1} \propto \left(\frac{DN}{h}\right)^{n-1} \tag{8.10}$$

Substituting in eqn (8.6) gives:

$$E \propto \frac{D^{n+2}N^{n+1}L}{h^n} \tag{8.11}$$

and:

$$dT \propto \frac{E}{Q} \propto \frac{D^n N^n L}{h^{n+1}} \tag{8.12}$$

Taking first the Newtonian case, if channel depth h is maintained constant, output only increases as D^2, i.e. a tenfold increase in output would require an extruder more than three times the diameter, with considerable increase in capital cost. However, the power input (and motor rating!) would increase as D^3 and the specific power would increase in proportion to diameter, speed and length, resulting in very much higher temperatures or dramatic changes in heating/cooling. In this case it is unnecessary to increase length to maintain mixing or die pressure, so L may be regarded as constant (decreasing the number of turns). However, if speed is decreased ($N \propto 1/D$) to maintain E/Q constant, the output and power increase only linearly with diameter D, which is evidently uneconomic. Scale-up at constant depth is thus extremely limited, by either excessive temperature rise or small increase in output.

A more common strategy is to increase all dimensions in the same proportion, i.e. geometric similarity; length L is proportional to diameter D and thus the number of pitches (turns) remains constant. Since channel depth h is also proportional to diameter, output Q increases as D^3N and mechanical power E increases as D^3N^2. The adiabatic temperature represented by E/Q is thus independent of D but increases with speed N as before (Fig. 6.11). Shear rate, residence time and mixing are independent of diameter, and mixing is also independent of speed. The pressure gradient decreases with increase of diameter, giving the same final pressure due to the greater length, both gradient and pressure being proportional to speed. At first this appears very satisfactory, requiring only just over twice the diameter to give tenfold increase in output, with temperatures, pressures and mixing remaining constant. However, the heat energy required to melt ten times the output must be transferred through only 4·5 times the barrel surface ($Q \propto D^3$, surface $\propto D^2$) so that some limitation in melting rate, at least at the same screw speed, may be encountered. The same situation applies in the melt pumping section if melt temperatures appreciably above or below the adiabatic value are required, and in addition this heat energy must be conducted through a greater mean distance due to the deeper channel; thus heating will be slower and probably give rise to additional variations in final melt temperature. The overall consequences at constant speed are likely to be less complete melting, less homogeneous product and limitations on melt temperature above or below the adiabatic point. These in turn will set additional limits to the screw speed and output achievable before the output becomes too inhomogeneous or unstable for satisfactory production. The position can be ameliorated by a reduction in speed, with consequent reduction in output and pressure; however the accompanying reduction in adiabatic temperature may exacerbate the problem if high melt temperatures are required. Another possibility is to increase the number of turns in the melt pumping section (L increases more than in proportion to D), giving higher power E and adiabatic temperature as well as higher pressure and mixing—see Fig. 6.5.

This case illustrates the ramifications of the scale-up problem and the difficulty of reproducing all aspects of small-scale performance; the disproportion between surface and volume is inherent.

Of the third case, where peripheral speed is constant, it is only necessary to point out that it implies that increased diameter is accompanied by decreased speed so that increases in output are very modest in relation to size and capital cost. By manipulation of length and channel depth it is, however, possible to hold adiabatic temperature, mixing and pressure approximately constant (but see Case 5).

Since shear rate is significant in both power absorption and mixing, Case 4 considers constant shear rate; this implies speed $N \propto h/D$, which, if h/D is constant is similar to Case 2a, geometric similarity at constant speed, but with L/D ratio unspecified. If L/h is constant, adiabatic temperature, residence time, mixing and pressure will remain constant. Other factors then depend very strongly on the choice of channel depth h; if $h \propto \sqrt{D}$, speed is reduced as $1/\sqrt{D}$ and output is proportional to D^2, but L is also proportional to $\sqrt{D}$ so that heat ($\propto D^2$) is transferred through a barrel surface proportional to $D^{1.5}$, but still through an increased channel depth. If channel depth h is increased more than in proportion to D, say as $D^{1.5}$, speed increases as $\sqrt{D}$ and output as D^4. Length is also increased as $D^{1.5}$, giving barrel surface proportional to $D^{2.5}$, but with severe limitation on heating or cooling.

A compromise frequently followed in practice approximates to Case

5, where the number of turns is constant ($L \propto D$), and $h \propto \sqrt{D}$. At constant speed, output increases as $D^{2.5}$ and E/Q, representing adiabatic temperature, is proportional to D. Mixing increases as $\sqrt{D}$ and pressure as D. Barrel surface increases as D^2, so that heat conduction problems will be slightly worse. If however output is sacrificed by a modest reduction in screw speed—say $N \propto 1/\sqrt{D}$, shear rate becomes constant and E/Q, residence time, mixing and pressure all increase as $\sqrt{D}$. Output is now proportional to D^2, as is barrel surface, so that the only factor in heat conduction is a slight ($\propto \sqrt{D}$) increase in the mean radial distance for conduction through the polymer.

Since the shear rates at laboratory and production scales are the same in Cases 2a, 4 and 5a, the same results will obtain for pseudoplastic polymers. In other cases eqns (8.10), (8.11) and (8.12) apply with the general result that increases in pressure and adiabatic temperature will be less than for Newtonian behaviour, making these factors less critical. As observed on p. 183 the effect of speed on adiabatic temperature will also be less, so that the scale-up problem is less critical with highly pseudoplastic materials.

The conclusion from Table 8.2 is that a strategy for scale-up similar to Cases 2a and 4b(i) represents a good compromise of all the factors considered. In specific cases the relative importance of these factors may be changed so that the strategy may be modified to suit the more important factors. It will be noted that increase of scale generally leads to higher specific power input and higher temperatures, as does increase of screw speed. Similar effects of diameter and speed are also anticipated in respect of difficulty in obtaining temperatures much above (or below) the adiabatic value, temperature uniformity and output stability. The last three, together with mixing, can be improved by increasing the length of the melt pumping section (compare Case 5a with 4b(i)), though at the expense of greater power input, shear history and mean residence time (for polymers liable to degradation) and higher adiabatic temperatures (where processing temperatures are low). For reasons of heat transfer, melting may also be less complete in large machines, though the effect is likely to be less than that of increasing screw speed; this may be corrected by adjustments to channel depth, by increasing the melting length (e.g. by reducing appropriate barrel temperatures) or increasing the proportion of mechanical work input (again by reducing barrel temperatures).

Chapter 9

PRACTICAL EXTRUDER OPERATION

9.1. STEADY OPERATION

The principles of steady operation of single-screw extruders have been developed in Chapters 5 and 6 and their implications for operation discussed in Chapters 7 and 8. In this chapter an attempt is made to cover more practical aspects of operation to complement the theory, including methods of achieving and maintaining safe and steady operation, starting-up and changing running conditions, shut-down and cleaning, and scrap recovery.

Unlike injection moulding, extrusion is fundamentally a continuous process and successful economic production depends on keeping steady conditions for long periods. Since both material and energy are used during start-up and changes due to material, operation or product, it is important that these are made economically and usually as quickly as possible. As discussed in Section 8.4, the single-screw extruder combined with a die is to some extent a self-regulating system, though compensation for changes in pressure or temperature may produce undesired changes in mixing or stability. However, the downstream operations of sizing, drawdown and cutting directly affect the product, and equilibrium between these and the extruder itself must generally be controlled by the operator. Instrumentation, preferably automatically recording, is valuable as a guide to incipient changes in conditions, which are frequently due to changes in feed material properties. However, this often does not include direct indication of melt pressure or temperature, and instruments provided may have a delayed response, for instance due to the thermal mass of the machine part to which they are attached; hence the need for an understanding of the mechanism of extrusion to avoid over-hasty or inappropriate 'corrective' action. It is thus wise to be cautious about deliberate alterations to the machine controls, and when the process is running steadily unnecessary stops or changes in speed or temperature should be avoided. It should also be appreciated that the reaction times of the process differ widely; a change in pressure will have an almost instantaneous effect on output and power input, but due to heat capacity of machine and polymer, temperature will respond more slowly, bringing consequent further changes to output, power, etc. Speed changes also produce rapid changes in output and power input, but more delayed changes in temperature, and as seen in Chapter 6, profound changes in energy balance may occur. Changes in set temperature profile will usually affect melt temperatures quite slowly, but with complex effects on energy balance, pressure and the melting mechanism. If changes in settings of pressure, speed, etc., must be made, they should generally be done one at a time and adequate time allowed for the system to come to near-equilibrium before further adjustments are attempted. Modern three-term temperature controllers are designed to respond automatically to wide changes of conditions, but the speed of their response must be a compromise with the avoidance of large overshoot or instability in the heat and flow mechanisms within the extruder. Sometimes the heat will shut off, even though the temperature is falling—they are usually programmed to do this if a thermocouple or connection breaks, as a protection against overheating the polymer. However, they can only act on the signal received—if a thermocouple is connected the wrong way round, when the machine is too hot, the controller will sense it is too cold and put more heat on, so that it gets even hotter—until something burns or breaks! If the wrong thermocouple is connected, the controller will sense it is too cold, and respond by putting more heat into the zone it controls, without any feedback to limit the temperature rise. The same will happen if the thermocouple does not touch the metal of the heater or extruder, or if it slips out. A simple check is to place the thermocouple tip on the floor or bench—the controller should read about room temperature; then if held in the hand it should gradually change to read 35–40°C, if it does not it is defective or wrongly connected. Sometimes repeated changes in output (surging) or temperature will take place over periods of several minutes—this is probably due to the temperature controllers being incorrectly adjusted, requiring expert attention.

With experience the operator may be able to anticipate the result of deliberate alterations and so save time and material, especially in setting temperatures, but it is always wise to watch for signs of overload or rapid fluctuations of the motor ammeter and pressure gauge (if fitted). The main requirement for steady operation of the extruder itself is a continuous feed of polymer—possible problems in the feeding mechanism have been dealt with in Section 5.10; however, it is easy when pre-occupied with problems in the downstream operations to forget to refill the hopper. A temporary reduction in output will upset downstream operations, possibly requiring rethreading to restore stable conditions, but if the screw has substantially emptied, care must be exercised in recommencing feeding to avoid pressure or power surges—see Start-up in Section 9.3. If the feed is piped or conveyed from a central store or compounding section, the operator only needs to watch the hopper level indicator for signs of overfilling or bridging. However, extruders are frequently fed direct from bins or sacks, and colour masterbatch or recycled scrap may also be under the operator's control. In filling the hopper, care should be taken to avoid including dust from the bottom of bags, dirt from the floor and solid objects dropped from clothing, all of which can contaminate the product. Care must also be taken to keep the feed material dry to avoid foaming and blistering in the product; moisture-sensitive polymers including ABS, PMMA, nylons and PET should be stored in closed containers as long as possible. Many polymers in powder form and dust will readily burn or even explode, so precautions such as 'no smoking', flameproof electrical gear and earthing strips for static electricity must be strictly observed, and spillage on exposed heaters avoided. Dust may also constitute a health hazard, so spillage should be minimised and cleaned up promptly by vacuum cleaner, not with brush or compressed air, and protective masks used where possible. Never reach into the hopper or use a metal rod, which may slip through the safety grid (which must always be in position) and damage the extruder screw. The main hazard during operation from the barrel is contact with hot surfaces, heaters or live electrics, including exposed terminals and wet cables. However, leakage from joints or fittings may indicate excessive pressures, due possibly to a blocked die or screen, leading to breakage of die fastenings and decomposition of polymer due to churning, giving gases which can violently eject hot polymer from the die. For this reason, or due to excessive temperatures or residence times, even when the screw is stationary, trapped gas, water vapour or degradation products may be ejected, so one should never stand in front of the die, and safety spectacles and, whenever possible, heat-resistant gloves should be worn when working near the die. Leakage may also lead to fire if the polymer touches electric heaters.

9.2. COLOUR AND GRADE CHANGING

A frequent cause of disturbance to steady running and good product is a change in the feed material; this may be due to variations in the raw material as supplied, to variations in compounding or rework, or feeding incorrect material. This can only be corrected by empirical and wasteful changes in operating conditions unless rigid quality control of the feed enables the cause to be discovered quickly. However, if the symptom is a sudden appearance of surface roughness or distortion in the product in an otherwise steady operation, it is worth checking the die temperature in case a die heater has failed.

If it is required to change from a low viscosity (high melt flow rate, MFR) to higher viscosity (low MFR) grade of the same polymer, this is readily achieved, since the higher viscosity usually ensures removal of the softer polymer. However, some mixing is likely to occur giving off-standard product, which may recur intermittently for some time. To minimise wastage, screw speed and output may be reduced temporarily. If the change of viscosity is considerable, this opportunity can be taken to increase barrel and die temperatures, so that drive power is not excessive when output is restored to its original value.

Changing from high to low viscosity is much more difficult, since the fresh material will tend to bypass the harder polymer, which is liable to break away intermittently when steady conditions have apparently been re-established. If it is not practicable to stop and remove the screw for cleaning, probably the best strategy is to remove any screw cooling, reduce speed and temperature to the minimum practicable to produce maximum shearing action for melting and mixing, stop feeding and allow the machine to empty as far as possible. Then the machine should be kept running for several minutes to allow the screw to warm up and soften any remaining polymer. Then feeding should be recommenced with the softer grade, slowly as in start-up, and the mixed polymer allowed to run to waste for say 5 min before attempting to thread up the haul-off. It may be desirable to repeat this procedure rather than keep running steadily, then watch for contamination when returning to desired operating conditions. The change may be assisted, though at some cost in material, by first feeding a blend of the original hard grade with much softer material in an attempt to match the viscosity of the new grade to be introduced.

Changing colour may follow a similar procedure, bearing in mind that different pigments and other additives will give different viscosities to their compounds. If detailed information on viscosity vs. temperature for each compound is available, it may be possible to programme the changes from softer to harder compounds, though for reasons of colour matching it is more usual to move from light to dark colours, through greens and blues to greys, or through oranges and reds to browns, before dismantling and cleaning for a return to white or pastel colours. Alternatively, since the change of viscosity with temperature may vary, it may be possible to operate temporarily in a temperature range in which the colour change is also from lower to higher viscosity.

When changing from one class of polymer to another, either temperature or shear rate (output) may be used to ensure the change is from lower to higher viscosity, especially for groups such as polyolefines, within which different polymers mix readily with each other. In changing between different groups such as polyethylene to PVC or vice versa, where little mixing occurs, factors such as adhesion to metal surfaces make generalisations impossible. Indeed these two polymers are frequently used in special formulations specifically for purging each other from extruders and other processing machines. However, as explained under 'partially filled screws' in Section 5.6, it is unlikely that the purging compound itself will be completely emptied and may cause contamination when production is resumed. Because of lack of mixing, such changes almost certainly lead to breaks in the haul-off line and scrap which cannot be re-used.

9.3. START-UP AND SHUT-DOWN

Steady operation requires a satisfactory start-up, and this requires care and often time. The operator should first check that there is a screw in the extruder and that it is pushed right back—the start of the thread should be in line with the back edge of the feed opening. Then he should check that the die and head bolts or clamps are correctly fitted and tightened, and that the thermocouples are all in position and connected to the control panel. If there is plastic in the hopper, the feed valve should be closed, the motor switch in the 'OFF' position and the speed set to the lowest position. The feed pocket cooling should be turned on and the screw cooling off. The heater controls should be set and switched on. Unless certain they are functioning correctly, it is a good idea to set them first all to 40 or 50°C; the controllers should then show they are supplying heat, the moving pointer should rise to the set point (40 or 50°C) in a few minutes and then the heaters should shut off. This checks that the thermocouples are connected, and the right way round, the heaters are functioning (sometimes they have separate power switches which must be 'ON') and that the controllers switch heat on and off as required; this only takes a few minutes but may save hours later and avoid damage due to overheating. It should then be safe to reset the controllers to the required temperatures and leave the machine to heat up. If temperatures near to the minimum for the polymer are required, it will reduce motor load at start-up if the controllers are first set 10 to 20°C *higher*, and then reduced to the proper setting when the extruder is running. On the other hand, if high temperatures are needed, or the polymer is very sensitive to heat (especially PVC), it is wise to set the controllers 10 to 20°C *lower*; when the temperatures have reached these settings, they may be raised to the proper temperatures until they are again steady before starting to run the extruder. When the temperatures have reached the correct point (which may take an hour or more for a large machine) the moving and fixed (indicating and set) pointers on all the controllers should be very close to the same temperature position. As a final check, and on controllers without an indicating pointer, the indicator light showing that the controller is supplying heat should either be flashing or coming on and off periodically. When quite sure that the extruder is at the correct temperatures, the haul-off should be in position and connected to power, water, air, etc., e.g. the wire threaded up in wire covering. With waterbaths, as in tube sizing, it is usually as well to move the bath a few inches from the die to give space for handling and avoid splashing water on the hot die. A tin with some water may be placed under the die for start-up material or the haul-off started running at minimum speed. While watching the motor ammeter for overload, and pressure gauge if fitted, the motor should be started at lowest speed *without* any polymer. If motor amperes or pressure get near the safe limit, stop the motor immediately and wait for it to heat further or investigate the cause. If molten polymer comes from the die, this indicates that it is clear and it is safe to feed polymer, but *slowly*. If no polymer comes from the die, the machine is probably empty and care must be taken until certain that the die is clear. The operator should drop powder or granules by hand a little at a time into the feed hopper, but never enough to fill the screw; this is because the polymer will travel quickly down an empty screw, without time to melt properly, and if then forced into the die when the screw fills, it may block the die with unmolten particles, bursting the joints or damaging the die. This is especially important with very narrow dies, e.g. for film, or where there is a large space between the screw and narrow part of the die, which may take several minutes to fill. When molten polymer is flowing steadily from the die, without excessive motor load, the feed may be gradually increased until the screw is full in the feed opening— it is then safe to fill the hopper or open the feed valve completely. To save waste, as soon as the polymer flows from the die, it can be fed through the haul-off—this is usually much easier when running slowly. Cooling water or air pumps and jets should then be started and the haul-off speed adjusted to give approximately the correct size of extrusion, waterbaths can be carefully moved closer to the die, and vacuum for tube sizing, air for tubular film blowing and cooling, etc., gradually supplied. When the whole line is running steadily, the extruder screw speed and the haul-off speeds should be increased to the running speed, approximately in step, while watching the main ammeter for overload. The speeds should then be adjusted to give correct dimensions as far as possible, and only then final adjustments made to the die settings—it is a waste of time to do it earlier since swelling and drawdown depend on temperatures and speeds. It is impossible in this volume to cover all the different types of haul-off equipment and end-product. The aims are first to get the right size and shape, then the right properties (opticals for film, wire adhesion for insulation, etc.), the right appearance and surface finish, and then to keep all these at the highest possible output.

To shut down the process in emergency only requires shutting off the feed and stopping the main motor. For a short break (say up to 2 h), feed pocket cooling and barrel heating should be left on. For heat sensitive materials such as PVC, company policy may be to reduce barrel temperatures while standing, or purge the barrel with a different polymer such as polyethylene. More usually, barrel and screw cooling, if used, should first be shut off and then speeds lowered to reduce the load on the motor. Barrel temperatures may then be reduced (say, 20 to 50°C) while checking on motor load, to delay decomposition of the polymer. Companies use different procedures and these should be followed carefully; usual methods are (i) feed through a purging compound and only when this has cleared all the working polymer from the die, the screw is stopped, (ii) close the feed valve and let the extruder pump itself out as far as possible, then stop the screw, (iii) close the feed valve and stop the screw immediately, leaving the extruder full of working polymer, which must be melted before restarting; this keeps air and moisture from rusting the inside of the machine and accelerating decomposition of the still-hot polymer. It is obviously unsuitable for polymers which will decompose under heat alone during lengthy heating up before the next run. Once the screw is stopped, the heat to the barrel may be shut off, but not if the machine is to be restarted in the next couple of hours or if it is to be dismantled (see Cleaning, Section 9.4). If there is still polymer in the feed end of the screw, feed pocket cooling should be left on until Zone 1 temperature has fallen below the melting/softening temperature.

Before leaving the machine, a prominent notice should be posted if electric supplies or the heaters are still on, or if parts of the machine are hot. Water supplies and pumps and air supplies should be turned off. Always *isolate* the motors and pumps so that they cannot be accidentally started.

9.4. DISMANTLING AND CLEANING

Generally extruders are best dismantled hot; it will save a lot of time and hard work if dismantling, and cleaning if necessary, is done at the end of a run, and with less risk of damage. If the machine is cold, the start-up and shut-down procedures should first be followed through. After shut-down (as above), the feed valve should be closed, cooling (except for feed pocket) shut off, and the motor turned off and *isolated* to prevent accidental starting. The operator should check that all heaters are 'ON' and temperatures high enough for the polymer to be quite soft, but not high enough for the polymer to decompose when exposed to the air for the time required for cleaning. Next he should make quite sure that no heaters have exposed terminals or bare wires. If covers or heaters have to be removed, say from the die, they should be switched off and unplugged before working on or near them; also any heaters should be switched off before removing thermocouples controlling them, to avoid overheating. Waterbaths, etc., should be moved to one side so that a space at least as long as the extruder screw is left clear. Safety gear, tools, cleaning materials and lifting gear should be placed handy and a bench, stout boards or other non-scratch, non-inflammable supports placed ready for hot metal parts. A metal bucket of cold water is useful to cool spanners, bolts and brushes. Gloves only delay the time when the heat gets through to the hands, and take as long to cool off, so if supporting a heavy die, or carrying it, use in addition a rubber-moulder's pad or a spare glove folded over. If hot plastic gets on the skin, it is instinctive to pull it off—this should be avoided if possible as the skin will stick to the plastic and cause an open wound. It will be very painful, but instead if possible plunge it in cold water and shake it about until cold—then get medical attention *immediately*.

The cooling pipes and any other connections not part of the heating system should now be disconnected. The aim is to keep each part hot as long as possible, so heaters should be disconnected one at a time, starting from the die exit. THIS IS THE MOST DANGEROUS PART OF EXTRUDER OPERATION, SO SAFETY PROCEDURES MUST BE FOLLOWED. When each heater or group is disconnected, the control thermocouple should be removed and tied back away from damage by tools or hot polymer. The heater cover should then be removed and *again* all exposed heaters checked for disconnection. Heaters are easily damaged by tools so should, generally, be removed first, with minimum flexing, using heat-resisting gloves. A mica-insulated heater (about 3 mm thick) need only be loosened by slackening screws in the joint and then sliding off endwise; if it cannot be slid off, it should not be flexed more than 6–12 mm (depending on size), but separated completely into halves or left until a later stage. A ceramic-insulated heater (about 12 mm thick) must not be flexed and must usually be separated into halves. The segmental ceramic type made up of separate pieces of rigid insulator surrounded by metal bands must be handled even more carefully—they will flex slightly, but the coiled heating element (like a radiant domestic fire) running between the insulators can be stretched and damaged. Cartridge heaters (rod type) should if possible be left in place. Heaters should not be handled by the cable, but by the joint flange or terminal box (not with tools). As soon as the heater is removed, the die should be dismantled one piece at a time. Before the next part is taken off, the first should be cleaned and placed carefully on a clean smooth surface, avoiding joint faces and parts which contact the polymer. Contact surfaces should be cleaned first and only with aluminium or brass tools, brass brushes or brass wool. Special care is needed with chromium-plated surfaces—if scratched they will spoil the product and be more difficult to clean next time. Rigid polymers like polystyrene and acrylic tend to crack off cold metal surfaces—others, including polypropylene, will peel off—the purpose in these cases of cleaning hot is to remove as much as possible while soft so that the remainder can be removed cold with as little force as possible—with accompanying reduction in damage to the die. Polypropylene sticks to hot metal while LD polyethylene sticks to both hot and cold metal—so it is better cleaned hot, while it is still soft. Rigid (unplasticised) PVC peels away quite cleanly from both hot and cold metal, provided the PVC has not degraded (decomposed) too far, and the metal is smooth and not rusty, but like polypropylene it is very stiff when cold, requiring more force and risk of damage. Plasticised PVC is similar, but being softer is probably better peeled off when cold. With most polymers, once cleaned and dry, the die only needs keeping dry or a light smear of grease. PVC is the exception—a clean, dry, warm die smeared with grease will (unless chromium-plated) rust under the grease overnight—because acids from the PVC soak into the pores of the metal, and these acids then pick up moisture from the air. The best remedy is that for cleaning army rifles and for the same reason; several litres of boiling water poured through the die, which is then allowed to dry and lightly greased—the water 'sweats out' and dissolves the acids and the heat ensures it dries quickly before being greased.

Tapered die entrances often hold a considerable amount of polymer; if this can be eased away from the metal at the wide end and then allowed to cool until it is slightly leathery, it can be gripped with a pair of pliers and *slowly* pulled out. Pulling makes it narrower, separating it gradually from the metal before it breaks off at the narrow end and releases a large piece; this is usually quicker and cleaner than poking or digging out the polymer. Any plug remaining at the narrow end can then be pushed through with a wooden, Tufnol or aluminium cleaning rod, preferably in the flow direction to avoid damage to the die lips. This should also be done slowly so that the polymer does not swell and wedge in the die. The die can then be brushed with a fine brass bottle-brush, preferably wetted with a little cold water to prevent clogging. If still hot enough, the die may be polished with a clean, strong cloth wrapped round the cleaning rod—if the die is too cold, the cloth will stick. If solvents are used, instructions should be followed, especially to avoid fire and breathing fumes. Solvent baths require special precautions, which should be followed carefully. On removal, die parts should be allowed to cool slowly out of draughts and supported away from cold metal to minimise distortion, but with fume removal. When cold, they should be kept dry and protected, e.g. with a film of grease, as the solvent leaves the surface very liable to corrosion. With dies having a separate core or mandrel (as for tube and tubular film) this should be cleaned as soon as removed since it will cool quickly and is difficult to re-heat out of the die. If the die's outer element has meanwhile cooled too much, it can be warmed on a laboratory hotplate, but quick work can often avoid this. Heating in an oven is not recommended; it is usually slow, produces fumes and drips of soft polymer in the oven, which may catch fire when the door is opened, and cleaning cannot continue while the die is heating up. Burning off in a furnace or with a gas torch should only be used as a last resort, e.g. if the polymer has already decomposed to a hard char—the burnt plastic residues are hard to remove, the metal surface will be roughened with 'scale' and die parts may be distorted by uneven heating or cooling and rendered useless.

When the die is fully dismantled and cleaned, it should whenever possible be left assembled mounted on the extruder or on a bench to minimise the possibility of accidental damage. The screenpack, breaker plate and die adapter can then be removed and cleaned in the same way, except that the screenpack will probably be replaced. Readymade packs with a soft metal binding are expensive, but much better than ones made from loose pieces—the soft edge makes a good seal, the finer gauzes cannot be displaced or trapped at one edge by the spigot, and correct combination of gauzes is ensured; they must still be replaced the right way round so that the fine gauzes are supported by the coarser ones from being pushed into the breaker-plate holes by the pressure from the extruder. If the screenpack is to be re-used, do not forget that it may contain 'gels' of hard plastic which will cause 'fish-eyes', etc., in the product, as well as dirt, fibres, burnt plastic, etc.

Next, the polymer at the end of the screw should be cleaned out and from the joint face ready for re-assembly of the adaptor and die. If the screw is also to be cleaned or changed, then a screw or hydraulic jack designed for the job should be used—a hammer is ineffective and liable to damage the machine. With twin-screw machines, maker's instructions must be followed, since the two screws usually interlock. If the screw is pulled out, the screw nose will usually be removed first so that the puller can be inserted. If the screw is pushed from the back, the screw-cooling rotary union and 'wand' (inner tube) must be removed and possibly the outer tube. This tube and the thread for it in the back of the main screw are much more easily damaged than the screw itself. Any locking ring or screw on the drive shaft should be released and the gland slackened at the back of the feed pocket. The screw should now be pulled or pushed forward, not more than to expose two turns. If the screw will not move or sticks at some point make sure (i) it is not locked back, e.g. by the screw cooling tube, (ii) it is not fouling, e.g. on a register ring or pressure transducer at the front, a drive key hitting the back of the feed pocket or solid polymer jammed in the feed pocket, (iii) temperatures are high enough to melt the polymer, (iv) it can be pushed back into its running position, and (v) it can be turned (by barring the motor by hand if still engaged with the drive shaft). The exposed turns of the screw should be stripped of polymer and thoroughly cleaned. Except on chromium-plated screws, steel scrapers can be used; a wallpaper or paint scraper with rounded corners and width to fit easily within the screw channel is convenient, but not harder than spring-temper. This may be frequently dipped in cold water to reduce sticking to the polymer and a steel wire handbrush similarly wetted will remove the remaining plastic without clogging. A strong clean cloth used like a strap may be used for a final polish, but first make sure that the corners of the channel are free from hard or decomposed plastic. The screw may then be pushed forward another two or three turns and the cleaning process repeated before it cools. The process is then repeated as necessary, ensuring that the polymer is removed while still soft and the part of the screw being cleaned is well supported by the barrel. This method is excellent for LD polyethylene but for polymers which do not stick to cold metal, it may be preferable to scrape off the bulk while hot, let the screw go quite cold and then peel off the thin film remaining. When the feed section is reached, the unmolten powder or granules should just brush off. Pull the screw right out and place it either flat on a wooden bench or vertically in a rack to cool naturally to avoid warping.

Only now should the barrel heaters be switched off, and while still hot the barrel should be cleaned with a steel wire flue brush welded to a handle at least 1 m longer than the barrel. It should be worked up and down with a screw action until the barrel is shiny throughout when viewed from the die end. Take care not to push the brush past the feed opening, where it may damage the rear seal. With a gentle blast of compressed air in the feed pocket, any loose particles of polymer or broken wire from the brush should be blown out. The barrel should be inspected again for cleanliness and finally the bore and face of the drive shaft checked for damage and that the drive key(s) is in position in the shaft or screw.

When re-assembling, it should be checked that keys are in position, that they are aligned with keyways in both shaft and screw, and that the screw is tight against the end of the shaft—this is important if mixing devices are fitted as otherwise the screw may foul or move back when the extruder is started. The correct position of the front of the screw relative to the barrel sealing face should also be checked. The feed pocket seal, if fitted, should be re-adjusted, the screw nose refitted and the screenpack (right way round) and the breaker plate. Before refitting the adaptor or die, it is as well to make sure there is a screw in the machine!

Rapid and effective cleaning can be aided by careful design. In one instance a production team required 32 h to strip, clean and re-assemble a 114 mm (4·5 in) extruder with a hot-melt cutter. An alternative design of cutter, incorporating a hinged door in line with the screw and several other aids including mountings for screw-puller gear, hooks for the dismantled heaters, flexible hydraulic hoses and plug connectors for heaters, was demonstrated to the production management in which two men completed the operation, from shutting down the extruder and cutter to operation at full output in 8·5 min, using procedures described in this chapter. This undoubtedly owed much to the experience and organisation of Mr J. Whitenstall as a Naval Chief ERA, but would have been impossible without the practical understanding and careful design of Mr B. Tunnicliffe.

9.5. WASTE RECOVERY

Scrap always represents a waste, usually in several ways, and often results in a waste of expensive raw material. Even if the material can be fully recovered, this requires extra energy, labour and cost. It also represents a waste of energy in the main process and a loss of production capacity—production which otherwise could have been sold at a profit. Scrap is of several types:

(i) Edge-trim, 'ends', etc., produced as part of the process.
(ii) Raw material spillage.
(iii) Clean undecomposed product with wrong dimensions, breaks,

incompletely melted or poor surface, etc. This results from wrongly compounded material, incorrect extrusion conditions (including running too fast—see Operational strategies, Section 7.5), unsteady operation, incorrect haul-off or cooling conditions.

(iv) Product containing slight contamination due to dirt, colour or grade changes, purge material, decomposition due to excessive temperatures, etc.

(v) Material in strips or long lengths produced during start-up or output changes, clean but of incorrect dimensions and surface, and possibly incorrectly melted or compounded.

(vi) Lumps and irregular pieces produced during start-up, shut-down, or jams in the haul-off, clean or dirty.

(vii) Purgings, cleanings, machine leakage, spillage on heaters and other hot parts, pieces 'lost' in water system, cutters, etc. (often grease or dirt-contaminated).

Scrap (i) should be totally recoverable for sale or re-use. Cleanliness is paramount and it is better to waste a little more than risk spoiling a whole batch. Sacks, bins and granulators must be scrupulously clean and it is better to run a trial first. If re-used, check operating instructions for permitted proportions as even a different particle shape may affect feeding or product properties, may be changed.

Scrap (ii) should be avoided or minimised at all costs as it is a direct addition to raw material cost. In this state it is easily contaminated and impossible to clean—only recover what is certainly clean and sweep up the remainder immediately—it is a hazard for falls and possibly fire.

Scrap (iii) should be recoverable as (i) if quantities and shape justify the cost of granulators, storage, etc. As in (i), check on quantities permitted for re-use.

Scrap (iv) should be stored separately and protected from further contamination if quantities are likely to justify recovery. It should only be regranulated if the granulator can be cleaned afterwards for use on (i), (iii) or (v). If so, it may be useful for trials, purging, start-up, etc., but contamination of the extruder must always be watched.

Scrap (v) can be treated as (iii) if of convenient shape and sufficient quantity.

Scrap (vi) and (vii) cannot easily be recovered, so should be kept separate from other scrap and disposed of as soon as possible, but not mixed with metal, paper, etc., because of risks of injury and fire.

Granulators are dangerous; watch should be kept for loose parts or insecure guards and instructions should be observed. If the machine jams and does not clear in 5–10 s, do not push in more scrap but switch off, wait for it to stop, and follow cleaning instructions. While cleaning or dismantled, pull the plug or make sure it cannot be switched on accidentally—do not rely on safety switches. Even without power, cutters are very sharp and have high momentum—they will cause severe injury even when turned by hand. The corners and crevices in the granulator are difficult to clean and some materials may smear on the blades or casing, only to break off later and spoil another batch. After cleaning, a trial run is always worthwhile to ensure correct working and clean product. Granulators produce dust and noise, so are often in a separate room. The dust attracts static electricity and so dirt, and is often highly inflammable—it also hinders refeeding to the extruder, so is best disposed of. Dust masks and/or ear muffs should be worn as instructed.

Compounding, of either powdered polymers or colours, etc., into granules is also often dangerous, dusty and noisy, and wherever possible should be carried out in a separate room; safety procedures for cleaning are especially important. Contamination is at least as great a hazard as in the extrusion operation, especially if scrap is being recovered. This may arise from spillage, dirty containers, cross-contamination of one batch or colour by another, by water or moisture in the atmosphere.

Chapter 10

APPLICATION TO THE INDIVIDUAL MACHINE

It would be easy to call for more data in every aspect of extruder operation; reliable data, especially on larger machines, are always valuable in verifying and extending generalised predictions. The trade processor, however, needs more immediate information to obtain the best quality product and then maximise the efficiency of production, usually by maximising output without significant loss of product quality. The following ten items are proposed to form a basis for satisfactory operation; most of them use his own equipment, and are to some extent specific to it. Some require instruments not commonly found on production machinery, but well worth installing for a limited period to obtain data important for product quality and uniformity.

1. Screw dimensions
As mentioned in Chapter 5, the basic screw dimensions of diameter, pitch, flight width, channel depth and length permit calculation of the volumetric drag flow at any speed by means of eqn (5.13). Comparison of the actual flow rate with (twice) the drag flow gives the value of Q/Wbh and hence the sign and approximate magnitude of the pressure gradient, which coupled with the (helical) length gives an indication of the total pressure. This dimensionless flow rate also gives an indication of the relative specific power input and the proportion of energy contributed by internal shearing. Most important, it indicates the general position in the operating field and so the likely consequences of any deliberate change in operating conditions.

2. Flow properties
As suggested in Section 8.4, unforeseen changes in flow properties of feed material are a common cause of unprogrammed changes in

operation and in product quality; equipment for routine testing of melt flow rate is therefore essential. This equipment can be used without modification to obtain the melt viscosity of the polymer at low shear rates and over a range of temperatures, representing conditions within the screw channel. This permits estimation of the effect of melt temperature on mechanical power input E, and with the specific heat (see 3 below), estimation of the change in adiabatic temperature or heater power input H (Fig. 6.8).

The standard equipment for determination of melt flow rate is restricted to a fairly narrow range of low shear rates, so gives little indication of viscosity at the high shear rates existing in the die and in the screw flight clearance. Capillary viscometers designed for these higher shear rates are more expensive, but can be used, with due precaution, to obtain melt flow rates as well as a wide range of much more informative data. Values of viscosity in the die permit estimation of die pressure drops and hence the balance between screw and die characteristics represented by the factor $(1 + B\eta_{die}/K\eta_{screw})$ on p. 103. Viscosity in the flight clearance permits estimation of the balance between mechanical power in the screw channel (eqn (6.22)), which varies with back pressure, and power in the flight clearance (eqn (6.26)), which is independent of pressure. The high shear viscometer also gives the variation of viscosity in the die with temperature—usually less than that at low shear rates applicable to the screw channel (p. 8). But the high shear viscometer not only gives the viscosities appropriate to different parts of the process, but by calculation of the pseudoplasticity index n gives the correct shape of the (isothermal) non-Newtonian flow curves for screw and die (Fig. 5.35) and also the curve of mechanical power E vs. speed (Fig. 6.10). The former gives more accurate estimates of the effects of speed change on output and pressure while the latter permits estimation of the effects of speed on mechanical power and energy balance or temperature change.

3. Thermal properties

Data on the specific volume of the polymer at processing temperature are essential to convert volumetric flow rates calculated in 1 above to mass flow rate and also in estimating energy requirements which are based on mass. The enthalpy content of the polymer at processing temperature I is a necessary factor in the energy balance (eqn (6.2)). As a simple guide, enthalpy multiplied by desired mass flow rate gives the mechanical power required at the adiabatic temperature, and an overestimate of that required at higher melt temperatures; this gives a rough guide to the maximum output achievable with a given drive motor, disregarding limitations of product quality. Enthalpy of the polymer is also an important factor in deducing the changes in mechanical power and heating/cooling requirements consequent on deliberate changes in temperature, screw speed or back pressure. However, for small changes from an established datum, the mean specific heat in the melt state is a sufficient approximation.

4. Heat losses
In Appendix C the measurement of heat losses from the machine is shown to be a simple, once-for-all test for each physical set-up of barrel length, adaptor and die type and heaters. It indicates the standby losses and, with the polymer enthalpy (3 above), the total energy input required and its variation with temperature (Fig. 6.8). These permit construction of the energy balance and determination of the maximum heater energy required to maintain a specified combination of output (speed) and temperature. Heat losses are also significant in the difference between adiabatic and autogeneous temperatures (or speeds); Figs 6.8 and 6.10 illustrate how these losses permit a lower melt temperature or higher screw speed than in the absence of losses, before intentional barrel cooling must be employed.

In addition to the physical set-up, the usual feed pocket cooling rate and inlet temperature should be set, the screw fitted (for heat conduction along the screw) and ideally the screw channel should be filled with polymer (for heat transfer to screw). Screw cooling should not be used and ambient temperature will have a slight effect on the general level of losses. The main variables are the (set) temperature level (say at the end of the barrel, approximating the desired final melt temperature) and the temperature profile along the barrel.

5. Mechanical power input
Measurement of output Q and mechanical power input E vs. speed N for fixed temperatures gives the basic performance of the screw with the polymer used. The output and screw dimensions (see 1 above) give the value of Q/Wbh and any changes with speed. The power input indicates whether operation is possible at the given temperature and, since power is non-linear with speed, whether the rating of the drive motor limits the maximum screw speed. Power measurement will also indicate (by extrapolation if necessary) the adiabatic and autogeneous speeds for the given temperature (Fig. 6.10). Since mechanical power

E increases more than in direct proportion to speed N, the specific power E/Q will increase with speed. However, the *rate* of increase will indicate how closely the viscosity approaches Newtonian behaviour (Table 6.2 for ηN), including the effect of power absorbed in the flight clearance. The trial may be carried out either at fixed back pressure or with fixed die dimensions. The former is perhaps more useful for academic investigations, while the latter is closer to the usual industrial situation. As shown in Section 5.5, in this case output is approximately proportional to screw speed and thus Q/Wbh is nearly constant. As suggested previously, this is a useful basis for comparing performance, both in respect of velocity and shear rate distribution in the channel, and hence of mixing and shear heating, and also of longitudinal pressure profiles (see Figs 5.11, 5.12 and 5.13). Constant Q/Wbh also gives wall shear rate proportional to speed (eqn (5.98)) and a constant value to the square bracket in eqn (6.22). Thus the change in power is due only to the direct effects of speed and viscosity, and for a power-law approximation gives a constant proportion between the power in the channel and in the flight clearance (the two terms of eqn (6.27)). The test at fixed die dimensions also avoids the use of a pressure transducer and manual adjustment of back pressure.

6. *Power versus temperature*

The preceding test would naturally be carried out at the desired operating temperature and need only be of sufficient duration to allow steady-state temperatures to be re-established at a few speed settings. It may be that the desired final melt temperature is unknown or that a range of temperatures or polymer viscosities may be used. This then requires the tests to be repeated at, say, three temperatures, though the changes are likely to be such as to permit linear interpolation for other temperatures within the range. This also gives an indication of the change in mechanical power input with temperature at a given speed, from which the heating/cooling power for other temperatures can be deduced (Fig. 6.8). It will also indicate the sensitivity of the system to small deviations in temperature and help fix the allowable tolerances for satisfactory process control (Table 8.1). Possibly even more important would be measurement of mechanical power at a fixed speed and final melt temperature, but varying the temperatures in the melting section only. As mentioned in Table 7.2, the relative contributions of conducted and shear heating will affect the mechanism of melting and possibly the position and temperature variation at break-up of the solid bed, with the consequences for subsequent elimination of the remaining solid particles.

If direct measurement of power is impossible, ammeter readings will give qualitative indications and a basis for resetting conditions—with the provisos for older types of drive mentioned in Section 6.2.

7. Limiting outputs

In the preceding tests, product quality has been largely neglected. As a first step, the maximum screw speed and corresponding output should be found for each final melt temperature and for several practicable temperature profiles (see Fig. 7.8). This may be on the basis of visually distorted or unmolten particles at the die or on an accepted criterion of acceptable product dimensions or quality; in the latter case assessment may be confused by consequential changes in downstream processes, such as cooling and drawing. The effect of a barrel temperature profile is particularly informative, since a profile giving a higher limiting output on any of the above criteria is also likely to permit higher output on any more severe criterion such as melt temperature variation, which cannot be so easily determined.

8. Space variation of melt temperature

This requires an array of melt thermocouples and/or a travelling thermocouple beyond the end of the screw and at various radial positions (see Section 8.1). Because of possible damage, it is suggested that these are only installed for the duration of these tests, but that they are of fairly robust construction (say, 3 mm diameter) since a time-averaged temperature is required. As described in Section 8.1, the profile appears to be characteristic of the screw type, but to amplify with screw speed. It is not clear to what extent it is dependent on the polymer used, or on the barrel temperature profile. The former may be of no consequence to a processor using a single polymer, but the latter would extend the information from 7 above. The two significant factors appear to be the average temperature in the melting section, and hence subdivision between conducted and shear heating (see 6 above), and the temperature rise in the melt pumping section, i.e. the excess of the final melt temperature over the softening/crystalline melting point.

The data from this series of tests would give direct evidence of lack of uniformity in the melt at the die, give warning of incipient incomplete melting and contribute to understanding of non-uniform flow in the die, drawdown problems and non-uniform properties in the product. It is basic to maintenance of product quality and could be combined with 6 above.

9. Movement of 'gel' point

This requires initial provision of at least three tappings in the barrel for pressure transducers, the latter being fitted only for these tests. The movement of the position of initial pressure rise with speed would be useful in any calculations of performance requiring a value for melt length z, and especially in modifying the effects of screw speed discussed in Section 6.6, where melt length was assumed constant. It will also indicate the melt length available for reducing temperature variations originating in the melting process and possibly the onset of incomplete melting.

Movement of the gel point with changes in temperatures in the melting section would help explain the effects on final temperature variation found in 8 above.

This test would be important in finding the correct temperature settings for the melting section, as well as its location, and might usefully be combined with 6 above.

10. Effect of back pressure on melting

This requires the same instrumentation as the preceding test, with the addition of an adjustable restrictor at the die. The aim would be to find the movement of the 'gel' point with increase of back pressure at constant speed and temperature. A movement towards the die might imply a delay in break-up of the solid bed, but also an increase in final melt temperature variations. The decrease in melt output rate should ease melting requirements and might permit a higher speed before surging or unacceptable temperature variations set a limit to output. Thus increased back pressure by means of screen packs might improve performance, or it might indicate improved performance from a shallower screw (see Table 7.2).

The last three of these tests require special equipment and may be considered as 'research' by industrial operators. They are included as giving direct indication of factors affecting product quality. The whole series of tests has been deliberately arranged in order of effort required, recognising that many processors will not have the wish or opportunity to complete them all. However, the author believes that each will contribute valuable information and together they will form a sound basis for improved operation and control; indeed they will provide much of the process model necessary for the most sophisticated computer control. Much of the foregoing is specific to a particular machine, screw, polymer, and die arrangement and lacks the generality for extension of the principles of operation and control. The author would select just five areas for more fundamental study as likely to make the greatest contribution to understanding these principles:

(1) Factors affecting the melting process, including the position of break-up of the solid bed, and the resulting non-uniformity of temperature.
(2) Actual comparative performance including 'quality' factors of different screw lengths, depths and profiles for several commercial polymers over a range of speeds, temperatures and back pressures.
(3) Effects of barrel and screw cooling on energy balance, mixing and final temperature variations.
(4) Factors affecting time fluctuations of final temperature as an indication of origins of instability and poor quality.
(5) Development of a *general* process model for single screw extrusion, including 'quality' factors, for use in improved overall process control.

Appendix A

ALTERNATIVE DERIVATION OF FLOW EQUATION

In a system of rectangular coordinates x, y, z, where the local velocities are respectively u, v, w, if there is a pressure gradient $\partial P/\partial z$ in the z direction only, the general flow equation for an incompressible fluid reduces to the Navier–Stokes equation:

$$\eta \frac{\partial^2 w}{\partial x^2} + \eta \frac{\partial^2 w}{\partial y^2} = \frac{\partial P}{\partial z} \qquad (A1)$$

where η is the local dynamic viscosity.

If the width b of the channel in the x direction is much greater than depth h in the y direction, the shear stress and shear rate $\partial w/\partial x$ in the x direction are constant and:

$$\frac{\partial^2 w}{\partial x^2} \simeq 0 \qquad (A2)$$

Then approximately:

$$\eta \frac{\partial^2 w}{\partial y^2} - \frac{\partial P}{\partial z} = 0 \qquad (A3)$$

If viscosity η is independent of shear rate $\partial w/\partial y$ (as for a Newtonian fluid) and integrating with respect to y:

$$\eta \frac{\partial w}{\partial y} - y \frac{\partial P}{\partial z} = C \qquad (A4)$$

and integrating again:

$$\eta w - \frac{y^2}{2} \cdot \frac{\partial P}{\partial z} = Cy + D \qquad (A5)$$

Note that if viscosity is dependent on shear rate, this cannot be simply integrated.

In an extruder screw, taking velocities relative to the screw, the boundary conditions are:

$$\text{at } y = 0, \quad w = 0$$
$$\text{and } y = h, \quad w = W$$

Substituting in eqn (A5) gives $D = 0$ and:

$$C = \eta \frac{W}{h} - \frac{h^2}{2h} \cdot \frac{\partial P}{\partial z} \tag{A6}$$

Putting this value of C in eqn (A4) gives:

$$\frac{\partial w}{\partial y} = \frac{W}{h} + \frac{(2y - h)}{2\eta} \frac{\partial P}{\partial z} \tag{A7}$$

and eqn (A5) becomes:

$$\eta w - \frac{y^2}{2} \frac{\partial P}{\partial z} = \eta \frac{Wy}{h} - \frac{yh}{2} \frac{\partial P}{\partial z}$$

or:

$$w = \frac{Wy}{h} + \frac{y^2 - yh}{2\eta} \frac{\partial P}{\partial z} \tag{A8}$$

The volumetric flow rate is given by:

$$Q_{\text{Tot}} = \int_0^h wb \, dy$$
$$= \left[\frac{Wby^2}{2h} + \frac{b}{2\eta} \cdot \frac{\partial P}{\partial z} \left(\frac{y^3}{3} - \frac{y^2 h}{2} \right) \right]_0^h$$
$$= \frac{Wbh}{2} - \frac{bh^3}{12\eta} \frac{\partial P}{\partial z} \tag{A9}$$

which is identical with eqn (5.17).

If the pressure gradient $\partial P/\partial x$ in the x direction is considered, eqn (A1) is transformed by change of coordinates as:

$$\eta \frac{\partial^2 u}{\partial y^2} + \eta \frac{\partial^2 u}{\partial z^2} = \frac{\partial P}{\partial x} \tag{A10}$$

eqn (A7) becomes:

$$\frac{\partial u}{\partial y} = \frac{U}{h} + \frac{(2y-h)}{2\eta} \cdot \frac{\partial P}{\partial x} \qquad (A11)$$

eqn (A8) becomes:

$$u = \frac{Uy}{h} + \frac{y^2 - yh}{2\eta} \cdot \frac{\partial P}{\partial x} \qquad (A12)$$

and eqn (A9) becomes:

$$Q_x = \frac{Uzh}{2} - \frac{zh^3}{12\eta} \cdot \frac{\partial P}{\partial x} \qquad (A13)$$

Appendix B

PRESSURE GRADIENTS IN A STEPPED SCREW

In a stepped screw, having sections of channel depth h_1 and h_2, the pressure gradient dP/dz will generally be different in the two sections. Since the flow equation (eqn (5.17) and (A9)) is a cubic in h, it is not possible to state unambiguously which is the greater, dP_1/dz or dP_2/dz. The objective here is to determine the conditions in which dP_1/dz is greater or less than dP_2/dz. Since downstream velocity W and channel width b are commonly the same in both sections, and volumetric flow rate Q is the same by continuity, pressure gradients can be represented by the (local) dimensionless flow rates Q/Wbh_1 and Q/Wbh_2, which are inversely proportional to the channel depth. When $Q/Wbh_2 > 0.5$, the pressure gradient dP_2/dz is negative and when $Q/Wbh_2 = 0.5$, $dP_2/dz = 0$; in both cases the whole of the pressure is raised in the first section, where if $h_1/h_2 > 1$, $Q/Wbh_1 < 0.5$. The effective melt lengths z_1 and z_2 will differ in different screw designs, and the former also with the position of the point of initial pressure rise (gel point). However, this study shows that, except over a very small range of flow rates ($Q/Wbh_2 \simeq 0.5$), the pressure gradient in the shallow second section is greater than in the deeper first section and thus unless the second section is very short, most of the pressure before the die will be generated in this section. In Table B1 the values of Q/Wbh and dP/dz are calculated for three ratios of h_1/h_2, viz. 2, 3 and 4, and in Fig. B1 the ratio $dP_2/dz : dP_1/dz$ is plotted against the respective dimensionless flow rates. If eqn (5.18) is rearranged, it gives:

$$\frac{1}{2} - \frac{Q}{Wbh} = \frac{h^2}{12\eta W} \cdot \frac{dP}{dz} \tag{B1}$$

TABLE B1
Pressure Gradients in a Stepped Screw

$\dfrac{h_1}{h_2}$	$\dfrac{Q}{Wbh_1}$	$\dfrac{Q}{Wbh_2}$	$\dfrac{dP_1/dz}{\times \dfrac{12\eta W}{h_2^2}}$	$\dfrac{dP_1/dz}{dP_1/dz_{(P_2=0)}}$ Fig. B2	$\dfrac{dP_2/dz}{\times \dfrac{12\eta W}{h_2^2}}$	$\dfrac{dP_2/dz}{dP_1/dz}$ Fig. B1
2	0·25	0·50	0·0625	1·00	0	0
2	0·22	0·44	0·070	1·12	0·060	0·858
2	0·21	0·42	0·0725	1·16	0·080	1·103
2	0·20	0·40	0·075	1·20	0·100	1·333
2	0·15	0·30	0·0875	1·40	0·200	2·285
2	0·10	0·20	0·100	1·60	0·300	3·00
2	0·05	0·10	0·1125	1·80	0·400	3·56
2	0	0	0·125	2·00	0·500	4·00
3	0·167	0·50	0·0370	1·0	0	0
3	0·160	0·480	0·0378	1·02	0·020	0·530
3	0·155	0·465	0·0383	1·036	0·035	0·914
3	0·150	0·450	0·0389	1·050	0·050	1·286
3	0·120	0·360	0·0422	1·14	0·140	3·32
3	0·100	0·300	0·0444	1·20	0·200	4·50
3	0·050	0·150	0·050	1·35	0·350	7·00
3	0	0	0·0556	1·50	0·500	9·00
4	0·125	0·500	0·0234	1·0	0	0
4	0·120	0·480	0·0238	1·014	0·020	0·842
4	0·115	0·460	0·0241	1·027	0·040	1·661
4	0·110	0·440	0·0244	1·04	0·060	2·462
4	0·100	0·400	0·0250	1·068	0·100	4·00
4	0·070	0·280	0·0269	1·15	0·220	8·17
4	0	0	0·0312	1·33	0·50	16·0

Thus

$$\frac{dP_1}{dz} = \left(\frac{1}{2} - \frac{Q}{Wbh_1}\right) \frac{12\eta W}{h_1^2}$$

$$= \left(\frac{1}{2} - \frac{Q}{Wbh_1}\right) \cdot \frac{12\eta W}{h_2^2} \cdot \frac{h_2^2}{h_1^2} \quad \text{(B2)}$$

and:

$$\frac{dP_2}{dz} = \left(\frac{1}{2} - \frac{Q}{Wbh_2}\right) \frac{12\eta W}{h_2^2} \quad \text{(B3)}$$

FIG. B1. Stepped screw—ratio of pressure gradients in the two sections. ——, Q/Wbh_2; – – –, Q/Wbh_1.

FIG. B2. Stepped screw—pressure gradient in first section dP_1/dz in terms of that when $dP_2/dz = 0$ ($Q/Wbh_2 = 0·5$). ——, Q/Wbh_2; – – –, Q/Wbh_1.

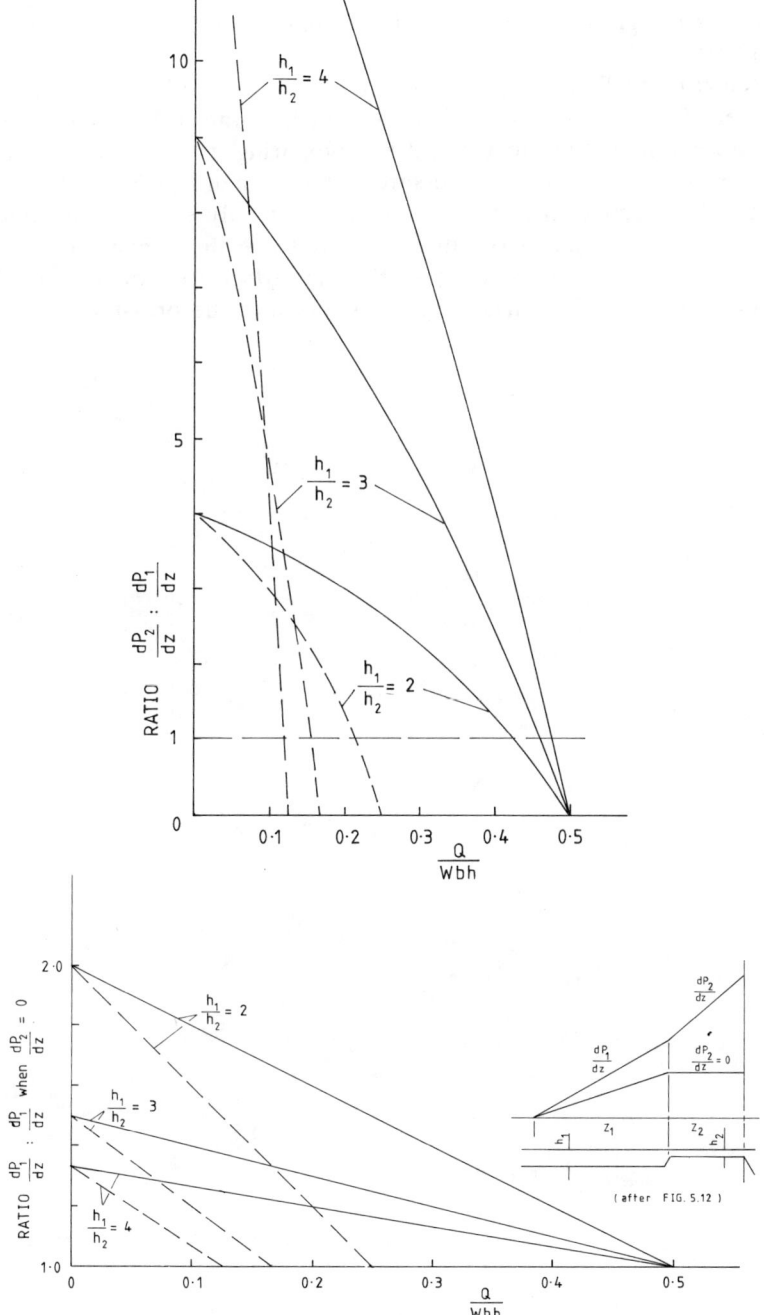

The pressure gradients are calculated from eqns (B2) and (B3) in terms of $12\eta W/h_2^2$.

Referring to Fig. 5.12, it is also interesting to see how the pressure gradient in the first section changes with flow rate (Q/Wbh_1) and die pressure; this is done in Fig. B2 by comparing dP_1/dz with its value when it represents the total pressure rise, i.e. when $Q/Wbh_2 = 0.5$. This shows that even with a depth ratio h_1/h_2 of only 2:1, the maximum pressure gradient (at zero output) is only twice that when there is no pressure change in the shallow section, i.e. dP_1/dz is not only usually smaller than dP_2/dz, but also changes less with die pressure.

Appendix C

EXPERIMENTAL DETERMINATION OF HEAT LOSSES

These trials were carried out on a 37 mm (1·5 in) ×20:1 L/D extruder having a barrel of 127 mm (5 in) outside diameter, approximately 600 mm (24 in) long. Three barrel heaters, approximately 200 mm (8 in) diameter by 150 mm (6 in) long, each rated at 3 kW, were controlled by three-term time-proportioning controllers acting through relays. The periphery of the adaptor flange was heated by a 2·2 kW heater of 280 mm (11 in) diameter by 63 mm (2·5 in) long and the die had a heater of 82 mm (3·25 in) diameter by 89 mm (3·5 in) long, rated at 0·6 kW. Adaptor and die heaters were controlled by similar time-proportioning controllers. Each control thermocouple was close to the corresponding heater element. In addition, at longitudinal positions between the heaters, were three thermocouples extending to within 3 mm of the inner barrel surface. Auxiliary contacts on the relays were connected to separate pens on a multichannel event recorder, so that 'on' and 'off' times were recorded for each heater over a 15 min period. The 'on' times were summed manually to give the average percentage 'on' time, which multiplied by the continuous rating (as above) gave the actual power input to each heater. Cold water was circulated continuously through the feed pocket jacket as in normal operation, but unfortunately flow rate and temperature rise were not recorded.

Table C1 gives the temperatures set on the controllers and the corresponding heater power inputs while Fig. C1 gives an example of the event recorder chart. Table C2 gives the recorded temperatures at various positions on the outer surfaces and of the 'deep' thermocouples. Also included in Table C1 are calculated values for settings 2 and 4 of heat loss by radiation and natural convection based on the

TABLE C1
Static Heat Losses

	Set temperatures (°C)					Power consumption (W)					
	Barrel zone			Flange	Die	Barrel zone			Flange	Die	Total
Heater rating	1	2	3	4	5	1 3000	2 3000	3 3000	4 2200	5 600	8800
Setting											
1	100	125	150	150	150	276	289	789	342	105	1801
2	150	150	150	150	150	996	0	498	356	123	1973
3	150	175	200	200	200	619	182	1056	626	182	2665
4	200	200	200	200	200	1189	324	757	562	185	3017
5	200	225	250	250	250	990	339	1396	867	241	3833
6	250	250	250	250	250	2015	444	1074	821	246	4600
Calculated heat losses by radiation and convection											
2	150	150	150	150	150	352	171	235	295	48	1101
4	200	200	200	200	200	588	342	400	497	99	1926

recorded surface temperatures. Table C1 shows that the total heat loss increases progressively as expected with set temperatures, for both 'ramp' and 'level' profiles, the latter also being greater than the former for a given final temperature. The distribution of energy between heaters has some unexpected features; Zone 3 is high even with 'level' temperature profiles, though this might be expected with 'ramp' profiles as heat will be conducted away to the cooler zones. The high contribution from Zone 1 is no doubt largely due to longitudinal heat conduction to the feed pocket, especially with the 'level' profiles, where conduction from Zones 2 and 3 will be minimal. A possible explanation of the low energies from Zone 2 is that the demand on Zone 1 causes this to be 'on' for most of the time and so its temperature rises above that of Zone 2, to which it then conducts heat; this is borne out to some extent by the surface temperatures in Table C2. The calculated values for settings 2 and 4 are considerably less than the experimental totals; the difference being attributed to heat conducted to the feed pocket. However, this difference closely approximates the energy from Zone 1, suggesting that this heater is supplying energy to the feed pocket while the remainder are supplying the radiation and convection losses. Whether this is coincidence, the heat to the feed pocket is substantial and measures such as a thermal barrier

Appendix C

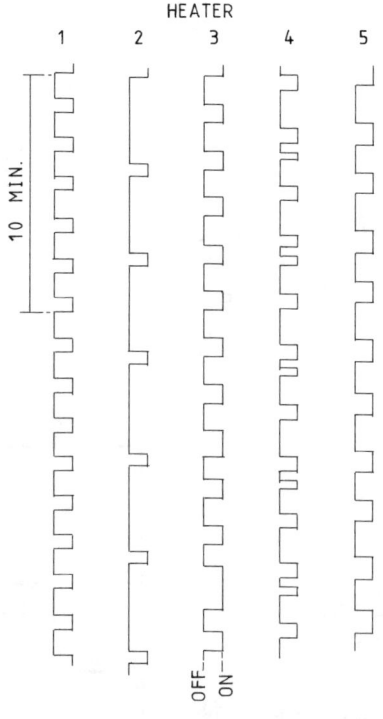

Fig. C1. Event recorder—heater switching.

between feed pocket and barrel would appear worthwhile not only in reducing losses and the need for feed pocket cooling, but also in permitting higher temperatures in the first barrel zones. The data in Table C1 make clear the importance of longitudinal conduction in the barrel in limiting temperature differences between adjacent zones; but a calculation based on the temperatures of the 'deep' thermocouples in settings 1, 3 and 5 shows that in this machine the longitudinal heat flow may not exceed 100/150 W, which is small compared with the energy consumptions of the heaters.

It will be noted that, especially at the higher set temperatures, the energy losses are a substantial proportion of the rated input of the heaters—8·8 kW total. Since losses increase with temperature, this not only sets an upper limit to achievable barrel temperatures but will also

TABLE C2
Surface Temperatures in Heat Loss Tests (°C)

Setting	K	F	A	G	L	B	H	M	C	I	N	D	J	E
1	57	74	125	105	110	105	134	145	203	154	145	137	—	130
2	65	90	240	131	145	128	144	155	175	150	155	150	146	150
3	74	111	210	152	160	157	183	190	250	193	190	210	173	216
4	100	137	322	188	200	221	201	210	248	207	215	207	200	233
5	104	137	286	205	210	212	240	250	231	262	250	255	250	265
6	115	172	390	240	245	242	255	255	315	260	255	252	250	285

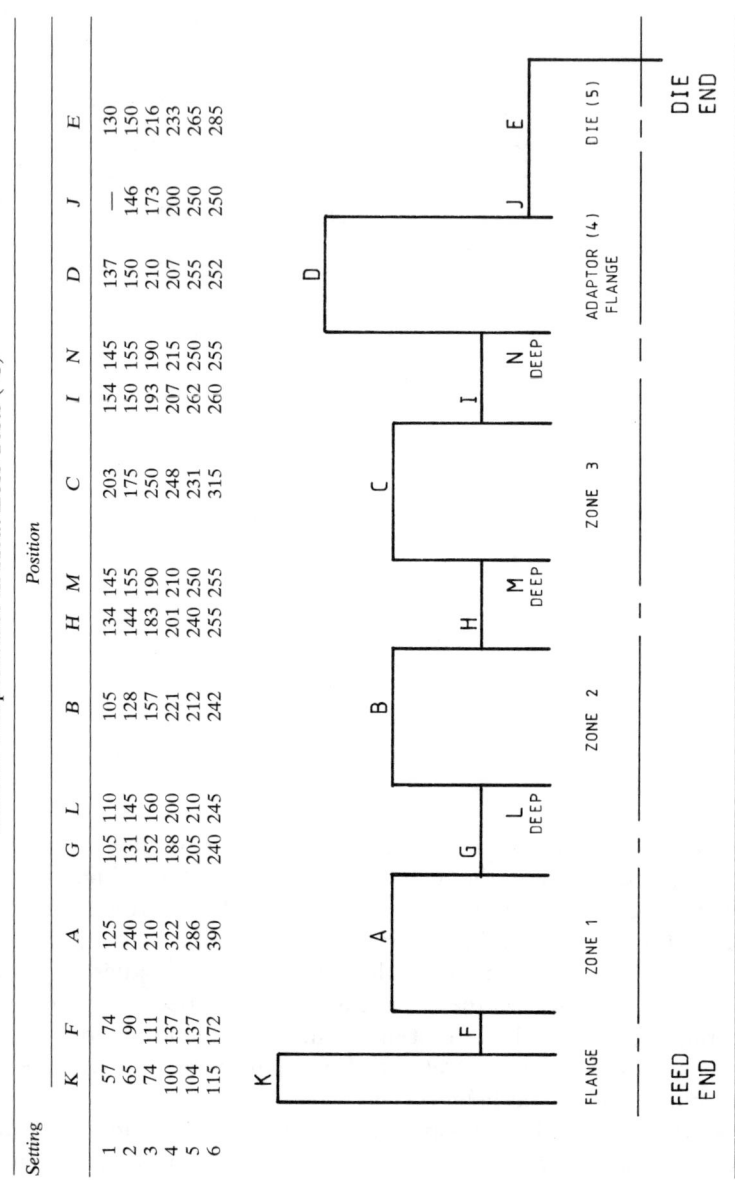

limit the rate of temperature rise and extend the time to re-establish steady conditions after a change of temperature settings. A type of heater suitable for continuous operation at (element) temperatures around 350°C or for use with external insulation, e.g. mains-frequency induction heaters, could allow substantial reduction of these losses and improvement in energy utilisation as well as more rapid response to intentional changes in set temperatures.

The data presented are strictly only applicable to the external temperatures given in Table C2, which do not necessarily relate directly to temperatures at the internal surface of the barrel. The distribution of energy between heaters, which largely determines these external temperatures, is governed primarily by the profile of set temperatures, especially near the feed end. For most practical purposes the energy losses can therefore be regarded as dependent only on the set temperatures and the physical arrangement, i.e. heater positions and ratings, barrel length, adaptor and die type, etc., and with these provisos need only be measured in a single series of experiments as outlined here.

Appendix D

STABILITY OF MELT PUMPING SECTION

In Section 5.5 the total output was expressed in terms of speed N and pressure P by eqn (5.78):

$$Q_{\text{Tot}} = AN - \frac{BP}{\eta} \qquad (5.78)$$

where A and B are constant depending on the dimensions of the screw and die (eqns (5.79) and (5.80)). For the present purpose it is convenient to define:

$$B' = BZ = \frac{bh^3}{12} \qquad (D1)$$

Equation (5.78) becomes:

$$Q_{\text{Tot}} = AN - \frac{B'P}{\eta Z} \qquad (D2)$$

and eqn (5.86) becomes:

$$Q_{\text{Tot}}\left(1 + \frac{B'}{KZ}\right) = AN \qquad (D3)$$

or:

$$Q = \frac{ANKZ}{KZ + B'} \qquad (D4)$$

If, due to an external cause such as a rearward movement of the 'gel' point, the melt length increases to $Z + \mathrm{d}z$, with the same die constant K, the output will change to $Q + \mathrm{d}q$ and pressure to $P + \mathrm{d}p$. Then:

$$Q + \mathrm{d}q = \frac{ANK(Z + \mathrm{d}z)}{K(Z + \mathrm{d}z) + B'} \qquad (D5)$$

Instantaneously, the rate of melting will remain at Q and the change of output represents an additional volume $bh\,\mathrm{d}z$. Therefore stable operation requires that the change of output with respect to melt length should be minimised; this is represented by $\mathrm{d}q/Q\,\mathrm{d}t$.

Substituting from eqn (D4) in (D5) gives:

$$\frac{\mathrm{d}q}{Q} = \frac{KZ+B'}{KZ} \cdot \frac{K(Z+\mathrm{d}z)}{K(Z+\mathrm{d}z)+B'} - 1 \tag{D6}$$

Then:

$$\frac{1}{Q}\frac{\mathrm{d}q}{\mathrm{d}t} = \frac{\mathrm{d}}{\mathrm{d}t}\left(\frac{KZ+B'}{KZ} \cdot \frac{K(Z+\mathrm{d}z)}{K(Z+\mathrm{d}z)+B'}\right) \tag{D7}$$

$$\frac{1}{Q} \cdot \frac{\mathrm{d}q}{\mathrm{d}t} = \frac{KZ+B'}{KZ} \cdot \frac{\mathrm{d}}{\mathrm{d}t}\left(\frac{K(Z+\mathrm{d}z)}{K(Z+\mathrm{d}z)+B'}\right)$$

$$= \frac{KZ+B'}{KZ} \cdot \frac{\mathrm{d}}{\mathrm{d}t}\left(1 - \frac{B'}{K(Z+\mathrm{d}z)+B'}\right)$$

$$= \frac{KZ+B'}{KZ}\left[\frac{-B'K}{(K(Z+\mathrm{d}z)+B')^2}\right]\frac{\mathrm{d}z}{\mathrm{d}t} \tag{D8}$$

For small values of $\mathrm{d}z/Z$:

$$\frac{1}{Q} \cdot \frac{\mathrm{d}q}{\mathrm{d}t} \simeq \frac{-B'}{Z(KZ+B')} \cdot \frac{\mathrm{d}z}{\mathrm{d}t} = \frac{-1}{Z\left(\dfrac{KZ}{B'}+1\right)} \cdot \frac{\mathrm{d}z}{\mathrm{d}t} \tag{D9}$$

Substituting for B':

$$\frac{1}{Q} \cdot \frac{\mathrm{d}q}{\mathrm{d}t} \simeq \frac{-1}{Z\left(\dfrac{12KZ}{bh^3}+1\right)} \cdot \frac{\mathrm{d}z}{\mathrm{d}t} \tag{D10}$$

From eqn (D10) it is seen that if channel depth h is decreased, $\mathrm{d}q/\mathrm{d}t$ is reduced. Similarly, if channel length Z is increased, $\mathrm{d}q/\mathrm{d}t$ is reduced; both of these will tend to make the melt pumping section produce smaller amplitude oscillations of output and pressure (and also of power input) when subjected to disturbance, e.g. due to an imbalance between melting and pumping rates.

Appendix E

PROPERTIES OF POLYMERS FOR HEAT AND FLOW

The physical properties of polymers which are important in melt processing have been discussed in Chapters 2 and 3. They include density (or its reciprocal, specific volume), specific heat, thermal conductivity and melt viscosity. These are dependent to a greater or lesser extent on polymer type, molecular weight, distribution of molecular weight, pressure, temperature, and flow conditions. They may also be influenced by previous history of temperature or shear, causing changes in crystallinity, orientation or residual strain. These in turn may lead to anisotropy of particular properties, e.g. thermal conductivity increases (in all directions) with crystallinity; it is also increased (anisotropically) in the direction of orientation in both amorphous and semi-crystalline states, though probably not to the same extent in each state. There is also some evidence that thermal conductivity is anisotropic in strong shear flows, possibly due to melt orientation.

Some of these properties and interactions are extremely difficult to measure with accuracy and reliability; consequently there is a scarcity of data, especially in the melt state, and in some cases lack of agreement. For these reasons the thermal data given here (Table E1), which are only intended as a guide, largely refer to the solid state and room temperature. These provide an approximate basis for comparisons between the processing behaviour of widely differing polymers; data for the melt state and a range of temperatures are given where possible (Figs E1–5), but must not be assumed to be valid for other polymers or grades. As pointed out on p. 24, for the polymers given, specific volume will depend on the reference density (molecular weight) while changes in specific volume and enthalpy from room temperature also depend on the initial crystallinity.

TABLE E1
Thermal Properties in the Solid State (Room Temperature)

Polymer (Table 4.1)	Glass transition temperature (°C)	Crystalline melting point (°C)	Density (kg/m^3)	Specific heat (J/kg K)	Thermal conductivity (W/m K)
ABS	—	—	{990, 1150}	{1380, 1680}	{0·19, 0·36}
PMMA	105	—	{1170, 1200}	1470	{0·19, 0·21}
CAB	—	—	{1150, 1220}	{1260, 1680}	{0·17, 0·33}
PC	160	—	1200	1260	0·19
PS	105	—	{1040, 1065}	1340	{0·10, 0·14}
PVAC	—	—	{1350, 1450}	{840, 1170}	{0·13, 0·29}
SAN	105	—	{1075, 1100}	{1340, 1420}	0·12
UPVC	{67, 80}	220	{1350, 1450}	{840, 1170}	{0·15, 0·17}
PPVC	—	—	{1160, 1350}	{1260, 2100}	{0·13, 0·17}
HIPS	—	—	{980, 1100}	{1340, 1470}	{0·04, 0·13}
PPO[b]	195	—	1060	—	0·21
PES[c]	195	—	1240	1260	0·26
POM	−70	180	1410	1470	0·23[a]
EVAC	—	—	{920, 950}	2300	—
PA 6-6	−25	{250, 267}	{1090, 1140}	{2300, 2350}	{0·21, 0·25}
PA 6	−30	{215, 225}	{1120, 1140}	—	0·24
PA 6-10	—	220	1090	—	0·21
PA 11	—	190	1040	2430	0·29
PETP	+70	{260, 267}	1350	—	{0·23, 0·29}
PBTP	—	—	—	—	—
LDPE	−60	{110, 117}	{910, 925}	2300	{0·33, 0·36}
HDPE	−80	{135, 138}	{941, 965}	2300	{0·46, 0·52}
PP	0	{165, 175}	{895, 910}	1930	{0·12, 0·14}
TPX	—	250	830	2180	0·17
PTFE	—	327	{2100, 2200}	1050	0·25

[a] POM homopolymer. Copolymer $k = 0·07$ approx.
[b] PPO polyphenylene oxide
[c] PES polyethersulphone
Data for density, specific heat and conductivity primarily from Ref. A

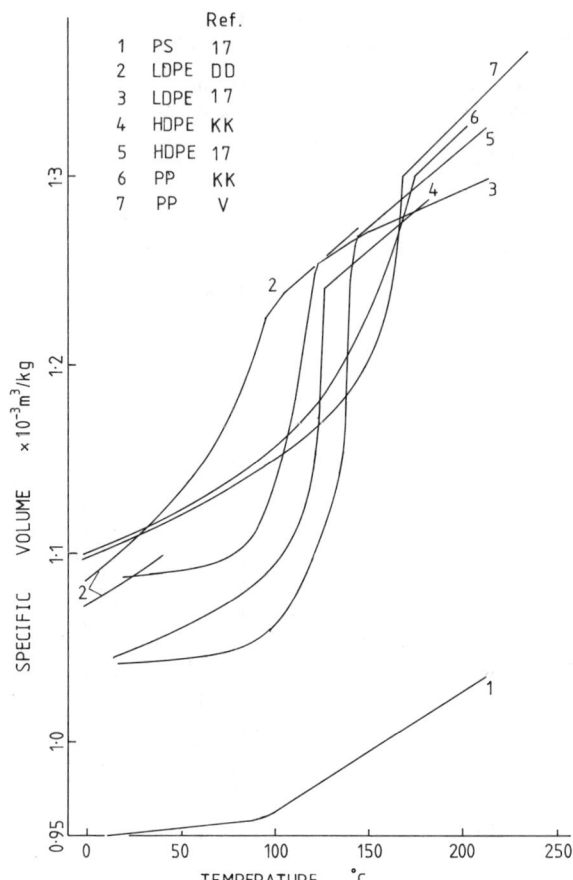

FIG. E1. Specific volume vs. temperature.

For data on melt viscosity the reader is recommended to Ref. 70. This may be taken as representative of each class of polymer, though the variation with shear rate is strongly influenced by distribution of molecular weight. However, the absolute values are liable to considerable variation with molecular weight, manufacturing variables, and compounding additives; since viscosity enters directly into calculations of die pressure, mechanical energy, etc., an in-house measurement of viscosity at a minimum of one known temperature and two (widely spaced) shear rates is highly desirable.

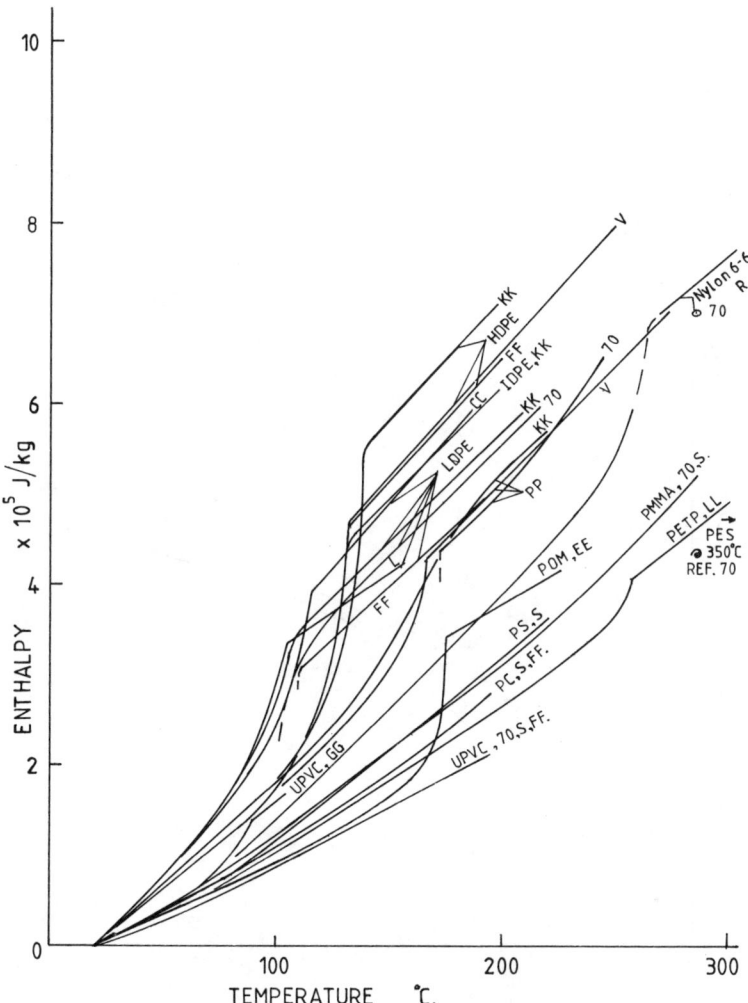

FIG. E2. Heat capacity—enthalpy above 20°C.

For certain purposes, such as calculating heat transfer coefficients and transient conduction, combinations of properties are required, e.g. thermal diffusivity ($\alpha = k/\rho c_p$), Prandtl number ($\Pr = c_p \eta / k$), Graetz number ($\mathrm{Gz} = W c_p / kL$); these may be calculated from individual values, but in some instances may be more reliably determined by other methods (e.g. thermal diffusivity from cooling experiments).

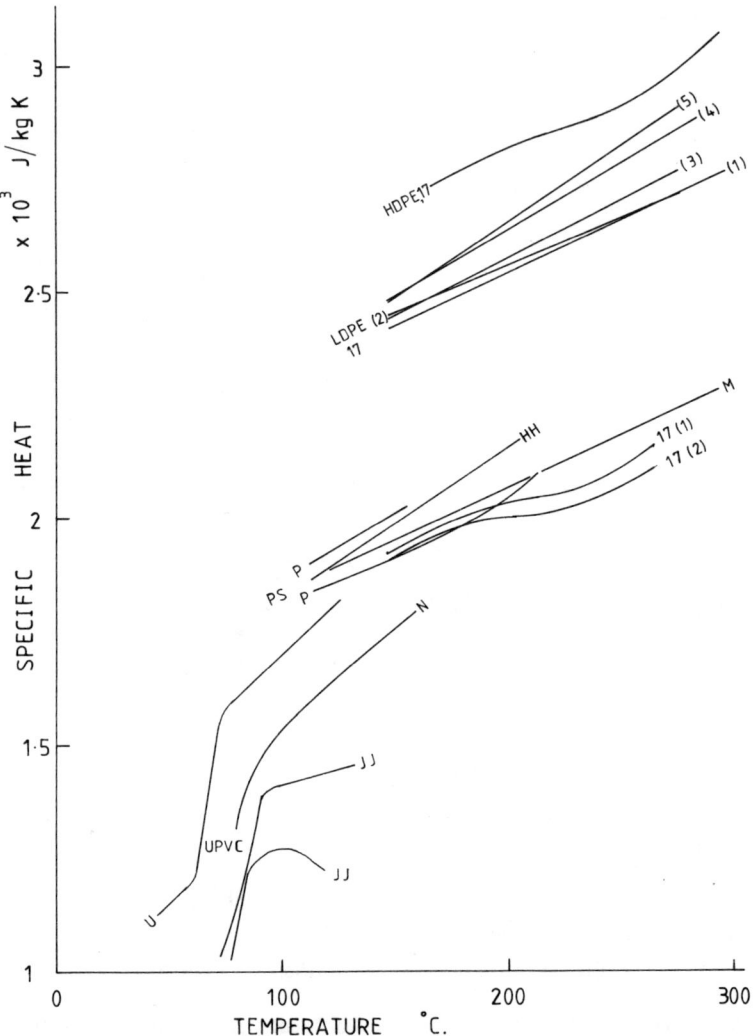

Fig. E3. Specific heats in the melt.

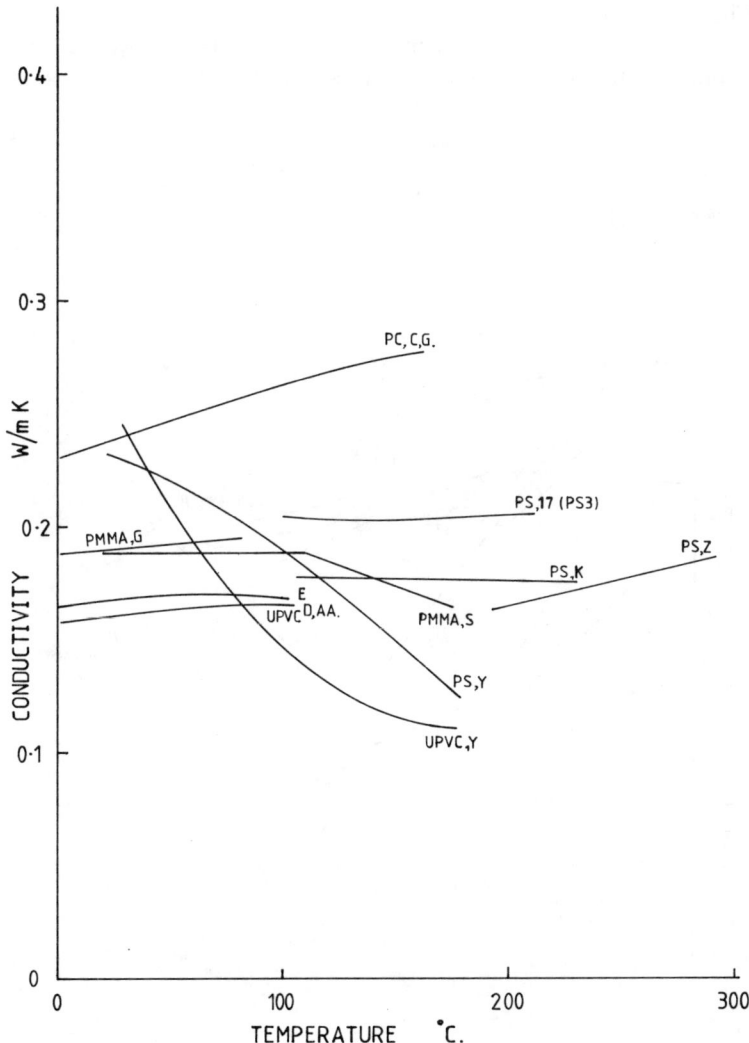

Fig. E4. Thermal conductivity—amorphous polymers.

314 *Extruder Principles and Operation*

Some typical data for LD polyethylene, air, water and steel are given in Table 3.1. An example of swell ratio data (for polystyrene) is given in Figs 2.10 and 2.11.

The reference letters given in the following list are quoted against the appropriate data in Figs E1–5. In some cases the same experimen-

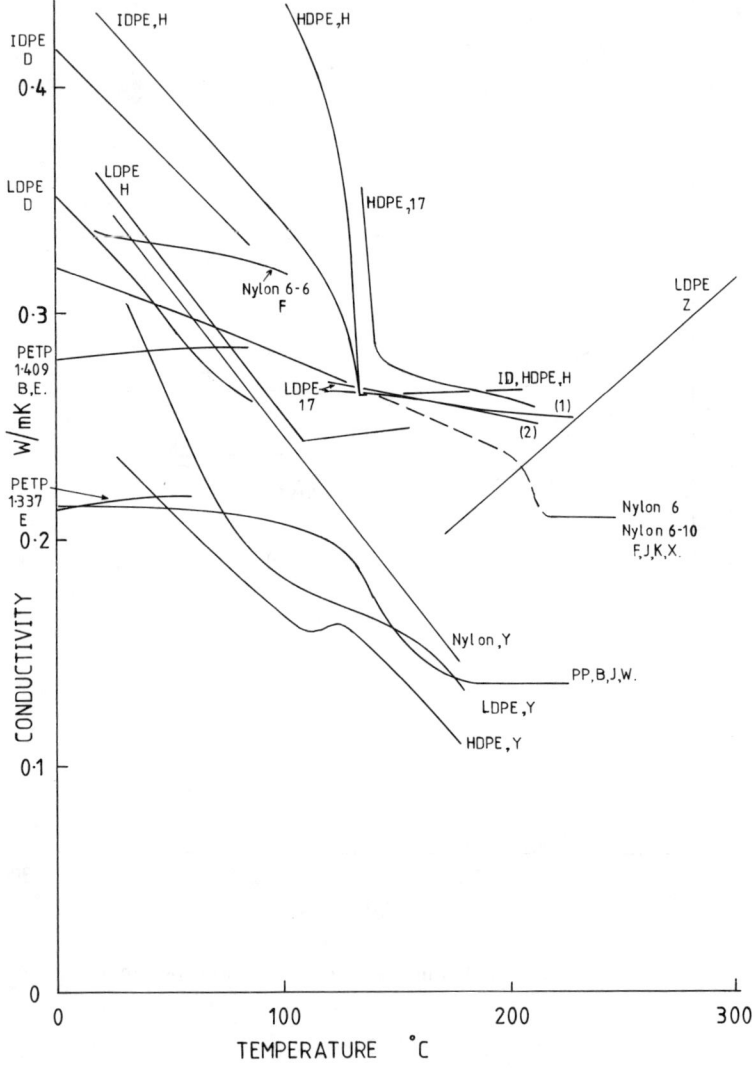

Fig. E5. Thermal conductivity—semi-crystalline polymers.

tal data appears to have been published in more than one journal and under different combinations of authors' names; duplication has been avoided in this Appendix where possible. The present author has endeavoured to reproduce the data as accurately as possible, but problems of scale and units may have led inadvertently to some averaging; the reader is directed to the original publications for exact values, individual data points and values outside the scope of this book. Reference back should also be made for precise materials and pretreatments used, which are here lumped under generic names for clarity. The materials tested in Ref. 17 were supplied by the UK Science and Engineering Research Council's Polymer Supply and Characterisation Centre (PSCC) at the Rubber and Plastics Research Association and coded by them LDPE4, PS2, etc., as indicated in the figures.

REFERENCES

A   Anon. *Plastiques Modernes et Elastomères*, **18**(6) (1966) 128–49.
B   Eiermann, K., *Koll. Zeit.*, **201** (1965) 3–15.
C   Eiermann, K., *Kunststoffe*, **55** (1965) 335.
D   Eiermann, K. and Hellwege, K-H., *J. Poly. Sci.*, **57** (1962) 99–106.
E   Eiermann, K., Hellwege, K-H. and Knappe, W., *Koll. Zeit.*, **174**(2) (1961) 134–42.
F   Hellwege, K-H., Hoffmann, R. and Knappe, W., *Koll. Zeit.*, **226** (1968) 109–15.
G   Hellwege, K-H., Hennig, J. and Knappe, W., *Koll. Zeit.*, **188** (1963) 121–7.
H   Hennig, J., Knappe, W. and Lohe, P., *Koll. Zeit.*, **189** (1963) 114–16.
J   Knappe, W. et al., *Kunststoffe*, **51** (1961) 707.
K   Lohe, P. et al., *Koll. Zeit.*, **203** (1965) 115–19.
L   Dole, M. et al., *J. Chem. Phys.*, **20** (1952) 781–90.
M   Abu-Isa, I. and Dole, M., *J. Phys. Chem.*, **69**(8) (1965) 2668–75.
N   Alford, S. and Dole, M., *J. Amer. Chem. Soc.*, **77** (1955) 4774.
P   Currie, J. A. and Dole, M., *J. Phys. Chem.*, **73**(10) (1969) 3384–9.
R   Wilhoit, R. C. and Dole, M., *J. Phys. Chem.*, **57** (1953) 14–21.
S   Wunderlich, B. and Baur, H., *Adv. Poly. Sci.*, **7** (1970) 151–368.
T   Bares, V. and Wunderlich, B., *J. Poly. Sci. A2*, **11** (1973) 861–73.
U   Wilski, H., *Koll. Zeit.*, **210**(1) (1966) 37–45.
V   Wilski, H., *Kunststoffe*, **50**(6) (1960) 335–6.
W   Tomlinson, J. N. and Kline, D. E., *J. Appl. Poly. Sci.*, **11** (1967) 1931–40.
X   Holzmüller, W. and Münx, M., *Koll. Zeit.*, **159** (1958) 26–8.
Y   Holzmüller, W. and Lorenz, J., *Plaste u. Kautschuk*, **8** (1961) 351–2.
Z   Fuller, T. R. and Fricke, A. L., *J. Appl. Poly. Sci.*, **15** (1971) 1729–36.

AA   Sheldon, R. P. and Lane, S. K., *Polymer*, **6** (1965) 77–83.
BB   Shoulberg, R. H. and Shetter, J. A., *J. Appl. Poly. Sci.*, **6**(23) (1962) S32–3.
CC   Raine, H. C., Richards, R. B. and Ryder, H., *Trans. Farad. Soc.*, **41** (1945) 56–64.
DD   Hunter, E. and Oakes, W. G., *Trans. Farad. Soc.*, **41** (1945) 49–56.
EE   Linton, W. H. and Goodman, H. H., *J. Appl. Poly. Sci.*, **1**(2) (1959) 179–84.
FF   Cogswell, F. N. and Couzens, D. C., unpublished data.
GG   Griskey, R. G. et al., *Mod. Plast.*, **43**(11) (1966) 119–23.
HH   Karasz, F. E., Bair, H. E. and O'Reilly, J. M., *J. Phys. Chem.* **69**(8) (1965) 2657–67.
JJ   Dunlap, L. H., *J. Poly. Sci.* A2, **4** (1966) 673–84.
KK   Hanna, R. D. and Lomax, J. Y., *Mod. Plast.*, **36**(6) (1959) 111–22.
LL   Smith, C. W. and Dole, M., *J. Poly. Sci.*, **20** (1956) 37–56.

Note. References 17 (Adamski) and 70 (Cogswell) referred to in Figs E1–5 are given in detail in the main reference list.

Appendix F

TEMPERATURE VARIATION IN THE FLIGHT CLEARANCE

The following example is taken to estimate the order of temperature variation due to energy dissipation in the flight clearance. The assumptions and conclusions are discussed in Section 6.5.

Diameter	D	100 mm
Channel depth	h	10 mm
Pitch	p	100 mm
Flight width	t	10 mm
Speed	N	1, 2, 4 rps
Viscosity	η	see below
Flight clearance	δ	0·15 mm

Then

Channel width	$b = (p - t)\cos\phi = 0{\cdot}08576$ m
Helix angle	$\phi = \tan^{-1} 1/\pi = 17{\cdot}567°$
Helical length	$dz = 0{\cdot}1\pi/\cos\phi = 0{\cdot}3297$ m (one turn)
Longitudinal velocity	$W = \pi 0{\cdot}1 \times 1 \cos\phi = 0{\cdot}2994$ m/s
Drag flow	$Q_D = 1{\cdot}283 \times 10^{-4}$ m³/s
Drag flow shear rate	$\dot{\gamma}_D = W/h = 0{\cdot}2994/0{\cdot}010 = 29{\cdot}94$ s^{-1}
Normal flight width	$= t \cdot \cos\phi = 0{\cdot}009529$ m
Tangential flight width	$= t/\tan\phi = 0{\cdot}01/0{\cdot}3183 = 0{\cdot}03142$ m
Circumference	$\pi D = 0{\cdot}3142$ m
Tangential velocity	$\pi DN = 0{\cdot}3142$ m/s at 1 rps
Flight shear rate	$\pi DN/\delta = 0{\cdot}3142/0{\cdot}00015 = 2094{\cdot}7$ s^{-1}
Transit time of flight	$= 0{\cdot}1$ s (1 rps)
	$= 0{\cdot}05$ s (2 rps)
	$= 0{\cdot}025$ s (4 rps)

'Rest' time between transits = 0·9 s (1 rps)
                              = 0·45 s (2 rps)
                              = 0·225 s (4 rps)
Channel power (150°C)   = 783 W/turn at 1 rps
Channel power (200°C)   = 259·8 W/turn at 1 rps

Thermal properties (from Table 3.1)

Density of steel	7800 kg/m³
Density of polyethylene	760 kg/m³ at melt temperature
Specific heat, steel	503 J/kg K
Specific heat, polyethylene	2300 J/kg K
Thermal conductivity, steel	50 W/m K
Thermal conductivity, polyethylene	0·50 W/m K
Thermal diffusivity, steel	$1·3 \times 10^{-5}$ m²/s
Thermal diffusivity, polyethylene	$1·5 \times 10^{-7}$ m²/s

Viscosity data for MFR 2·0 LD polyethylene

Shear rate (s^{-1})	Temperature (°C)	Viscosity (Ns/m²)	Temperature coefficient (per °C)
30	150	2200	$\beta = 0·0134$
30	200	730	
2000	150	124	$\beta = 0·0102$
2000	200	61	

Then

pseudoplasticity index n at 150°C, $n = 0·3152$
pseudoplasticity index n at 200°C, $n = 0·4090$

Clearance volume per turn $= 0·3297 \times 0·009529 \times 0·00015 = 471·26 \times 10^{-9}$ m³/turn.

In Table F1 are tabulated for several temperatures in the required range, the viscosity in the flight clearance, the shear power ($\eta(\dot{\gamma})^2$ W/m³) and the energy (J per turn) at 1 rps. The temperature rise is also tabulated assuming the flight clearance was insulated, i.e. all the shear heat was absorbed in heating the polymer.

The heat flows in the barrel are illustrated in Fig. F1, showing the approximation of one-dimensional radial heat flow in the barrel wall. The barrel is assumed initially at a uniform temperature T_0, equal to

Appendix F

TABLE F1
Shear Heating in Flight Clearance

Mean temperature (°C)	$\beta(T-T_0)$	Viscosity at $2094\cdot7\,s^{-1}$ $(N\,s/m^2)$	Shear energy/ unit volume $(W/m^3 \times 10^6)$	Shear work per turn $(J/turn)$	Adiabatic (insulated) temperature rise (°C)
140	−0·102	132·38	580·85	27·375	36·48
150	0·0546	120·13	527·10	24·840	33·10
151	0·0102	118·90	521·70	24·586	32·77
152	0·0204	117·68	516·35	24·334	32·43
154	0·0408	115·23	505·60	23·827	31·75
156	0·0612	112·78	494·85	23·320	31·08
158	0·0816	110·33	484·10	22·814	30·40
160	0·102	107·88	473·35	22·307	29·73
170	0·204	95·62	419·56	19·772	26·35
180	0·306	83·37	365·81	17·239	22·97
190	0·408	71·12	312·06	14·706	19·60
200	0·510	58·86	258·26	12·171	16·22

Screw speed 1 rps
Viscosity at $30\,s^{-1}$ shear rate and $150°C = 2200\,N\,s/m^2$
Clearance volume per turn $= 471\cdot26 \times 10^{-9}\,m^3/turn$
Temperature rise at end of transit $(T_a - T_0) = 1\cdot54°C$ ⎫
Residual temperature rise after 'rest' $= 0\cdot239°C$ ⎭ See text p. 323.

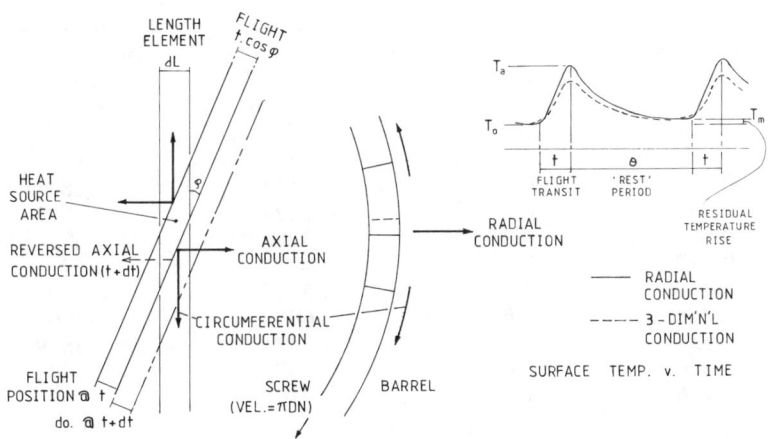

FIG. F1. Local heat flows in barrel from flight clearance.

that of the polymer in the channel, and it is in contact with the polymer in the flight clearance at a temperature T_a for a time $t = 0 \cdot 1$ s at 1 rps. Referring to Ref. 19 p. 39, for one-dimensional transient heat conduction, resistance ratio $m = k/hx = 0$ (negligible surface resistance).

Dimensionless group Z (related to Fourier number $\alpha t/r_m^2$)

$$= x/2\sqrt{(\alpha t)}$$
$$= x/2\sqrt{1 \cdot 3 \times 10^{-5} \times 0 \cdot 1}$$
$$= 438 \cdot 5x$$

where t is time and x is distance below the barrel surface.

Unaccomplished temperature change $Y = \dfrac{T_a - T}{T_a - T_0}$

where T is the temperature at position x and time t.

Reference 19, p. 39, tabulates values of Z and the corresponding Gauss 'error integral' Y, from which Table F2 is calculated and Fig. F2 is plotted. The unique values of Y imply that the temperatures distributed within the barrel wall at a given time are directly proportional to the surface temperature difference $T_a - T_0$, and eqn (3.4) of Ref. 19 shows that the total heat absorbed in this time is also directly

TABLE F2
Radial Temperature Distribution in Barrel when Flight Passes

Depth below inner surface x mm	Dimension-less depth $Z = x/2\sqrt{(\alpha t)}$	Unaccomplished temperature change $Y = \dfrac{T_a - T}{T_a - T_0}$	Fractional temperature rise $\dfrac{T - T_0}{T_a - T_0}$
0	0	0	1·0
0·228	0·1	0·1125	0·8875
0·456	0·2	0·2227	0·7773
0·684	0·3	0·3286	0·6714
0·912	0·4	0·4284	0·5716
1·140	0·5	0·5205	0·4795
1·368	0·6	0·6039	0·3961
1·824	0·8	0·7421	0·2579
2·280	1·0	0·8427	0·1573
2·736	1·2	0·9103	0·0897
3·192	1·4	0·9523	0·0477
4·561	2·0	0·9953	0·0047

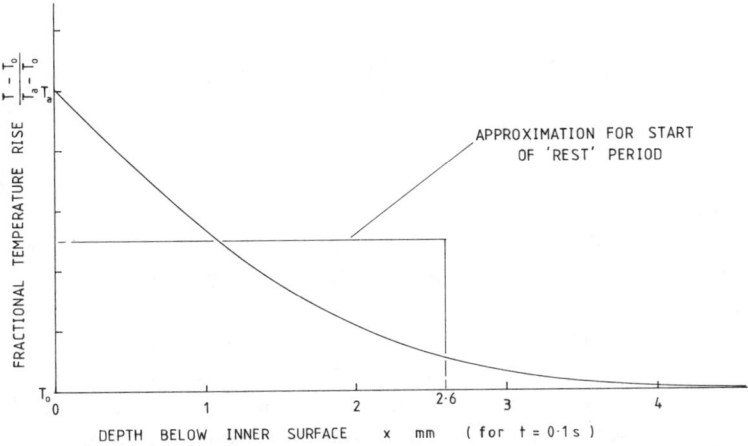

FIG. F2. Radial temperature distribution in barrel when flight passes.

proportional to $T_a - T_0$. Thus by equating this heat to that produced by shearing, a common value of T_a can be found which satisfies Table F1 and eqn (3.4) of Ref. 19. For this example, the left-hand side of eqn (3.4) of Ref. 19 is:

$$\frac{Q}{(2)r_m A \rho c_p (T_a - T_0)} = \frac{1}{0{\cdot}3297 \times 0{\cdot}009529 \times 7800 \times 503} \cdot \frac{Q}{r_m(T_a - T_0)}$$

The factor (2) is omitted since the barrel is semi-infinite, i.e. receives heat from one face only. The right-hand side is:

$$1 - \frac{8}{\pi^2}(\exp(-a_1 X) + \tfrac{1}{9}\exp(-9a_1 X) + \tfrac{1}{25}\exp(-25a_1 X) \cdots)$$

where

$$a_1 = \left(\frac{\pi}{2}\right)^2$$

and

$$X = \alpha t/r_m^2 = 1{\cdot}3 \times 10^{-5} \times 0{\cdot}1/r_m^2$$

Trial calculations show that for the present conditions the values obtained by taking r_m as $0{\cdot}005$ or $0{\cdot}010$ m differ by less than $0{\cdot}2\%$. Taking $r_m = 0{\cdot}010$ m, $X = 0{\cdot}013$ and the right-hand side equals

0·128817. Then:

$$\frac{Q}{T_a - T_0} = 0{\cdot}128817 \times 0{\cdot}01 \times 0{\cdot}3297 \times 0{\cdot}009529 \times 7800 \times 503$$

$$= 15{\cdot}878 \text{ J/K}$$

Interpolation from Table F1 gives $T_a - T_0 = 1{\cdot}54°C$ for $Q = 24{\cdot}45$ J/turn.

For the remaining 0·9 s of the revolution, this portion of the barrel is in contact with the polymer in the channel, and may be treated as thermally insulated, so that the heat in the barrel represented by the area under the temperature distribution in Fig. F2 will diffuse deeper into the barrel with accompanying drop of surface temperature. By inspection this area is approximated by a uniform temperature $T_b = (T_0 + T_a)/2$ over a depth of 2·6 mm from the surface. Treating this as a semi-infinite slab of thickness $r_m = 2{\cdot}6$ mm in contact with the remainder of the barrel at T_0, the resistance ratio $m = 1$ and from Fig. 3.3 of Ref. 19 for the 'mid-plane' temperature T_m (i.e. the barrel surface):

$$X = \alpha\theta/r_m^2 = 1{\cdot}3 \times 10^{-5}\theta/(0{\cdot}0026)^2 = 1{\cdot}923\theta$$

For $\theta = 0{\cdot}9$ s, $X = 1{\cdot}7307$ and for $m = 1$, $Y_m = 0{\cdot}31 = (T_0 - T_m)/(T_0 - T_b)$. Then for $T_0 - T_b = 1{\cdot}54/2$, $T_m = T_0 + 0{\cdot}239°C$. If, with the same polymer characteristics, speed is increased to $N = 2$ and 4 rps, viscosity will be reduced in proportion to the $(0{\cdot}315 - 1)$ index of speed, but power (shear heating) increases as the $+1{\cdot}315$ index of speed.

On p. 39 of Ref. 19, a unique relation is shown between Y, representing the local temperature and the heat stored, and Z representing the time and distance from the surface. The profile of temperature in Fig. F2 is therefore a function of $Z = x/2\sqrt{(\alpha t)}$. But for a given value of Z, distance x is proportional to the square root of time t of transit of the flight (time of shearing $\propto 1/N$). Therefore the heat stored by the barrel is proportional to $\sqrt{t}$ and $1/\sqrt{N}$. During the 'rest' period, the Fourier number $X = \alpha\theta/r_m^2$ where θ is the time of rest. But, r_m is proportional to distance x (see Fig. F2) and thus $\sqrt{t}$. Therefore X is proportional to $\theta_{\text{rest}}/t_{\text{transit}}$, i.e. is a constant for a fixed ratio of transit time to 'rest' time, which is determined by the geometry of the flight and independent of screw speed N. The value of Y_m from Fig. 3.3 is therefore also constant at 0·31. The values for 2 and 4 rps are tabulated in Table F3.

TABLE F3
Shear Heating and Temperature Rises in Flight at Increased Speeds

Mean temperature (°C)	Viscosity in flight (N s/m²)	Shear energy/ unit volume (W/m³ × 10⁶)	Shear work per turn (J/turn)	Temperature rise at end of transit $(T_a - T_0)$ (°C)	Residual temperature rise after 'rest' (°C)
140	82·35	1445·3	34·05		
150	74·73	1311·6	30·90	2·65[a]	0·41
160	67·11	1177·8	27·75		

[a] Heat content of barrel for $(T_a - T_0) = 1°C$ is 11·34 J.
Screw speed 2 rps; transit time 0·05 s; 'Rest' time 0·45 s
Drag flow shear rate in channel = 59·86 s⁻¹
Viscosity in channel = 1369 N s/m²
Power consumption in channel = 1949 W/turn
Shear rate in flight clearance = 4189·4 s⁻¹

Mean temperature (°C)	Viscosity in flight (N s/m²)	Shear energy/ unit volume (W/m³ × 10⁶)	Shear work per turn (J/turn)	Temperature rise at end of transit $(T_a - T_0)$ (°C)	Residual temperature rise after 'rest' (°C)
140	51·23	3596·6	42·37		
150	46·49	3263·8	38·45	4·58[b]	0·71
160	41·75	2931·0	34·53		

[b] Heat content of barrel for $(T_a - T_0) = 1°C$ is 8·01 J.
Screw speed 4 rps; transit time 0·025 s; 'Rest' time 0·225 s
Drag flow shear rate in channel = 119·72 s⁻¹
Viscosity in channel = 851·4 N s/m²
Power consumption in channel = 4848 W/turn
Shear rate in flight clearance = 8378·8 s⁻¹

REFERENCES

1. Janssen, L. P. B. M. *Twin screw extrusion*, Elsevier, Amsterdam, 1978.
2. Fisher, E. G. *Extrusion of plastics* (3rd ed.), Newnes–Butterworth, London, 1976.
3. Ibid, p. 48.
4. Ibid, p. 76.
5. Ibid, p. 73.
6. Ibid, p. 83.
7. Schenkel, G. *Plastics extrusion technology and theory*, Iliffe, London, 1966.
8. *Plastics and Rubber Weekly*, Maclaren & Sons Ltd, London.
9. Fenner, R. T. *Extruder screw design*, Iliffe–Butterworth, London, 1970.
10. Ibid, pp. 56, 82, etc.
11. Brydson, J. A. *Flow properties of polymer melts* (2nd ed.), Geo. Godwin, London, 1981.
12. Ibid, p. 60.
13. Weeks, D. J. Some aids to the design of dies for plastics extrusion, *Brit. Plast.* **31** (Apr/May 1958), 156–70.
14. Powell, P. C. Design of extruder dies using thermoplastics melt properties data, *Polym. Engng. Sci.*, **14** (Apr 1974), 298–307.
15. Westover, R. F. Effect of hydrostatic pressure on polyethylene melt rheology, *S.P.E. Trans.*, **1** (1961), 14–20.
16. Westover, R. F. The significance of slip in polymer melt flow, *Polym. Engng. Sci.*, **6** (1966), 83–89.
17. Adamski, T. Unpublished work, Loughborough University of Technology, Loughborough, 1974.
18. Kay, J. M. and Nedderman, R. M. *An introduction to fluid mechanics and heat transfer* (3rd ed.), Cambridge University Press, Cambridge, 1974.
19. McAdams, W. H. *Heat transmission* (3rd ed.), McGraw–Hill, New York, 1954.
20. Ibid, p. 171.
21. Ibid, pp. 235–7.
22. Ibid, p. 140.
23. Forrest, G. and Wilkinson, W. L. Laminar heat transfer to power law fluids, *Trans. Inst. Chem. Engrs.*, **51** (1973), 331–8; **52** (1974), 10–16.

24. Abedi, P. Energy transport to molten flowing polymer systems, Doctoral thesis, Loughborough University of Technology, Loughborough, 1975.
25. Briston, J. H. *Plastic films*, Iliffe–Butterworth, London, 1974, p. 81.
26. The Farrel Corporation, Ansonia, Connecticut.
27. Francis Shaw & Co. Ltd, Manchester.
28. Frieseke u. Hoepfner G.m.b.H., Erlangen-Bruck.
29. McKelvey, J. M. *Polymer processing*, John Wiley, New York, 1962, p. 255, Fig. 10.11.
30. Ibid, p. 234.
31. Ibid, p. 236, Fig. 10.6.
32. Ibid, p. 247.
33. Welding Engineers Inc., Norristown, Pennsylvania.
34. Klein, J. H. Aspects of process design: mechanics and economics of large extrusion lines, *Plast. Polym.*, **41** (1973), 286–91.
35. Weeks, D. J. and Allen, W. J. Screw extrusion of plastics, *J. Mech. Engng. Sci.*, **4** (4) (1962), 380–400.
36. Bernhardt, E. C. (Ed.). *Processing of thermoplastic materials*, Reinhold, New York, 1959, p. 170.
37. Ibid, p. 180–1, Figs 4.17, 4.18.
38. Ibid, p. 195.
39. Ibid, p. 211, eqn 107.
40. Ibid, p. 189.
41. Ibid, p. 200, eqn 84.
42. Mohr, W. D., Saxton, R. L. and Jepson, C. H. Theory of mixing in the single screw extruder, *Ind. Engng. Chem.*, **49** (1957), 1857–62.
43. Eccher, S. and Valentinotti, A. Experimental determination of velocity profiles in an extruder screw, *Ind. Engng. Chem.*, **50** (1958), 829–36.
44. Kennaway, A. Some recent developments in extrusion, in *Plastics progress*, Iliffe, London, 1957 (referred to in Ref. 35).
45. Bernhardt, E. C. The vacuum extruder screw, *S.P.E. J.*, **12** (1956), 40.
46. Grant, D. Developments in extrusion machinery: valved and vented extrusion, *Trans. Plast. Inst.*, **29** (1961), 130.
47. Marshall, D. I., Klein, I. and Uhl, R. H. Metering screw performance with temperature gradients, *S.P.E. J.*, **21** (1965), 1192–202.
48. Barnett, R. D., Klein, I., Mallory, R. D. and Marshall, D. I. *15th Annual wire and cable symposium*, Atlantic City, New Jersey, 1966.
49. Tadmor, Z. and Klein, I. *Engineering principles of plasticating extrusion*, Van Nostrand Reinhold, New York, 1970, Figs 5.7, 5.10, 5.23, 5.24, 5.25.
50. Ibid, Figs 5.1–5.5, 5.13–5.22.
51. Ibid, Fig. 5.57.
52. Ibid, p. 163.
53. Marshall, D. E. Private communication, Loughborough University of Technology, Loughborough, 1983.
54. Gale, G. M. Dry-blend extrusion of rigid PVC, *Plast. Polym.*, **38** (1970), 183–91.
55. Stevens, M. J. The melting process in screw extruders, *Plast. Polym.*, **38** (1970), 107–13.
56. Halmos, A. L., Pearson, J. R. A. and Shapiro, J. Melting in single screw

extruders, *Polymer Rheology and Plastics Processing Conference*, Loughborough University of Technology, Loughborough (Sept 1975), 124–131.
57. Maddock, B. H. An improved mixing-screw design, *S.P.E. J.*, **23** (1967).
58. Maillefer, C. Analytical study of the single screw extruder, *Brit. Plast.*, **27** (1954), 394–7, 437–40.
59. Hayward, I. F. The design of a model screw extruder for the examination of the factors affecting the feed characteristics, Project for B. Tech., Dept. of Mech. Eng., Loughborough University of Technology, Loughborough, 1977.
60. Darnell, W. H. and Mol, E. A. J. Solids conveying in extruders, *S.P.E. J.*, **12** (1956), 20.
61. Kosel, U. M. A novel concept of single-screw extrusion, *Plast. Polym.*, **39** (1971), 319–27.
62. ASEA Ltd, Earl Road, Cheadle Hulme, Cheshire SK8 6QP, England.
63. Allen, W. J. and Hillman, D. A. Torque measuring in an extruder, *Plastics*, **22** (1957), 423.
64. B.I.C.C. Pyrotenax Ltd, Hebburn on Tyne.
65. Kirby, R. B. Process dynamics of screw extruders, *S.P.E. J.*, **18** (1962), 1273.
66. Tadmor, Z. and Klein, I. *Engineering principles of plasticating extrusion*, Van Nostrand Reinhold, New York, 1970, p. 410.
67. Bowyer Engineering Ltd, Andover.
68. Teh, J. W., Rudin, A. and Schreiber, H. P. Characterisation of linear low density polyethylene by capillary rheometry, *Plast. Rubb. Proc. Appl.*, **4** (2) (1984), 149–56.
69. Lolley, R. A. and Wykes, H. The measurement of melt pressure, *Plast. Polym.*, **34** (1966), 187–8.
70. Cogswell, F. N. *Polymer melt rheology*, Geo. Godwin, London, 1981.

INDEX

Adiabatic operation, 179
Adiabatic temperature, 179, 265
Ammeter, drive, 256, 291
Analysis
  hydrodynamics, 68
  variables, of
    leakage, 88
    output equation, 67
    parallel screw, 89
    power equation, 161, 163
Application of single-screw extruder, 34
Approximations
  flow and power, for
    Newtonian isothermal, 107, 155, 265
    pseudoplastic isothermal, 124, 163
Assumptions, simplified flow, 71
Autogeneous (autothermal) operation, 245
Autogeneous speed, 246
Autogeneous temperature, 179, 246

Back pressure, 39, 67, 193, 203
Bagley end correction, 22
Barrel
  cooling, 174, 180, 197, 241, 243, 252, 293
  heat capacity, 245
  temperature (set), 194, 197, 203, 206, 207, 242, 243, 256
  thermal (heat) conduction, 145, 165–7, 196, 241, 242, 317

Barrier screw, 134, 213
Bearing, thrust, 203
Bingham fluids, 33
Blending
  feed, 262
  melt, 53
Blow moulding, 55, 56
Blown film, *see* Tubular film (blown)
Breaker plate, 216, 228, 282

Capillary
  flow, 11
  viscometer, 21
Channel (screw)
  curvature, 72, 120
  depth, 103, 107, 177, 210
    energy balance, effect on, 177
    output, effect on, 103, 107
  partial filling, 110, 112, 121, 246
  power input, 158, 175–87, 215–35
  velocities in, 74, 78, 80, 83, 190, 191, 242
  width correction (shape factor), 63, 70, 72, 120, 189
  see also Screw channel depth
Characteristics
  output/pressure, 67, 107, 109, 115
  output/speed, 102, 106
Circular capillary, 11
Cleaning, 279
  dies, from, 281
  granulators, of, 286
  screws, of, 283
  tools, 281

Clearance, flight, 20, 52, 70, 85
  power, 159, 197, 241, 317
  temperature variations, 165, 317
Coating
  paper, 45, 58, 110
  supported, 45
Colour (pigments)
  changing, 275
  compounding, 36, 61, 286
  dispersion of, 67, 201
Combined (pressure and drag) flow, 80, 83, 124, 169, 294
Comparison
  machine types, of, 60, 65
  requirements for processes, of, 58, 60
  thermal properties, of, 26
Compounding, 36, 51, 58, 61, 286
  twin-screw, 66
Compression ratio, 104, 210, 212
  twin-screw, 66
Computer control, *see* Control
Condition
  changes, 213, 237, 250, 272, 278
  operating, 258
  uniformity, 58; *see also* Fluctuation (with time); Variations
Conduction
  circumferential
    barrel wall, in, 167, 317
  longitudinal, 145
    barrel wall, in, 166, 196, 241, 242, 317
    polymer, in, 196
  radial, 28
    barrel wall, in, 165, 241, 317
    polymer, in, 242
  steady, 25
  transient, 25, 31, 167
  twin-screw, 66
Conductivity, *see* Thermal conductivity
Constant depth screw, 91
Contamination, 274–6, 285, 286
Continuous operation, *see* Operation
Control, 253
  accuracy, 258
  action (temperature), 150, 152, 154

Control—*contd.*
  computer, 258, 262, 293
  external, 254
  medium, 255
  melt length, of, 258
  methods, 237
  power, 259
  signal, 255
  specific processes, for, 258
  speed, 254, 259
  temperature, 195, 206, 245, 259, 273
Controlled variables, effect of, 201
Conversion factors (units), 26
Conveying, solids, 68, 139, 189
Cooling, 37, 38, 40, 43, 44, 48, 52
  barrel, 174, 180, 197, 241, 252, 293
  air, 243
  feed pocket, 139, 145, 193, 289
  screw, 145, 199, 211, 241, 258, 283, 289, 293
Corrections
  end (Bagley), 22
  heat transfer, 30
  non-isothermal, 67, 129, 164
  output, 67, 73, 120, 126, 129
  power, 163, 164
  pseudoplastic, 67, 103, 124, 125, 129, 163, 183
  shape factor, 63, 70, 72, 120, 189
Corrosion, from PVC, 281
Covering (wire), 39, 58
Crystallinity, 24, 44, 46, 49, 153, 162

Data
  thermal properties, 26, 288, 308
  viscosity, 10, 262, 308
Defects of flow, *see* Flow
Definition
  efficiency, 146
  energy terms, 146
  product quality, 238
Degassing, 54, 58
Degradation
  shear, 54
  thermal, 53, 199, 208, 213

Degradation—*contd.*
 thermal—*contd.*
  dismantling and cleaning, during, 279, 282
Density, 24, 308
Derivation
 flow equation
  dies, 11, 13
  extruder, 70, 294
 power equations, 155
 stability, 306
 strain energy, 15, 169
 temperature variation over flight, 165, 317
 wall shear rate, 123
Design
 cleaning, for, 284
 dies, of, 4, 38, 192, 209
 feed pocket, of, 136, 262
 machines, of, 2, 201, 265
 screw, of, 200, 209, 257, 264
Dewatering, 55
Die
 adjustment, 39, 41, 278
 cleaning, 281
 crosshead, 38, 39, 48
 design, 4, 38, 192, 209
 flow distribution, 41, 192
 heating, 196, 275
 in-line, 48
 lips, 39, 50, 192, 203
 multi-hole (spinaret), 41, 46
 oscillating, 47, 48
 quality at, 238, 251, 258, 265, 291
 rotating, 47
 swell, 19, 23, 37, 42, 192, 209
 temperature distribution at, 42, 196, 200, 202, 206, 226–30, 240, 243, 255, 275, 291, 293
Dimensionless output (Q/Wbh), *see* Output
Dismantling, 279
 screw, of, 282, 284
Dispersive mixing, 33, 61, 67, 201
Distribution
 energy (dissipation), of, 64, 169
 flow in die, of, 41, 192

Distribution—*contd.*
 mixing, of, 64
 temperature at die, of, 42, 196, 200, 202, 206, 226–30, 240, 243, 255, 275, 291, 293
 velocity, of, *see* Velocity
Distributive (mixing), 32, 62, 200, 252
Double-parallel screw (stepped), 93, 115, 209, 211, 247, 297
Drag flow, 34, 40, 73, 101
 power, 187
 shear
  heating, 173
  rate, 75
 velocity distribution, 75
Drawdown, 42, 43, 44, 46
Drawing, 19, 40, 44
Drive
 ammeter, 256, 291
 efficiency, 147, 149
 motor, 147, 149
  characteristic, 246, 248, 253
  regulation, 253
 torque measurement, 147, 256
Drying
 feed, of, 274
 process, 54

Efficiency, 35, 142
 definition, 146
 drive, of, 147, 149
 pumping, 62
Elastic effects, 19, 56, 63, 238, 239, 263
 measurement of, 20
Elastic melter, 63
Energy
 balance, 142, 144, 265
 determination, 147, 287
 heating, effect of, 203
 pressure, effect of, 184, 203
 pseudoplastic, 183
 speed, effect of, 181, 202
 variables, effect of, 174, 201
 wall shear rate, effect of, 183
 dissipation (absorption), 64, 165, 169, 196

Energy—contd.
  feed, in, 143
  heater, 143, 150
  input (specific), 142, 174–87, 222–33, 265–71
  losses, 143, 145, 154, 181, 289, 301
  mechanical, 143, 147, 155, 223–6, 289
  pressure, 145
  properties, 24
  strain, 15, 30, 169
    distribution of, 17, 31, 169
  transfer, 196; *see also* Energy balance
  see also Power
Enthalpy, 25
  increase, 143, 145, 161, 177–82
  feed, of, 143
  product, of, 143, 152
Equations for
  die flow, 101, 106
  drag flow, 76
  flow, 4, 16, 67, 70, 78, 294
  output, 67, 78, 93, 96, 294
  power, 158, 159
  pressure flow, 77
  shear rate, 16
  stability, 248, 306
  transverse flow, 83
Experimental determination of
  energy balance, 147, 287
  heat losses, 154, 301
  heater energy, 150
  mechanical power, 147
  melt length (gel point), 121, 292
  melting process, 131
  solids feeding, 136
  temperature variations, 243
  viscosity, 10, 20, 262, 264
Experimental screw nose, 200
Extensional flow, 19
Extraction, vacuum, 54, 64, 109, 119
Extruder
  operation, 237
  ram, 62
  single-screw, 1, 67, 142, 188
  smooth disc (Weissenberg), 63
  spiral disc, 62

Extruder—contd.
  twin-screw, 1, 64
  vented, 109
Extrusion, vii, 1, 3, 34, 67, 188, 272, 287
  processes, 34, 237

Feed
  blending, 262
  enthalpy, 143
  properties, 262
  variations, 275
Feed-pocket, 55, 136
  cooling, 139, 145, 193, 289
  design, 136, 262
  heating, 141
Feeding, 51, 54, 55, 68, 135, 188, 213, 274, 275
  deblending, 200, 262
  force-, 140
  melt-, 61, 115, 141, 144
  powder, 135, 140, 195
  preheating, 144, 153, 193
  pressure, effect of, 140, 193
  rubber, 136, 140
  scrap, 141, 262, 275, 286
  segregation, 200, 262
  starve-, 140
  temperature, effect of, 139, 206
  see also Solids conveying
Fibres, 46, 58, 118, 254
Filling, partial, of screw channel, 110, 112, 121, 246
Film, 43, 48, 58, 254
Filtering, 39, 44, 46, 55, 216, 228, 282; *see also* Contamination
Flat film, 43, 58
Flat sheet, 41, 58
Flight clearance, 20, 52, 70, 85
  power in, 159, 165, 186, 197, 241, 317
  temperature variations in, 165, 317
Flow
  behaviour, 4
  capillary, 11
  defects, 22, 40, 50, 192, 238, 239

Flow—contd.
  drag, 34, 40, 73, 101, 173, 187
  equations, 4
    dies, for, 16, 101, 106
    extruder, for, 67, 70, 78, 83, 294
  extensional, 19
  fluctuations, 123, 253
  infinite slit, 13
  isothermal, 107, 190
  leakage, 51, 84
  Newtonian, 11
  non-isothermal, 28, 129, 208
  non-Newtonian, see Flow, pseudoplastic
  solids, of, 68, 135, 189
  patterns, 67, 83
  pressure, 11, 76
  properties, 4, 8, 10, 20, 22, 262, 264, 287, 308
  pseudoplastic, 123, 263
  rate, 37, 67, 78
  transverse, 81, 87, 190, 242, 295
  see also Properties; Measurement; Shear rate; Viscosity
Fluctuation (with time), 67, 237, 262
  flow, of, 123, 253
  pressure, of, 118, 193, 248, 253
  speed, of, 118, 149, 254, 259
  temperature, of, 121, 293
    over flight, 165, 317
    melting, in, 245, 248
Foams, 50
Force feeding, 140
Fourier number (transient conduction), 320
Functions of the single-screw extruder, 34, 51, 67, 68

Gauss error integral (transient conduction), 320
Gear pump, 46, 117
Gearing, 147, 148, 264
'Gel point', see Melt pumping effective length
Grade of polymer, 178, 208
  changing, 275

Gradient, temperature, see Temperature
Graetz number, 311
Granulators, 286
Graphical representation
  energy balance 176–83
  output/pressure/speed, 67, 101–19
Grashof number (natural convection), 73

Haul-off, 38, 39, 43, 46, 48
  speed, 258
Hazards
  burns, 274, 279, 280
  contamination, 274, 285, 286
  decomposition, 274, 282
  electric shock, 274, 280
  excess pressure, 278
  leakage, 274
  powders, 274
  static electricity, 274, 286
Heat
  capacity of barrel, 245
  content (enthalpy), 143, 145, 152, 161, 177–82
  crystallisation, of, 24, 197
  flow, 27, 189
    barrel, in, 166, 196, 203, 241, 319
    screw, in, 165, 199, 317
  flux, 27
  losses, 143, 145, 154, 181, 289, 301
  specific, 24, 26, 153, 308
  transfer, 25, 29
    coefficient, 29
    through barrel, 149, 165, 196, 241, 317
    non-isothermal, 28
    screw, to, 165, 199, 241
  see also Thermal conduction; Thermal conductivity; Thermal degradation; Thermal diffusivity; Thermal properties
Heater(s)
  energy, 143, 150
    change with temperature, 151, 203

Heater(s)—*contd.*
  handling, 280
  induction, 152
Heating
  barrel, 166, 196, 203, 241, 319
  die, in, 196, 275
  feed, of, 193
  feed pocket, of, 141
  shear (viscous), 19, 30, 52
    channel, in, 169, 203, 241
    over flight, 165, 241, 317
    melting, in, 196
    space, 143
Hollow sections, 37, 58

Individual machine, 287
Injection moulding, 35, 55, 56, 261
Input, *see* Feed, Heater, Power
Instruments, 272, 287
  pressure, 255
  temperature, 243, 256, 291
  viscosity, 21
Isothermal flow, 190
  approximation, 107, 124, 265
Isothermal power, 155, 163

Leakage flow, 51, 84
  power in, 159, 165, 186, 197, 241, 317
  temperature variations in, 165, 317
  twin-screw, in, 64
  variables, 88
Limits
  (limiting) factors, 238
  motor load, 277
  output, vii, 238, 291
Lips, *see* Die
Longitudinal conduction, 145, 155, 166, 196, 241, 303, 317
Longitudinal insulation, 243
Longitudinal mixing, 242
Longitudinal pressure profile, 89, 92, 94, 98
Longitudinal temperature gradients, 139, 155, 161, 194, 207, 242, 303

Longitudinal velocities, 74, 78, 80, 190, 242
Losses
  drive motor, 149
  heat (energy), 143, 145, 154, 181, 289, 301

Machine
  comparison of types, 60, 65
  individual, 287
  twin, *see* Twin-screw
  variables, *see* Operational strategies, Operational variables
Material flow, 68, 135, 188; *see also* Feeding; Flow; Output
Mean temperature difference, 29
Measurement
  elastic effects, 20
  output, 152, 255
  power, 147, 256
  pressure, 255
  torque, 147, 256
  viscosity, 10, 20, 262, 264
Mechanical construction, 1
Mechanical power input, 143, 147, 155, 223–6, 289
Melt
  blending, 53
  feeding, 61, 115, 141, 144
  fracture, 22, 40, 50, 192, 239
  pool (stream), 132, 189, 194, 197, 248, 292
  properties, *see* Properties; Elastic effects; Viscosity
  pumping
    effective length, 120, 243, 258, 292
    effect on energy balance, 176
    variation, 161, 246
  equations for, 67, 70, 78, 190, 294
  temperature, 178, 207, 216–22, 243, 255
    measurement, 256
  viscosity, 4, 6, 7, 8, 10, 20, 53, 90, 166, 208, 262, 263

Melt flow rate, 20, 41, 54, 208, 275
Melting
  bed break-up, after, 133, 134, 195, 239
  mechanism, 132, 134, 189, 206, 207, 243
  mixing in, 200
  powder, 195
  power consumption, 160
  pressure, effect of, 133, 193
  process, 131, 189, 194, 197, 248, 292
    control, for, 258
    improvement, 230–2, 248, 293
    temperature variations, 246
  region, 69
  temperature (crystalline), 24, 208, 308
Mesh, 47
Metering, see Melt pumping
Mixing, 31, 35, 51, 199, 251, 265
  continuous mixer, 61
  dispersive, 33, 51, 61, 67, 201, 239
  distribution of, 64
  distributive, 32, 62, 200, 252
  improvement, 233–5
  'intensive' powder mixer, 61
  internal melt mixer, 61
  longitudinal, 242
  melting, in, 200
  pressure, effect of, 201
  radial, 53, 200
  scale of, 239
Modelling, process, 258
Molecular weight, 53, 208
  fluctuations, effect on, 262
  power, effect on, 178
  distribution, 54, 56, 208
    operation, effect on, 262
Monofil, 46
Motor, see Drive
Moulding
  blow, 55, 56, 261
  injection, 35, 55, 56, 261

Navier–Stokes equation, 294
Netting, 47

Newtonian approximation for flow, 107, 155, 265
Newtonian flow, 11
Newtonian power, 155
Newtonian viscosity, 6
Non-isothermal corrections, 67, 129, 164
Non-isothermal flow, 28, 129, 208
Non-isothermal heat transfer, 28
Non-isothermal power, 164
Non-isothermal shear rate, 28
Non-Newtonian flow, 7, 263; see also Pseudoplastic flow

Objectives, 1
  operational changes to achieve, 213
Operating conditions, 258
  changes to achieve objectives, 213
Operation, 2, 188, 201
  adiabatic, 179
  as part of a total process, 237
  continuous, 35, 51, 61, 272
  molecular weight distribution, effect of, 262
  practical, 272
  steady, 56, 57, 272
  viscosity, effect of, 263
Operational strategies, 201, 213
Operational variables, 201
Orientation, 44, 46, 47, 50
Output, 67, 215
  corrections, 67, 73, 120, 126, 129
  dimensionless (Q/Wbh), 78, 90, 105, 117, 265
  equations, 67, 78, 93, 96
  limit for quality, 238, 291
  measurement of, 152, 255
  output/pressure stability, 253
  screw/die combinations, 67, 107, 109, 115
  speed, effect of, 102, 106
  temperature, effect of, 203, 253
  viscosity, effect of, 90
Overall performance, 188, 238

Paper coating, 45, 58, 110

Parallel
  double-, screw (stepped), 93, 115, 209, 211, 247, 297
  screw (constant depth), 73, 79, 91, 159
Partial filling of screw channel, 110, 112, 121, 246
Performance
  overall, 188, 238
  variables, effect of, 2
Pigments, *see* Colour (pigments)
Pipe, *see* Tube; Hollow sections
Polymer
  changing, 276
  properties, 20, 24, 208, 308
    operation, effect on, 209, 262
    see also Properties
Polymer Supply and Characterisation Centre (SERC), 315
Polymerisation, 55
Post-forming, 3, 37, 43, 46, 48
Powder
  feeding, 135, 140, 195
  melting, 195
  seals, 135, 140, 283, 284
Power, 155
  consumption (input), 155, 289
    calculation, temperature for, 149, 161
    channel, in, 158, 175–87, 215–35
    equations, 158, 159
    isothermal, 155
    melting, in, 160
    over flight, 159, 165, 186, 197, 241, 317
    pressure, effect of, 164, 184
  law, 8
  measurement, 147, 256
  molecular weight, effect of, 178
  Newtonian, 155
  non-isothermal, 164
  pseudoplastic, 163
  specific, 175–87, 222–33, 265–71
  temperature, effect of, 162, 179, 290
Practical extrusion processes, 34
Practical operation, 272
Prandtl number, 311

Preheating of feed, 144, 153, 193
Pressure, 11, 76, 216
  back, 39, 67, 193, 203
  energy, 145
  energy balance, effect on, 184, 203
  feed, in, 140, 193
  flow, 11, 76
  fluctuation at die, 193, 248, 253
  gradient(s), 11, 13, 79, 127, 265, 297
  die, in, 192
  maximum, 97, 100
  maximum, 98
  measurement, 255
  melting, effect on, 133, 193
  mixing, effect on, 201
  power, effect on, 164, 184
  pressure/output stability, 253
  profile, 67
    longitudinal, 89, 92, 94, 98
    vented, 111, 113
  speed, effect of, 118
  venting, effect on, 112, 114
  viscosity, effect on, 10
Principles
  energy balance, 142
  single-screw, 67
  twin-screw, 1, 64
Process(es)
  control for specific, 258
  modelling, 258
  practical, 34
  requirements, 1, 3, 58, 60, 237, 258
    non-shaping, 51
    shaping, 37
    subsidiary, 3
    total, 2, 237
Product
  dimensions, *see* Die swell, Drawdown
  quality, *see* Quality of product
  uniformity, *see* Uniformity
Profile
  pressure, 67
    dies, in, 192
    longitudinal, 89, 92, 94, 98
    vented, 111, 113
  screw, 211; *see also* Screw types

Index

Profile—*contd.*
  temperature
    radial, experimental, at die, 244
    set, 195, 206, 207, 243, 256, 290
  velocity
    dies, in, 12, 14, 28
    screw, in, 80, 191
      barrel cooling, effect of, 197
    simple shear, in, 10
Properties
  energy, 24
  feed, 262
  flow, 4, 8, 10, 20, 22, 264, 287, 308
  melt, determination of, 262
  polymer, 20, 24, 208, 308
    operation, effect on, 209, 262
  thermal, 24, 26, 288, 308
Protection of metal surfaces, 280, 281, 283
Pseudoplastic flow, 7, 123
  corrections, 67, 103, 125, 129
  energy balance, 183
  power in, 163
  shear rate, 124
Pseudoplasticity index, 8, 208, 263
Pump, gear, 46, 117
Pumping efficiency, 35, 62, 131, 145; *see also* Pressure; Flow
Purging, 276, 278

Quality of product, vii, 3, 237
  definition, 238
  die, at, 238, 251, 258, 265, 291
  screw type, effect of, 293

Regulation
  drive, of, 253
  self-, of process, 253
Requirements, 1, 3, 58, 60
  complete processes, of, 237, 258
  non-shaping processes, of, 51
  shaping processes, of, 37
Residence, *see* Time, Mixing
Restrictor
  adjustable, 292
  bar (ring), 41, 203

Reynolds number, 73
Rheology, 4; *see also* Flow; Properties, flow
Rubber
  extrusion, 35, 68, 70, 120, 211, 252
  feeding, 136, 140
Rubber and Plastics Research Association (PSCC), 315

Safety
  compounding machines, of, 286
  dismantling/cleaning, in, 280
  dust masks, 286
  ear muffs, 286
  gloves, 275
  granulators, of, 286
  housekeeping, 274
  shut-down, in, 279
  spectacles, 275
  start-up, in, 278
Scale of mixing, 239
Scale-up of machines, 3, 264
Science and Engineering Research Council (SERC), 315
Scrap, 55, 276
  feeding, 141, 262, 275, 286
  recovery, 284
Screen, *see* Filtering
Screw
  barrier, 134, 213
  channel depth, 103, 107, 177, 210
  cleaning and storage, 283
  constant depth (parallel), 73, 79, 91, 159
  conveying of solids, 68, 139, 189
  cooling, 145, 199, 211, 241, 258, 283, 289, 293
  design, 209, 257, 264
  flight, *see* Flight clearance, Clearance, flight, Leakage flow
  length, 56, 104, 115, 121, 176, 209
    effective, 120, 161, 176, 243, 246, 258, 292
  nose (tip), 200
  output equations, 78, 89
  partial filling, 110, 112, 121, 246
  pressure profiles, 89

Screw—*contd.*
quality, effect on, 293
removal, 282, 284
sections, 69, 210, 212
specific processes, for, 258
speed, 63, 66, 67, 202
  alteration, 278
  control, 249, 254, 259
  energy balance, effect on, 181, 202
  measurement, 147
  output, effect on, 102, 106
  power, effect on, 163
  pressure, effect on, 118
  temperature variations, effect on, 202
stepped, 93, 115, 209, 211, 247, 297
taper, 96, 127, 191, 211
taper-parallel, 96
torsional strength, 63, 213
twin, *see* Twin-screw
types, 212
venting, for, 111, 113, 114
wear, 70, 89
Screw/die combination, 67
  output, 107, 109, 115
Seals, powder, 135, 140, 283, 284
Sections
  hollow, 37, 58
  screw, of, 69, 210, 212
  solid, 37, 58
Segregation of feed, 200, 262
Set temperature, *see* Temperature set
Shape factor, 63, 70, 72, 120, 189
Shaping processes, 37, 58
Shear
  flow, 10
    drag flow in screw, 67, 75
    simple shear, 10
  heating (viscous), 19, 30, 52
    channel, in, 169, 203, 241
    melting, in, 196
    over flight, 165, 241, 317
  history, 64, 67, 251
  rate, 6, 10, 128
    apparent, 21
    capillary, 13
    drag flow, 75

Shear—*contd.*
rate—*contd.*
    energy balance, effect on, 183
    equation, 6, 10, 12, 15, 16, 75, 124, 295, 296
    non-isothermal, 28
    pseudoplastic, 124
    slit, 15
    wall, 80, 124, 265
  strain, 5, 32, 268
  variation, 251
  stress, 5, 6, 11, 13, 33
Sheathing, *see* Wire covering
Sheet, 41, 58
Shut-down, 276
Simple shear, 10, 75
Simplified theory, *see* Theory, simplified
Single-screw extruder, 1, 3
  comparison with twin-screw, 64
  comparison with other machines, 60
  control, 253
  energy balance, 142
  operation, 188
  output, 67
  practical operation, 272
  principles, 67
  processes, 34
  quality, 237
  scale-up, 264
  stability, 245
Sizing, 37, 38; *see also* Die adjustment; Die swell; Drawdown
Slit flow, 13
Solid(s)
  conveying, 68, 139, 189
  sections, 37, 58
Specific energy input (power), 142, 174–87, 222–33, 265–71
Specific heat, 24, 26, 153, 308
Specific processes, 258
Specific volume, 24, 308
Speed, 63, 66, 67, 202
  alteration, 278
  control, 254, 259
  energy balance, effect on, 181, 202

Speed—*contd.*
  haul-off, 258
  measurement, 147
  output, effect on, 102, 106
  power, effect on, 163
  pressure, effect on, 118
  regulation of drive, 149
  temperature variations, effect on, 202
Spinning, 46, 58, 118, 254
Stability, 237
  condition changes, in, 250
  equation for, 248, 306
  output/pressure, 253
  output/temperature, 253
  start-up, in, 250
  surging, 123, 239, 245, 306
  venting, in, 112, 114
Start-up, 276
  stability, 250
  transients, 278
Steady operation, 56, 57, 272
Stepped, *see* Double-parallel screw (stepped)
Strain, 4, 32, 251, 268
  energy, 15, 30, 169
    distribution, 17, 31, 169
  rate, 4, 6, 46; *see also* Shear rate
Strategy
  operation, for, 201, 213
  scale-up, for, 265
Strength of screw, *see* Torsional strength of screw
Stress, 4, 6, 11, 13, 33, 157–9
Surface heat transfer, 25, 29
Surface protection, 280, 281, 283
Surging, 123, 239. 245, 306
  origin, 246
  venting, in, 112, 114
Swell ratio, 23, 42, 192, 209

Tape, 46, 58
Taper
  -parallel screw, 96
  screw, 96, 191, 211
    velocity in, 191
    wall shear rate, effect on, 127

Temperature, 193
  adiabatic, 179, 265
  autogeneous (autothermal), 179, 246
  control, 195, 206, 259
  controllers, 151, 178, 197, 256, 277
    adjustment, 245, 273
    stability, 250
    time-proportioning, 245
  difference, 27, 31, 63, 195
    between flight clearance and channel, 242
    mean, 29
    melt pumping, in, 195
    melting, in, 194, 195, 240, 243
  distribution from screw, 42, 196, 200, 202, 206
  feed, effect on, 139, 206
  fluctuations, 121, 245, 248, 293
  gradients, 27
    longitudinal, 139, 155, 161, 194, 207, 242, 303
    radial, 149, 172, 241, 245
  melt, 178, 207, 216–22, 243, 255
  output, effect on, 203, 253
  power calculation, for, 149, 161
  power, effect on, 162, 179, 290
  profile, 243, 256
    melt pumping, in, 206, 207
    melting, in, 195, 206, 207, 290
  set, 203, 206, 243, 256
    melt pumping, in, 197, 207, 242
    melting, in, 194, 207
  variations (space), 239
    at die, 42, 196, 200, 202, 203, 206, 226–30, 240, 243, 255, 275, 291, 293
    experimental determination of, 243
    injection moulding, in, 57
    length (screw), in, 121
    melting, from, 246
    over flight, 165, 318
    reduced by transverse flow, 242
  viscosity, effect on, 8, 166
Theory, simplified, 2, 67, 107, 155, 238, 265

Thermal conduction, 25
  barrel, in, 145, 165–7, 196, 241, 242, 317
  polymer, in, 196, 242
  radial (in capillary), 28
  transient, 25, 31, 167
  twin-screw, in, 66
  see also Conduction
Thermal conductivity, 25, 308
Thermal degradation, 53, 199, 208, 213, 279, 282
Thermal diffusivity, 25, 311
Thermal properties, 24, 288, 308
  conversion factors, 26
  typical values, 26
Thermocouple
  control, 257, 273
  experimental array, 243, 291
  melt, 256
  position, 149, 161
  Pyrotenax, 243
Thermoplastics
  abbreviations, 36
  extrusion, for, 35
Thrust
  bearing, 203
  load, 63, 203
Thyristor
  fast cycling, 152
  phase angle firing, 152
  speed control, 249, 254
Time
  fluctuations, of temperature, 121, 293
    melting, in, 245, 248
    over flight, 165, 317
  reaction, for process control, 273
  residence, 19, 41, 51, 56, 201, 209, 265
Tools, cleaning, 281
Torque measurement, 147, 256
  ASEA Torductor, 148
  belt tension, 148
  reaction, 147
  strain gauge, 147
Torsional strength of screw, 63, 213
Total process, 2, 237
Transient conduction in barrel, 25, 31, 167, 317

Transients, from condition changes, 237, 250, 257, 278
Transverse flow, 81, 87, 190, 242, 295
  die, in, 41, 192; see also Distribution
  reducing temperature variations, 242
Tube, 37, 58
Tubular film (blown), 48
Twin-screw, 1, 64
Two stage screw, see Screw, stepped
Types of screw, see Screw types

Uniformity
  conditions, of, 58
  product, of, 51, 258
  see also Fluctuation (with time); Variations
Units, 6
  conversion factors, 26
  shear strain, of, 251
Unmolten material (particles) at die, 239
  see also Die temperature distribution at; Fluctuation (with time); Melting; Mixing; Temperature

Vacuum extraction, 54, 64, 109, 119
  die pressure, effect of, 112, 114
  pressure profile, 111, 113
  see also Venting
Variables
  analysis of
    output equation, 67
    parallel screw, 89
    power equation, 161, 163
  controlled, effect of, 201
  energy balance, effect on, 174, 201
  leakage flow, in, 88
  machine (operational), 201
  performance, 2
Variations
  feed, in, 275
  melt properties, 262
  pressure with speed, in, 118

Variations—*contd.*
  temperature, in, *see* Temperature variations
  see also Fluctuation (with time)
Velocity
  distribution
    capillary, 12
    combined flow, in, 80, 83
    drag flow, in, 75
    non-isothermal, 28
    pressure flow, in, 77
    slit, in, 14
  screw channel, in, 74, 83, 190
    longitudinal, 74, 78, 80, 242
    tapered, 191
    transverse, 81
Venting, 54, 64, 109, 119
  die pressure, effect of, 112, 114
  pressure profile, 111, 113
  screw types, 111, 113, 114
  see also Vacuum extraction
Viscometer, capillary, 21

Viscosity, 4, 41, 53, 208, 275
  apparent, 21
  changing, 275
  experimental determination, 10, 20, 262, 264
  Newtonian, 6
  operation, effect on, 263
  output, effect on, 90
  pressure, effect of, 10
  pseudoplastic, 7
  temperature, effect of, 8, 166
Viscous heating, *see* Shear
Viscous history, *see* Shear

Wall shear rate, 80, 124, 265; *see also* Shear rate
Waste, 55, 276; *see also* Scrap
  heat, 143
  recovery, 284
Wear, 70, 89
Weissenberg extruder, 63
Wire covering, 39, 58